Power Prediction Modeling of Conventional High-Speed Craft

Dejan Radojčić · Milan Kalajdžić ·
Aleksandar Simić

Power Prediction Modeling of Conventional High-Speed Craft

Springer

Dejan Radojčić
Mechanical Engineering
Department of Naval Architecture
University of Belgrade
Belgrade, Serbia

Milan Kalajdžić
Mechanical Engineering
Department of Naval Architecture
University of Belgrade
Belgrade, Serbia

Aleksandar Simić
Mechanical Engineering
Department of Naval Architecture
University of Belgrade
Belgrade, Serbia

ISBN 978-3-030-30606-9 ISBN 978-3-030-30607-6 (eBook)
https://doi.org/10.1007/978-3-030-30607-6

This Springer imprint is published by the registered company Springer Nature Switzerland AG
The registered company address is: Gewerbestrasse 11, 6330 Cham, Switzerland

Foreword

This book is about evolution, taking note of the history of technological progress for mathematical models to predict calm water resistance, trim and propeller characteristics for high-performance marine vessels. Dr. Dejan Radojčić has been an activist from early in his academic career to develop mathematical models using emerging techniques with validated predictions using available experimental data.

His and his team's goals have focused on calculation procedures to improve naval architect's capability to reliably develop high-performance vessel hydrodynamic designs. With these improved resources, designers are able early on to optimize vessels with respect to their requirements to maximize performance in their operational environment.

I have been a friend and colleague of Dr. Radojčić for more than thirty years. I have learned much from him and encouraged this documentation of his passion for improving analytical prediction techniques. This book is one of many achievements of Dejan's life work and the work of his team, resulting in especially useful prediction methods developed with artificial neural networks (ANNs). Their recent ANN methods have been used to develop techniques sensitive to hull geometry input resulting in predictions of calm water resistance and trim near to that of towing tank quality.

This expanded version of his earlier publication with the addition of Part II brings increased ease of use for designers to make mathematical model predictions of resistance as well as trim for their high-performance craft. Data expansions include representative body plans of model test data which Dr. Radojčić and his team used for this collection of published mathematical models. The equations and coefficients of each method as well as their respective boundaries of application are included. Thus, naval architects and designers can readily develop software for resistance and trim predictions is in keeping with their specific design procedures.

Now retired after sixty-plus years as a designer of high-performance vessels, I believe you will find this book to be essential for your work and career.

Virginia Beach, VA, USA
June 2019

Donald L. Blount, P.E.
Founder of Donald L. Blount
and Associates, Inc.

Preface

This work was initially intended to be a review paper of the first author, but gradually grew and became a Springer Brief titled *Reflections on Power Prediction Modeling of Conventional High-Speed Craft (Radojčić 2018).* It was inspired by, and envisaged as an extension of, the seminal Blount (1993) paper titled "Reflections on Planing Hull Technology." Hence, the titles are intentionally similar. The Brief was a summation of author's insights in, and experiences with, high-speed craft (HSC) design and modeling, lovingly accumulated over a 35-year career in the industry and academia, and almost entirely focused on that specific topic.

Formatting errors in the Brief impelled the author to consider an expansion of the initially presented material and to add

1. Resistance and propulsion design data for high-speed craft
2. Excel Macro Codes with software for the mathematical models outlined
3. Worked examples.

Hence, the authorship of this book is expanded to include two additional contributors. The three authors of this enlarged work have been collaborating for years and have published several papers on high-speed craft. The result is this new and more usable two-part book on the subject of HSC. Part I retains the same name as the original Springer Brief, i.e., "Reflections on Power Prediction Modeling of Conventional High-Speed Craft," as it is essentially identical. Part II, titled "Resistance and Propulsor Design Data with Examples," contains tabulated data for resistance and propeller efficiency evaluations. The relevant Excel files can be provided to the reader as electronic supplementary material (ESM). The reader can access ESM via the Springer sites which are given in Chaps. 8–10.

The above-mentioned Blount (1993) paper was updated and upgraded to a book in 2014 and is still a very relevant HSC reference. The focus of that work is on the design of *conventional high-speed craft*, which covers by far the largest number of yachts and boats currently in existence or under construction. Another work that merits a mention here is the Molland et al. (2011) book *Ship Resistance and Propulsion—Practical Estimation of Ship Propulsive Power*, which treats resistance, propulsion and powering, but is mostly focused on the conventional ships.

Both of these books should be regarded as companion books to the work in hand. It is believed, however, that the present work fills in the gap on specific high-speed vessel powering procedures which are not treated by either of the above-mentioned or other sources.

The present book focuses on the practical and concrete topics of high-speed craft resistance and propeller efficiency prediction, and avoids unnecessary issues, mathematical derivations and similar rigor. It is assumed that the reader has the basic university-level knowledge of ship hydrodynamics and is intended primarily for the naval architects who design and develop various types of conventional high-speed vessels. It may also be of use to students attending various naval architecture and marine engineering courses who are interested in the design of fast vessels.

Belgrade, Serbia
June 2019

Dejan Radojčić
Milan Kalajdžić
Aleksandar Simić

References

Blount DL (1993) Reflections on planing hull technology. In: 5th power boat symposium, SNAME Southeast Section

Blount DL (2014) Performance by design. ISBN 0-978-9890837-1-3

Molland AF, Turnock SR, Hudson DA (2011) Ship resistance and propulsion—practical estimation of ship propulsive power. Cambridge University Press, ISBN 978-0-521-76052-2

Radojčić D (2018) Reflections on power prediction of conventional high-speed craft. Springer, ISBN 978-3-319-94899-7

About This Book

High-speed craft is very different from conventional ships. This dictates the need, from the very outset, for special treatment in designing high-speed vessels. The professional literature, which is mostly focused on conventional ships, leaves a gap in the documentation of best design practices for high-speed craft.

The various power prediction methods, a principal design objective for high-speed craft of displacement, semi-displacement and planing type are addressed. At the core of the power prediction methods are mathematical models, based on experimental data derived on models representing various high-speed hull and propeller series. Regression analysis and artificial neural network (ANN) methods are used as extraction tool for this kind of mathematical models. A variety of mathematical models of this type are discussed in the book.

The most significant factors for in-service power prediction are bare hull resistance, dynamic trim and the propeller's open-water efficiency. Therefore, mathematical modeling of these factors is a specific focus of the book.

This book is arranged in two parts. Part I, presented as a text and titled "Reflections on Power Prediction Modeling of Conventional High-Speed Craft," covers the mathematical modeling of resistance, dynamic trim and propeller hydrodynamic characteristics of high-speed craft. This is followed by brief treatment of additional resistance components, propulsive coefficients and power prediction.

Part II titled "Resistance and Propulsor Design Data with Examples" presents Excel Macro Codes for evaluation of resistance and propulsion characteristics, tabulated data for assessment of mathematical models of resistance and propeller efficiency as outlined in Part I, and worked example applications for a number of interesting cases.

This book is aimed at the high-speed craft community in general and particularly at the naval architects who design and develop various types of high-speed vessels. It may also be of use to students who are interested in the design of fast vessels.

About This Book

Contents

Part II Resistance and Propulsor Design Data with Examples

Abbreviations

AEW	Admiralty Experiment Works, Haslar
ANN	Artificial neural network
ATTC	American Towing Tank Conference
BSRA	British Ship Research Association
CFD	Computational fluid dynamics
CPP	Controllable pitch propeller
DSDS	Delft systematic deadrise series
DTMB (TMB)	David Taylor Model Basin
DTNSRDC (NSRDC)	David Taylor Naval Ship Research and Development Center
DUT	Delft University of Technology
ESM	Electronic supplementary material
HSC	High-speed craft
HSMV	High-speed marine vessel
IMO	International Maritime Organization
ITTC	International Towing Tank Conference
KCA	Kings College Admiralty (Newcastle)
MARIN (Wageningen)	Maritime Research Institute of the Netherlands
MM	Mathematical model
NPL	National Physical Laboratory
NSS	Naples Systematic Series
NTUA	National Technical University of Athens
PHF	Planing hull form
RINA (INA)	The Royal Institution of Naval Architects
SKLAD	Series tested in the Naval Institute in Zagreb
SNAJ	The Society of Naval Architects of Japan
SNAME	The Society of Naval Architects and Marine Engineers
SPP	Surface piercing propeller
SSPA	Swedish Maritime Research Centre

SVA	Potsdam Model Basin
SWATH	Small Waterplane Area Twin Hull
TUNS	Technical University of Nova Scotia
USCG	United States Coast Guard
VBA	Visual Basic Application
VTT	Technical Research Centre of Finland
VWS	Versuchsanstalt für Wasserbau und Schiffbau (Berlin)
WEGEMT	EU Marine University Association
WUMTIA (Wolfson Unit)	Wolfson Unit for Marine Technology and Industrial Aerodynamics

Symbols

A_D	Developed propeller blade area (m^2)
A_E	Expanded propeller blade area (m^2)
A_{FZ}	Area of zinc anodes (m^2)
A_O	Propeller disk area (m^2)
A_O’	Immersed area of SPP (m^2)
A_P	Projected propeller blade area (m^2)
A_P	Projected planing bottom area (m^2)
A_T	Transom area (m^2)
A_V	Above-water transverse area projection
A_W	Area exposed to wind (m^2)
A_X	Maximum section area (m^2)
Ax	Area of hull openings (catamaran) (m^2)
$A_P/\nabla^{2/3}$	Planing area coefficient
A_T/A_X	Transom area ratio
BAR	Blade area ratio
B_{EF}	Effective planing beam (m)
B_M	Beam at midship ($L_P/2$) (m)
$B_{PA} = A_P/L_P$	Mean beam over chines (m)
B_{PT}	Projected chine beam at transom (m)
B_{PX}	Maximal projected chine beam (m)
$B_{WL} = B = B_X$	Beam of hull on DWL (m)
B_{XDH}	Maximum beam of demihull (catamaran) (m)
c	Section chord (m)
$C = 30.1266 \cdot V/(L)^{1/4} \cdot (\Delta/2P_E)^{1/2}$	C-factor (note: V in kn; not non-dimensional coef.)
C_A	Correlation allowance
C_{AA}	Air resistance (allowance) coefficient
CA_P	Centroid of A_P forward of transom (m)
$C_B = \nabla/(L \cdot B \cdot T)$	Block coefficient
C_{DP}	Structural roughness allowance coefficient

C_F	Specific frictional resistance coefficient
C_R	Specific residuary resistance coefficient
$C_S = S/(\nabla \cdot L)^{1/2}$	Taylor wetted surface coefficient
$C_{T\nabla} = R_T/(\rho/2 \cdot v^2 \cdot \nabla^{2/3})$	Total resistance coefficient
$C_T{*}= T/(\rho/2 \cdot v_{0.7R}^2 \cdot A_O)$	Thrust index (coefficient) of propeller
$C_Q{*} = Q/(\rho/2 \cdot v_{0.7R}^2 \cdot A_O \cdot D)$	Torque index (coefficient) of propeller
$C_X = A_X/(B \cdot T)$	Maximum section area coefficient
$C_\Delta = \nabla/B_{PX}^3$ or ∇/B_X^3	Beam load coefficient
$C_\nabla = C_{DL} = \nabla/(0.1 \cdot L)^3$	Volume displacement coefficient (also in use ∇/L^3)
d	Shaft diameter (m)
D	Propeller diameter (m)
$DAR = A_D/A_O$	Developed area ratio
$EAR = A_E/A_O$	Expanded area ratio
DWL	Designed waterline at rest
$Fn_B = C_V = v/(g \cdot B_{PX})^{1/2}$	Beam Froude number
$Fn_h = v/(g \cdot h)^{1/2}$	Depth Froude number
$Fn_L = Fn = v/(g \cdot L_{WL})^{1/2}$	Length Froude number
$Fn_\nabla = v/(g \cdot \nabla^{1/3})^{1/2}$	Volumetric Froude number
$Fn_{\nabla/2} = v/(g \cdot (\nabla/2)^{1/3})^{1/2}$	Volumetric Froude number of demihull (catamaran)
g	Acceleration of gravity (m/s^2)
h	Water depth (m)
h	Immersion of SPP (m)
h/D	Immersion ratio of SPP
$H_{1/3}$	Significant wave height (m)
i_e	Half-angle of entrance of waterline at bow (deg)
$J = v_a/(n \cdot D)$	Advance coefficient
$J_\Psi = v_a \cdot \cos\Psi/(n \cdot D)$	Revised advance coefficient for SPP
$K_T = T/(\rho \cdot n^2 \cdot D^4)$	Thrust coefficient
$K_T' = T/(\rho \cdot n^2 \cdot D^2 \cdot A_O')$	Revised thrust coefficient for SPP
$K_Q = Q_O/(\rho \cdot n^2 \cdot D^5)$	Torque coefficient
$K_Q' = Q_O/(\rho \cdot n^2 \cdot D^3 \cdot A_O')$	Revised torque coefficient for SPP
l	Length of shaft or bossing (m)
$L = T_a \cdot \sin(\psi+\tau) + N \cdot \cos(\psi+\tau)$	Vertical propeller force (lift) (kN)
L_C	Chine wetted length (m)
L_K	Keel wetted length (m)
$L_M = (L_K + L_C)/2$	Mean wetted length (m)
L_{OA}	Length overall (m)
L_P	Projected chine length (m)
L_{PP}	Length between perpendiculars (m)
$L_{WL} = L$	Waterline length (m)
L_P/B_{PX} or L_{WL}/B_{WL}	Length–beam ratio
$L_P/\nabla^{1/3}$ or $L_{WL}/\nabla^{1/3} = (M)$	Slenderness ratio

$L_P/(\nabla/2)^{1/3}$	Slenderness ratio of demihull (catamaran)
LCB	Longitudinal center of buoyancy (m)
LCG	Longitudinal center of gravity forward of transom (m)
LCG/L_P	Longitudinal center of gravity relative to transom
$\%LCG = (CA_P - LCG) \cdot 100/L_P$	Longitudinal center of gravity aft of A_p centroid (%)
N	Force normal to propeller shaft line (kN)
n	Propeller rotational speed (1/sec)
P	Propeller pitch (m)
P/D	Pitch–diameter ratio
P_B	Brake power (kW)
P_D	Delivered power (kW)
P_E	Effective power (kW)
$P_E{}^*$	Effective in-service power (kW)
p_a	Atmospheric pressure (kPa)
$p_h = \rho \cdot g \cdot h$	Static water pressure (kPa)
p_v	Vapor pressure of water (kPa)
Q	Propeller torque behind the vessel (kNm)
Q_O	Propeller torque (kNm)
$Q_C = Q/(\rho/2 \cdot D \cdot A_P \cdot v_{0.7R}^2)$	Torque load coefficient
RCG	Rise of center of gravity
R_F	Frictional resistance (kN)
$Rn = v \cdot L/\nu$	Reynolds number
$Rn_d = v \cdot D/\nu$	Reynolds number based on shaft diameter
R_{AA}	Air and wind resistance (kN)
R_{AP} (R_{App} in some refer.)	Appendage resistance (kN)
R_{PAR}	Parasitic resistance (kN)
$R_R = R_{Rd}$	Residuary resistance in deep water (kN)
R_{Rudder}	Appendage resistance due to rudder (kN)
R_{SB}	Appendage resistance due to strut bossing (kN)
R_{Sh}	Appendage resistance due to propeller shaft (kN)
R_{Skeg}	Appendage resistance due to skeg (kN)
R_{St}	Appendage resistance due to strut (kN)
$R_T = R = R_{TBH}$	Total bare hull resistance (kN)
$R_T{}^*$	Total in-service resistance (kN)
$R/\Delta = (R_T/\Delta)_{100000}$	Resistance-to-weight ratio (for Δ=100,000 lb = 45.36 t)
R_W	Wave-making resistance (kN)
R_{Wh}/R_{Wd}	Shallow water resistance factor
S	Wetted surface area (m^2)

S_R	Profile area of rudder or strut as seen from one side (m^2)
$(S) = S/\nabla^{2/3}$	Wetted surface area coefficient
$WSC = S/(\nabla/2)^{2/3}$	Wetted surface area coefficient (catamaran)
t (t_X in some refer.)	Thrust deduction fraction
t	Section thickness (m)
$T = T_H$	Hull draught at DWL (m)
T	Propeller thrust (kN)
T_a	Axial propeller force (kN)
$T_h = T_a \cdot \cos(\psi+\tau) - N \cdot \sin(\psi+\tau)$	Horizontal propeller force (kN)
v	Velocity of craft (m/s)
V	Velocity of craft (kn)
$v_a = v \cdot (1 - w)$	Speed of advance of propeller (m/s)
v_w	Wind speed (m/s)
$v_{0.7R} = [v_a^2 + (0.7\pi nD)^2]^{1/2}$	Resultant water velocity at 0.7R (m/s)
w (w_T in some refer.)	Wake fraction
z	Number of propeller blades
$\beta = \arctan [v_a/(0.7 \cdot \pi \cdot n \cdot D)]$	Hydrodynamic pitch angle at 0.7R
β_{EF}	Effective deadrise angle (deg)
$\beta = \beta_M$	Deadrise angle at midship ($L_P/2$) (deg)
β_{Bpx}	Deadrise angle at B_{PX} (deg)
β_T	Deadrise angle at transom (deg)
γ	Buttock angle (average centerline angle from $L_P/2$ to transom) (deg)
δ_W	Angle of transom wedge (catamaran) (deg)
$\Delta = \nabla \cdot \rho$	Displacement, mass (tons)
ΔC_F	Roughness allowance
$\Delta K_T = K_{Tatm} - K_{Tcav}$	K_T Reduction for cavitating conditions
$\Delta K_Q = K_{Qatm} - K_{Qcav}$	K_Q Reduction for cavitating conditions
∇	Displacement volume (m^3)
$\varepsilon = \beta_M - \beta_T$	Warp angle (according to DUT terminology twist angle) (deg)
$\varepsilon_B = P_B/(\Delta \cdot g \cdot v)$	Specific break power (catamaran)
ε_R	Residual resistance–weight ratio (catamaran)
η_B	Propeller efficiency behind the vessel
$\eta_D = P_E/P_D$	Propulsive efficiency (quasi-propulsive efficiency)
$\eta_H = (1-t)/(1-w)$	Hull efficiency
η_M	Gear and shaft losses
$\eta_O = (K_T/K_Q) \cdot (J/(2\pi))$	Propeller open-water efficiency
$\eta_P = \eta_H \cdot \eta_R \cdot \eta_S \cdot \eta_O$	Overall (total) propulsive coefficient (OPC)
$\eta_R = \eta_B/\eta_O$	Relative rotative efficiency
η_S	Shaft efficiency (including gearing efficiency)
ν	Kinematic viscosity of water (m^2/s)
λ	Scale ratio

ρ	Mass density of water (kg/m^3)
ρ_A	Mass density of air (kg/m^3)
$\sigma = (p_A + p_H - p_V)/(\rho/2 \cdot v_a^2)$	Cavitation number based on advance velocity
$\sigma_{0.7R} = (p_A + p_H - p_V)/(\rho/2 \cdot v_{0.7R}^2)$	Cavitation number based on resultant water velocity at 0.7 radius
τ (θ in some refer.)	Dynamic (running) trim relative to its value at zero speed (deg)
$\tau_{BL} = \tau_O + \tau$	Baseline trim angle (deg)
$\tau_c = T/(\rho/2 \cdot A_P \cdot v_{0.7R}^2)$	Thrust load coefficient
τ_O	Initial static baseline trim (deg)
Ψ	Shaft inclination relative to buttock (deg)

Subscripts

h	Finite water depth (shallow water)
d	Infinite water depth (deep water)
TK	Tank conditions
TR	Trial conditions

Chapter 1
Introduction

1.1 Objectives

The main goals of this book are to:

- Review various statistically based Mathematical Models (MM) for power prediction
- Spotlight some very useful MMs
- Encourage the HSC designer to use existing MMs.

The reason for the abovementioned objectives is the authors' belief that a large number of recently published papers are too complex to be useful in everyday practice. As a consequence, practicing naval architects in need of a power prediction, are indirectly forced to rely on commercial software whose essence is often not properly understood. Moreover, the few decades-old experimental results, MMs, etc., are often considered to be archaic, particularly by the younger engineers, and are hence frequently marginalized, although they have not been replaced by better MMs or routines.

Moreover, MM development is an evolutionary process, i.e. MM developers should be familiar with what their predecessors have done. Thus, one of the objectives of this work is to aid new MM developers by reviewing the existing models along with their principal characteristics and tradeoffs.

The core of this work are statistically based MMs based on the results of model experiments of various HSC series. A variety of regression analysis and lately Artificial Neural Network (ANN) methods were used to develop these MMs. The resulting MMs can be easily programmed into software tools, thereby enabling the parametric analyses required for design optimization.

D. Radojčić et al., *Power Prediction Modeling of Conventional High-Speed Craft*,
https://doi.org/10.1007/978-3-030-30607-6_1

1.2 Conventional High-Speed Craft (HSC)[1]

The term *conventional HSC* is applied here to the high-speed craft of displacement, semi-displacement, and planing types which achieve speeds that include and/or exceed the main resistance hump. This corresponds roughly to the length Froude number $Fn_L > 0.4$, volume Froude number $Fn_\nabla > 1$, and beam Froude number $Fn_B > 0.5$, approximately resulting in the following classifications:

- IMO according to which "HSC is a craft capable of maximum speed equal to or exceeding $3.7 \cdot \nabla^{0.1667}$ m/s" (which is actually equivalent to $Fn_\nabla > 1.18$), and
- ITTC according to which "high-speed marine vehicles are defined to be vessels with a design speed corresponding to a Froude number above 0.45, and/or a speed above $3.7 \cdot \nabla^{0.1667}$ m/s, and/or where high trim angles are expected, or for dynamically supported vessels".

The lowest speed for the dry or fully-vented transom approximately corresponds to $Fn_L > 0.3$ or $Fn_\nabla > 1$ (Blount 2014).

The relatively wide speed range of HSC should be emphasized. Namely, a single vessel may travel in displacement ($Fn_L < 0.40$), semi-displacement ($0.40 < Fn_L < 0.65$), semi-planing[2] ($Fn_L > 0.65$ but $Fn_\nabla < 3.0$), and pure planing regimes ($Fn_\nabla > 3.0$). Note that the approximate Fn values given in parenthesis are typical for slenderness ratio $L/\nabla^{1/3} \approx 6.0$; see Blount (1995, 2014). Each regime has its peculiarities, so that different parameters are necessary to model the performance for each of them. For instance, for speeds below $Fn_L \approx 1$ it is better to use Fn_L than Fn_∇ and vice versa (see above). This makes modeling relatively difficult, as it is desirable to describe the performance over the entire sailing regime with the same parameters (input variables) and if possible, with a single continuous equation.

1.3 Resistance, Propulsion, and Power Prediction

Typically, one of the main design objectives is the minimization of power. To achieve this, optimization of the whole system—including resistance, propulsion, and engine (with the gearbox)—is necessary, since separate optimization of the components, in isolation to the rest of the system, does not necessarily result in the

[1]According to the 16th ITTC HSMV Panel, HSC are grouped and divided into following types: (a) Hydrofoils, (b) Hard chine planing craft, (c) Round bilge semi-planing, (d) SWATH ships, (e) Air Cushion vehicles (amphibious), (f) Surface Effect Ships (non-amphibious), and (g) Others. HSC belonging to groups (b) and (c) are treated here.

[2]According to Savitsky (2014) "Vessels operating in the speed range between hull-speed and planing inception speed are totally supported by buoyant forces, hence the use of the terms 'Semi-Planing' or 'Semi-Displacement' hulls is inappropriate". Nevertheless, the authors of present book decided to continue using those well-established terms, for reasons of continuity and tradition.

same answer. The holistic system optimization, i.e. *complex* or *integrated approach*, is much more important for the HSC than for conventional displacement vessels. However, in order to achieve some clarity, the current work presents the subject in the conventional way, i.e. MMs for resistance and propulsion predictions are presented separately, while the integrated approach is elaborated in Chap. 6. Separate treatment of resistance and propulsion enables independent investigation of the influence of hull and of propeller parameters on hydrodynamic performance. With an integrated approach these influences have to be examined simultaneously, which significantly complicates the evaluation.

One of the key lessons learned is that power prediction for a desired speed (or vice versa, speed prediction for installed power) must be considered from the very early design phases. Moreover, initial predictions of speed should be rechecked whenever any design changes or modifications that affect performance are initiated. Thus, neglecting power prediction during the various design phases often results in degraded performance, and the actual achieved speed is almost certainly below the predicted speed.

For power prediction (P_B) evaluation of in-service total resistance (R_T*) and overall propulsive efficiency (η_P) are necessary as

$$P_B = P_E^*/\eta_P = R_T^* \cdot v/\eta_P$$

where P_E and P_B are effective and brake powers respectively. Note that routines that model bare hull resistance are usually valid for ideal test conditions in calm and deep water (denoted here as R_T). Various additional factors should be taken into account to predict actual in-service performance (e.g. appendages, air resistance, waves, restricted waterways, etc.) and "*" is used after P_E and R_T to denote this. Detailed subdivision of HSC total resistance, as well as various components that form R_T* are given, for instance, in Müller-Graf (1997a, b) and Molland et al. (2011), and will be discussed later.

The most significant portion of the overall propulsive efficiency (η_P) is the open water efficiency (η_O) of the selected, presumably optimal, propeller. Assessment of a propeller's open water efficiency, however, is not a straight forward procedure; it is actually a separate task to determine the best (i.e. optimal) propeller for a given set of requirements. The optimal propeller should produce thrust that overwhelms in-service resistance for both design and off-design conditions. Consequently, the open water efficiency of a propeller is a result of a complete propulsion analysis.

As is well known, the dynamic trim angle (τ) is very important for HSC resistance, propulsion, and performance in general. In a way, hull resistance mirrors the dynamic trim angle, and hence dynamic trim angle evaluations usually go hand-in-hand with the resistance evaluations. The relationship between hull resistance and dynamic trim is particularly pronounced for the hump speeds, which HSC cannot avoid, so that modeling this range is of the utmost importance.

Thus, the most significant factors that must be reliably evaluated for power prediction are:

- Bare hull resistance (R_T) and Dynamic trim (τ), and
- Propeller's open water efficiency (η_O).

Hence, this work is focussed on the mathematical modeling of these factors.

Other quantities in the abovementioned equation are also important, but are typically less significant. They are essential parts of power prediction routines, but are not the subject of this work per se. For the sake of completeness, these components are briefly discussed in Chap. 5.

1.4 Not Treated Topics

1.4.1 Resistance Evaluation Using Empirical Methods

Not addressed in this work is the Savitsky (1964) method. It is based on equations for prismatic hull forms and is by far the most frequently used amongst various empirical methods. The other planing hull resistance prediction methods are mentioned in Almeter (1993). Savitsky's method is applicable for higher planing speeds where hydrodynamic forces are dominant. Note that Savitsky's method was modified a few times (see for instance Blount and Fox 1976; Savitsky 2012).

1.4.2 Resistance Evaluation Using Computational Fluid Dynamics (CFD)

It is believed that the CFD-based methods will become common everyday tools in the future. At this time however, CFD still depends very much on interpretation of the simulated results by the user. Therefore, these methods are typically not yet sufficiently mature to be used by regular engineers in everyday engineering practice. CFD's subjective nature (Molland et al. 2011; Almeter 2008) is also not yet practical for application within broader numerical optimization tools (typically nonlinear multi-criterion optimization with constraints), where evaluations of resistance and propeller efficiency are just segments of an integrated approach. CFD applied to HSC is given for instance in Savander et al. (2010), Brizzolara and Villa (2010), Garo et al. (2012), De Luca et al. (2016).

Actually, CFD and the MMs treated here are complementary methods, although they are fundamentally different techniques. Namely, MMs based on experimental data provide a low-cost and reasonably accurate preliminary design tool. If needed, further analysis, improvements and hull/propeller adjustments should be done using CFD and/or tow tank tests. Furthermore, designers typically do not do their own

CFD evaluations; these are usually done by the CFD specialists, thereby resulting in 'once-removed' relationships that are similar to experimental facilities (tow tank, cavitation tunnel etc.). This may change in the future as CFD tools become more practical and useful to designers.

The state-of-the-art viewpoint on statistical power performance predictions and CFD, is given by Van Hees (2017):

- Statistical methods are fast, while CFD (including 3D hull modeling etc.) require time.
- In practice, CFD is used to supplement statistical methods, with the objective to further optimize the hull form.

1.4.3 Other Excluded Topics

1.4.3.1 High-Speed Ships

MMs based on the experimental series with hull forms that resemble displacement ships more than HSC are not treated here, although they are valid for relatively high speeds (Fn_L approaching 1 or so). In other words MMs for high-speed ships are not the subject of this work (for instance, Fung and Leibman 1993; Bojović 1997, etc.).

1.4.3.2 Not Released Mathematical Models

MMs for hull/propeller series for which complete and usable MMs have not been released are excluded. For instance, the following high-speed hull and propeller series are valuable, but have not been addressed here:

- MARIN systematic series of fast displacement hulls consisting of no less than 33 models, see Kapsenberg (2012), or
- SVA high-speed, 3-bladed, inclined shaft propeller series consisting of 12 models; see Heinke et al. (2009).

1.4.3.3 Commercial Software

MMs found in commercial software programs are also excluded. Note, however, that most of them are based on the routines that are discussed here.

1.4.3.4 Waterjets

Waterjets in general, as for their sizing cooperation with the waterjet manufacturer is usually required.

1.5 Common Mistakes

The most common power prediction mistakes of the so called statistical methods which are treated here are:

- An incorrect prediction model (MM) is selected, i.e. the vessel under analysis has different characteristics than those upon which the MM is based.
- Violation of the boundaries of applicability of MM.

Therefore, it is important to know how a particular MM was developed, the constraints and assumptions it used in formulation. This information is often missing, particularly when commercial power prediction software tools are used. For example, several very important hull or propeller characteristics may be "masked" (i.e. MMs are inherently valid for a particular hull form or propulsory type). These hidden characteristics, which may not be required by a MM should be regarded as additional and prescribed quantities that limit the applicability of a given MM. In other words, wise usage of readymade computer programs requires that the designers are familiar with the characteristics of the hull- and propeller-series the MM is based on, as well as with the technique used for its derivation.

MacPherson (1993) gives a good review of dos and don'ts concerning numerical prediction techniques. Some recommendations deserve to be cited:

- Not all methods are appropriate for all problems.
- Know your prediction method. The numerical procedure must be fully understood.
- Complete and reliable program cannot ignore the user. The human interface is very important.
- Numerical methods cannot eliminate model testing.

1.6 Book Arrangement

This book is arranged in two parts. The first part covers mathematical modeling of resistance, dynamic trim and propeller hydrodynamic characteristics of HSC (Chaps. 3 and 4). This is followed by brief treatment of additional resistance components, propulsive coefficients and power prediction (Chaps. 5 and 6).

Concluding remarks are given in Chap. 7. The first part (Chaps. 2–7) is entitled "Reflections on Power Prediction Modeling of Conventional High-Speed Craft" and is essentially the only textual part of the book.

The second part (Chaps. 8–10), entitled "Resistance and Propulsor Design Data with Examples" presents Excel Macro Codes for evaluation of resistance and propulsion characteristics. It provides tabulated data for assessment of mathematical models of resistance and propeller efficiency (Chaps. 8 and 9 respectively) which are outlined in the first part. MMs for evaluation of additional resistance components and propulsive coefficients are given in Chap. 10. Worked example applications are given for number of interesting cases.

The relevant Excel files can be provided to the reader as Electronic Supplementary Material (ESM). The reader can access ESM via the Springer sites which are given in Chaps. 8, 9 and 10.

References

Almeter JM (1993) Resistance prediction of planing hulls: state of the art. Marine Technol 30(4)

Alemeter JM (2008) Avoiding common errors in high-speed craft powering predictions. In: 6th international conference on high performance marine vehicles, Naples

Blount DL (1995) Factors influencing the selection of hard chine or round bilge hull for high froude numbers. In: Proceedings of the 3rd international conference on fast sea transportation (FAST'95), Lubeck-Travemunde

Blount DL (2014) Performance by design. ISBN 0-978-9890837-1-3

Blount DL, Fox DL (1976) Small craft power prediction. Marine Technol 13(1)

Bojović P (1997) Resistance of AMECRC systematic series of high-speed displacement hull forms. In: High Speed Marine Vehicles Conference on (HSMV 1997), Sorrento

Brizzolara S, Villa D (2010) CFD simulation of planing hulls. In: 7th international conference on high performance marine vehicles, Melbourne Florida

De Luca F, Mancini S, Miranda S, Pensa C (2016) An extended verification and validation study of CFD simulations for planing craft. Ship Res 60(2)

Fung SC, Leibman L (1993) Statistically-based speed-dependent powering predictions for high-speed transom stern hull forms. Chesapeake Section of SNAME

Garo R, Datla R, Imas L (2012) Numerical simulation of planing hull hydrodynamics. In: SNAME's 3rd chesapeake power boat symposium, Annapolis

Heinke HJ, Schulze R, Steinwand M (2009) SVA high speed propeller series. In: Proceedings of 10th international conference on fast sea transportation (FAST 2009), Athens

Kapsenberg G (2012) The MARIN systematic series fast displacement hulls. In: 22nd international HISWA symposium on yacht design and yacht construction, Amsterdam

MacPherson DM (1993) Reliable performance prediction techniques using a personal computer. Mar Technol

Molland AF, Turnock SR, Hudson DA (2011) Ship resistance and propulsion—practical estimation of ship propulsive power. Cambridge University Press. ISBN 978-0-521-76052-2

Müller-Graf B (1997a) Part I: resistance components of high speed small craft. In: 25th WEGEMT School, Small Craft Technology, NTUA, Athens. ISBN Number: I 900 453 053

Müller-Graf B (1997b) Part II: powering performance prediction of high speed small craft. In: 25th WEGEMT School, Small Craft Technology, NTUA, Athens. ISBN Number: I 900 453 053

Savander BR, Maki KJ, Land J (2010) The effects of deadrise variation on steady planing hull performance. In: SNAME's 2nd Chesapeake Power Boat Symposium, Annapolis

Savitsky D (1964) Hydrodynamic design of planing hulls. Marine Technol 1(1)
Savitsky D (2012) The effect of bottom warp on the performance of planing hulls. In: SNAME's 3rd Chesapeake Power Boat Symposium, Annapolis
Savitsky D (2014) Semi-Displacement Hulls – A Misnomer?. SNAME's 4th Chesapeake Power Boat Symposium, Annapolis
Van Hees MT (2017) Statistical and theoretical prediction methods. In: Encyclopedia of maritime and offshore engineering. Wiley

Part I
Reflections on Power Prediction Modeling of Conventional High-Speed Craft

Part I

[illegible]
[illegible]

Chapter 2
Mathematical Modeling

2.1 Statistical Modeling

Mathematical modeling which is of interest for present work belongs to the *predictive modeling* class, as opposed to *explanatory* or *descriptive* modeling. According to Shmueli (2010): "Predictive modeling is a process of applying a statistical model or data mining algorithm to data for the purpose of predicting new or future observations.... The goal is to predict the output value (Y) for new observations given their input values (X)". The modeling process segregated into a set of steps is described in Fig. 2.1 (Shmueli 2010).

Note that the first three steps are usually performed by one team and the rest by another, although it would be desirable if the entire process was executed by a single multidisciplinary team. In addition, the entire modeling process (Steps 1–8) is usually performed solely by the engineers, although knowledge of subject-matter mathematics, specifically statistics, is desirable (see for instance Weisberg 1980; Draper and Smith 1981).

The basic steps given in Fig. 2.1 are clear and logical, and MM developers naturally follow them. Note that Step 4, abbreviated EDA (Exploratory Data Analysis), usually requires transformation of the available data into a format suitable for mathematical modeling. EDA is a very important step because various variables, and their eventual transformations, should be considered at this point in the process. The choice of dependent (target), and most influential independent (input) variables (Step 5); statistical data modeling tool, i.e. MM extraction methods (Step 6); evaluation and selection of final MM amongst several considered (Step 7); and use of recommended MM together with reporting (Step 8); all follow Step 4 and are therefore affected by the decisions made there.

D. Radojčić et al., *Power Prediction Modeling of Conventional High-Speed Craft*,
https://doi.org/10.1007/978-3-030-30607-6_2

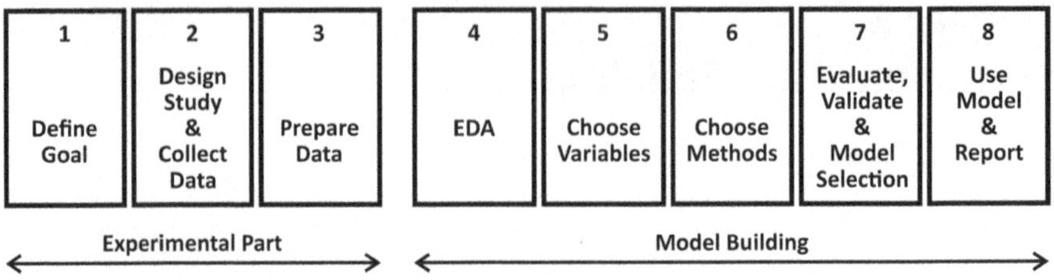

Fig. 2.1 Basic steps in statistical modeling

Datasets as treated here are usually scarce, so that data partitioning[1] is often avoided and all available data is used for building of the MM. This, however, requires thorough MM validation (Step 7), particularly stability checking (possibility of waving between the tested values). Moreover, the use of the entire dataset produces repeatable results, as the holdout samples are usually randomly selected.

The reliability of a MM (*predictive accuracy* or *predictive power* according to Shmueli 2010) is very important, particularly when MM is applied to everyday design problems when the correct value is not known, and the user relies on the MM's predictive accuracy. Therefore, during the development phase, the following should be checked in order to verify the derived model:

- Statistics of the accuracy of the model versus the data set used to develop the model
- Discrepancies between evaluated and measured values
- Behaviour of the model between the data points, where there are no measurements (naturally within the applicability boundaries).

2.1.1 *Statistical Modeling Applied to Ship Data*

The authors of this work never considered parameters of no physical significance or no physical meaning to be primary input variables, despite their possible high statistical correlation. Correlation analysis among the independent variables and versus the dependent variable, was always performed in order to ensure the validity of the selection. When methodical series data is used, the input variables are usually the same as the parameters varied during the model-based series' testing.

Note that the main disadvantage of the statistical data modeling tools is that only a limited number of variables are used to adequately describe the HSC's hull form and loading over a relatively wide speed range. For this reason, the secondary hull

[1]Data partitioning, to data that are used for MM development (often called the training set), and to the holdout sample which MM "did not see" (expression from Shmueli 2010) is not unusual. The purpose of the holdout sample is to validate the MM, thus it is also named the validation set.

form parameters are important and should be regarded as a kind of supplement to the MM.

To clarify, the input parameters for a HSC could for example be $L/\nabla^{1/3}$, L/B, and B/T. These hull and loading parameters, although most significant, do not reflect the hull form, i.e. whether it is a hard chine or a round bilge type, and if hard chine then whether it is wide- or narrow-transom, etc. Additional hull description is obviously necessary. This is provided through the secondary hull parameters. However, the secondary hull parameters are often not explicitly specified, but the MM is instead described as being valid for, for instance, the NPL series. This therefore means that the MM is based on the semi-displacement round bilge hulls whose secondary parameters are $C_B = 0.397$, LCB = 6.4% L aft. amidship, $A_T/A_X = 0.52$, etc. (see Table 3.1). Thus, the additional information typically given in the form of a comment such as “MM is valid for (or is based on) the NPL series” is a very important supplement of the MM. Similarly, the secondary parameters for the propellers would include information about the blade shape and section etc. Naturally, the MM user must be aware of this fact.

Statistical power performance predictions for conventional ships, with a focus on the developmental philosophy of prediction methods, are discussed in a related paper (Van Hees 2017). Amongst the observations is that the statistical methods should be “refreshed” when some new data is available.

2.2 Model Extraction Tools

The authors used two methods—statistical data modeling tools—to extract (i.e. develop) the mathematical models for prediction of resistance and propulsive coefficients:

- Regression analysis, and
- Artificial Neural Networks (ANN).

Note that there are several types of regression and ANN methods, but further elaboration on this topic is beyond the scope of the present text.

2.2.1 Regression Analysis

With regression analysis, and in particular with the linear multiple regression analysis (e.g. Weisberg 1980), the independent variables consist of two sets of input data:

- Basic independent variables, and
- Various transformations of basic independent variables.

For the simplest case, with only two basic independent variables X_1 and X_2 (e.g. assuming $X_1 = L/\nabla^{1/3}$, $X_2 = Fn_\nabla$ and dependent variable $Y = R/\Delta$), the additional transformed variables could then be $X_3 = X_1X_2$, $X_4 = X_1^2$, …. Thus the number of terms in the initial polynomial equation rapidly increases i.e.

$$Y = a_0 + a_1X_1 + a_2X_2 + a_3X_3 + a_4X_4 + \cdots + a_nX_n$$

where a_0, a_1, … a_n are the coefficients determined by the regression analysis. Note that this is a linear equation, although the basic variables can be transformed for the purpose of simulating nonlinear relationships.

Many different transformations of both independent and dependent variables have been tested by the authors, forming equations of, for instance, logarithmic, exponential, or reciprocal types that produced different curves. However, it has been found that when the number of equation terms is relatively high, and the cross-products and different powers of independent variables are used, then there are no great differences in the results. That is, the original polynomial form is sufficient.

Therefore, despite of the fact that the number of basic parameters is typically five or less, the initial polynomial equation, produced by standard automated polynomial fitting, may have 100 or more terms. Consequently, if a certain hull characteristic is not represented directly through the basic hull parameters, then it will most likely be represented indirectly, through one of the many polynomial terms that appear in that initial equation.

By applying a step-by-step procedure and statistical analysis, a subset of significant terms is chosen and less significant variables are eliminated, resulting in the final equation. This final equation, often called "the best equation", comprises considerably fewer independent variables than the initial polynomial equation. The weakness of this approach is that it assumes that there is a single optimal subset of terms in the equation; whereas in fact, usually there may be several sets of terms that work equally well.

In general, more terms in the equation enable better fitting to the data that the MM is based upon, but the interpolated results (those between the data points) may be poorer. That is, a larger number of terms results in more waving in the function, i.e. a less-smooth curve or surface.

The first author's experience spans the period from early custom-made programs written in BASIC through to the commercial PC software now in common use, with the regression routines evolving from the so called *backward elimination*, to *forward selection*, and to the more sophisticated *multiple stepwise methods* (combination of previous). This experience indicates that successful application of any of the approaches requires the user to have sufficient subject matter knowledge to choose the best subset (i.e. to obtain the "best equation"). Implementation of the abovementioned process actually does require an understanding of statistics and statistical methods.

2.2.2 *Artificial Neural Network (ANN)*

The Artificial Neural Network is a nonlinear statistical data modeling technique that can be used to determine the complex relationships between dependent and independent variables. A brief discussion of the potentials of ANN techniques, and probably some of the first applications of the novel ANN tools in marine design and modeling, are given in Mesbahi and Atlar (2000), Mesbahi and Bertram (2000), Koushan (2001), etc. Amongst the conclusions drawn was that ANN can be successfully used as an alternative to regression analysis.

There is a family of ANN methods that may be used for the derivation of the MMs. The authors used a software tool aNETka 2.0 (see Zurek 2007) based on a feed-forward type of ANN routine with a back-propagation algorithm; see Rojas (1996) for instance.

With ANN more attention is paid to the selection of independent variables than with regression analysis (see Radojčić et al. 2014). Specifically, the independent variables must be carefully chosen at the very onset of the fitting process, because the final model is based on the selected input parameters (which form the input layers for ANN). Use of incorrect, or insufficient, independent variables may result in an erroneous MM, with for example, the dependent variable being insensitive to the variations in a poorly-chosen set of input variables. On the other hand, if too many independent variables are assumed, validation of the model stability becomes considerably more complex.

With ANN, in order to find a 'good' solution, in terms of accuracy, reliability, or applicability, the number of layers and number of neurons in each layer must be selected by the user (MM model developer) in advance; see Fig. 2.2. The number of layers and type of activation function are usually constrained by the software used.

Activation or transfer function convert the input signal of a node to an output signal, which then becomes the input signal for another node in the next layer. The nonlinear activation function enables nonlinear transformation between input and output. It also facilitates neural network learning. Without it, the neural network would be a kind of linear regression model. Various activation functions, including linear, sigmoid, and hyperbolic tangent functions were tested, and the sigmoid function ($sig = \frac{1}{1+e^{-x}}$, S-shaped curve) was found to produce the best results and was hence adopted by the authors.

An illustration of ANN application to catamaran resistance is given in two related papers by the same authors (Couser et al. 2004; Mason et al. 2005). Different ANN network architecture was investigated, i.e. number of layers and number of neurons were systematically varied. ANN was implemented directly in treatment of the test data, so that multidimensional data smoothing was obtained as a side effect.

ANN technique with multiple outputs (see Fig. 2.2), as opposed to a single output, were applied lately for generating MMs for R/Δ and τ, with very encouraging results (see Radojčić et al. 2017; Radojčić and Kalajdžić 2018).

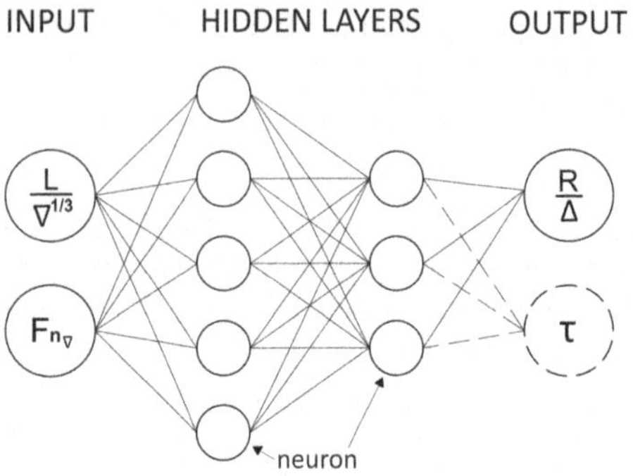

Fig. 2.2 ANN with 2-5-3-1 structure for single-output, or 2-5-3-2 structure (*dashed lines*) for multiple-output. Input variables are $L/\nabla^{1/3}$ and Fn_∇, while output variables are R/Δ, or R/Δ and τ, for single- and multiple-output, respectively

2.3 Hardware

Statistically based power prediction methods emerged with the proliferation of the use of computers in everyday engineering practice. In the early days of MM development computer power and its peripherals were an important factor, primarily because in practice they constrained the complexity of the MMs. For example, the first MMs developed by the first author in the 1980s, were done using successively, SHARP MZ80K (RAM 32 kb, with an audio-tape as external memory), Apple II + (RAM 48 kb, floppy drive), and Olivetti M21 (RAM 640 kb, hard disk). The MMs produced were relatively simple compared to the modern models, and those hardware platforms would be more than insufficient nowadays[2]. Note however that despite the fact that the computational capability has subsequently evolved exponentially, model development is still a lengthy process since the complexity of MMs has also grown. Thus, the selection of the "best equation" amongst many good options, stability checking, validation etc. is still a complex and a time-consuming job, regardless of the virtually unlimited computational power available today.

2.4 Conclusions on Mathematical Modeling[3]

2.4.1 General Conclusions

Both statistical and physical perspectives are important for correct modeling. Consequently, formation of good MMs requires the interdisciplinary approach, i.e. specific knowledge of:

[2]Incidentally, only ten or so years earlier, the 1969 Moon-landing Apollo 11 Guidance Computer (AGC) had a RAM of 2 kb running at 1.024 MHz!

[3]These are the summary conclusions obtained during the derivation of the mathematical models discussed here, hence some of these will also be mentioned later.

- Physics of HSC hydrodynamics.
- Statistics, curve fitting, regression analysis, ANN technique etc.
- Procedures for developing and checking MM (choosing “the best equation” out of the many derived candidates).
- Optimization techniques (i.e. what type of MM can be used in a larger numerical optimization tool—type of numerical expression suitable for computer evaluation, numerical boundaries of applicability etc.).

2.4.2 Conclusions on Application of Regression Analysis and ANN

- For extraction of MMs, ANN requires less of the user’s manual interference than regression-based methods.
- Regression Analysis seems to be more convenient than ANN for modeling simpler relationships with the lesser number of input variables and vice versa.
- Regression analysis (stepwise method) allows screening and rejection of less significant polynomial terms, which is not the case with ANN, where the number of terms is defined at the very beginning.
- Regression based MMs were stiff, at least the MMs of polynomial form. ANN based MMs, with complex relationship between dependent and independent variables and with many equation terms, proved to be more elastic. Moreover, the interpolated data (those between the data points) did not show instability.
- ANN with a larger number of hidden layers usually produced MMs which could better fit the data, but in return those MMs were usually not stable between the data points.
- MMs developed by ANN, as opposed to those developed by regression analysis, allowed limited extrapolation[4], although as a rule bounds of applicability should not be violated. This is due to the complexity of MMs derived by ANN which may not show an abrupt change of character beyond the applicability zones.
- Once the MM is developed and validated, no further knowledge of ANN or regression technique is needed. Although this sounds logical, the authors’ experience is that potential MM users tend to be hesitant whenever ANN is mentioned, probably because ANN, in general, has broad applicability and requires specific expertise.

[4]Similar conclusion was derived in Couser et al. 2004.

References

Couser P, Mason A, Mason G, Smith CR, Konsky BR von (2004) Artificial neural network for hull resistance prediction. In: 3rd international conference on computer and IT applications in the maritime industries (COMPIT' 04), Siguenza

Draper N, Smith H (1981) Applied regression analysis, 2nd edn. Willey

Koushan K (2001) Empirical prediction of ship resistance and wetted surface area using artificial neural networks. In: Cui W, Zhou G, Wu Y (eds) Practical Design of Ships and Other Floating Structures. Elsevier Science Ltd, Amsterdam

Mason A, Couser P, Mason G, Smith CR, Konsky BR von (2005) Optimisation of vessel resistance using genetic algorithms and artificial neural networks. In: 4th International conference on computer and IT applications in the maritime industries (COMPIT' 05), Hamburg

Mesbahi E, Atlar M (2000) Artificial neural networks: applications in marine design and modelling. In: 1st international conference on computer and IT applications in the maritime industries (COMPIT' 2000), Potsdam

Mesbahi E, Bertram V (2000) Empirical design formulae using artificial neural nets. In: 1st international conference on computer and IT applications in the maritime industries (COMPIT' 2000), Potsdam

Radojčić D, Kalajdžić M (2018) Resistance and trim modeling of naples hard chine systematic series. RINA Trans Int J Small Craft Technol. https://doi.org/10.3940/rina.ijsct.2018.b1.211

Radojčić D, Zgradić A, Kalajdžić M, Simić A (2014) Resistance prediction for hard chine hulls in the pre-planing regime. Polish Maritime Res 21(2)(82). Gdansk

Radojčić DV, Kalajdžić MD, Zgradić AB, Simić AP (2017) Resistance and trim modeling of systematic planing hull series 62 (With 12.5, 25 and 30 Degrees Deadrise Angles) using artificial neural networks, part 2: mathematical Models. J Ship Prod Design 33(4)

Rojas R (1996) Neural networks—a systematic introduction. Springer-Velrag, Berlin

Shmueli G (2010) To Explain or to predict. Statisti Sci 25(3)

Van Hees MT (2017) Statistical and theoretical prediction methods. In: Encyclopedia of maritime and offshore engineering. Wiley

Weisberg S (1980) Applied linear regression. Wiley, New York

Zurek S (2007) LabVIEW as a tool for measurements, batch data manipulations and artificial neural network predictions. National Instruments, Curriculum Paper Contest, Przeglad Elektrotechniczny, Nr 4/2007

Chapter 3
Resistance and Dynamic Trim Modeling

3.1 An Overview of Early Resistance Prediction Mathematical Models

The first application of regression analysis (actually of the 200 year old Gauss least square method) for prediction of ship resistance is believed to have been used for the design of trawlers (few papers were published by a single author, e.g. Doust 1960). Model-test data of residuary resistance were curves fitted for various speeds; six hull form and loading parameters important for the trawlers were chosen as the independent variables, but an estimation of their relative significance was not reported. In that work, the number of equation terms for resistance prediction was relatively large—typically 30 and in some cases even 86—depending on the speed-length ratio. Accuracy, reported as the differences between measured and calculated values, was acceptable at around 3% for 95% of cases, though it was lower for the hump speeds, where residuary resistance fluctuated more.

Sabit performed separate regression analyses of resistance data for various merchant ships series—BSRA, 60, SSPA (for instance Sabit 1971), and special attention was paid to the correlation amongst the polynomial terms. Evaluated regression coefficients consisted of only up to 16 terms for each Froude number. From the statistical viewpoint, Sabit's approach is considered as more advanced than Doust's.

Van Oortmerssen (1971) produced a single equation for a range of speeds, by using Havelock wave-making resistance theory from 1909. Despite of the fact that it included about 50 polynomial terms, this single equation MM, applicable for trawlers and tugs, is less accurate than Doust's. A similar approach, but for different vessel types, was later followed by other authors from MARIN including a well-known Holtrop and Mennen method (Holtrop and Mennen 1982).

Farlie-Clarke (1975) focused on the use of statistical methods in interpretation and evaluation of ship data. Linear and nonlinear least-square methods were applied, resulting in much smaller MMs, with fewer than 10 terms. The first author

D. Radojčić et al., *Power Prediction Modeling of Conventional High-Speed Craft*,
https://doi.org/10.1007/978-3-030-30607-6_3

has adopted the Farlie-Clarke (1975) approach, and has leveraged it in development of multiple MMs throughout his career.

Fung published a series of papers in the 1990s (e.g. Fung 1991; Fung and Leibman 1993), reporting on the application of multiple linear regression analyses to very large databases of transom-stern ships (Fn_L up to 0.9), covering in some cases even 700 ships (10,000 data points). An extensive overview of the application of statistically based regression analysis to ship performance data, and of statistics as a modeling tool, is given in Fung (1991). In the abovementioned references the in-depth analyses and validations of resistance predictions, as well as the limitations of the use of statistically based MMs for resistance predictions are given, hence this may be regarded as a turning point in the MM derivation practices used in naval architecture.

3.2 Types of Mathematical Models for Resistance Prediction

3.2.1 Random Hull Forms Versus Systematical Hull Forms

From the previous overview it follows that regression analysis has been successfully used to analyze the resistance data for random hull forms (e.g. Holtrop and Mennen 1982; Fung 1991; Fung and Liebman 1993) and methodical series (Sabit 1971).

If several random hull forms are considered, then the individual characteristics of each hull or series (i.e. the secondary hull form parameters) cannot be taken into consideration. Namely, the secondary hull form parameters may be lost among the primary parameters, even if many independent (explanatory) variables are introduced into the MM. Therefore, it is often better to narrow down the applicability of the MM to a specific methodical series, and thus to increase the reliability of the model. With this approach, only the primary hull form parameters are modeled explicitly, and it is up to the user to consider that the subject hull must correspond to the series' hull form, and hence to factor in the characteristics which are not explicitly encompassed. The disadvantage of this approach however, is that multiple MMs are required, each for a given methodical series or a group of similar hull forms, instead of having a single MM needed for random hull form approach.

The authors of the present work have used both approaches. In the first case (random hull forms), special attention was paid to formation of a database, i.e. a database consisting of multiple but similar hull forms was assembled (hence was not so random), so that the secondary hull form parameters, assumed to be similar, were also factored in. In general, with this approach, it is very important to choose representative hull parameters for the whole database (see Radojčić et al. 2014a). Concerning planing hulls for instance, choosing *effective beam* and *effective deadrise* is of extreme importance; see Blount and Fox (1976) and Savitsky (2012).

Depending on the specific problem, in some cases it may be more convenient to rely upon a MM based on random hulls, in other cases on a methodical series, or often on both.

3.2.2 *Speed-Independent Versus Speed-Dependent*

There are two general types of equations for resistance evaluation that have evolved over time (see Fung 1991), each having some advantages and disadvantages:

- Speed-independent models (e.g. Sabit 1971; Fung 1991) where separate equations are generated for each discrete speed, since the speed is not included as an independent variable.
- Speed-dependent models (e.g. Van Oortmerssen 1971; Fung and Leibman 1993) with vessel's speed included as an independent variable.

The advocates of speed-dependent models claim that the predicted resistance in speed-independent models often does not vary properly with speed, since the resistance computed at one speed is not directly linked to that at another speed. This is because the speed variable is not explicitly included in the regression with this approach. The accuracy of the speed-independent models, however, is believed to be somewhat better at given characterized speeds, since independent equations are developed at each speed point.

Speed-dependent MMs can be further segregated into those which are based on some wave-making theory (as in van Oortmerssen 1971; Fung and Leibman 1993 for instance), and those which are not based on the wave-making theory. MMs which are based on a wave-making theory should realistically represent the position of humps and hollows (which are probably only important at lower speeds). These MMs, however, are typically rather difficult to extract. Speed-dependent MMs which are not based on wave-making theory, typically do describe continuous dependence of resistance on speed, but do not always accurately predict for the HSC-important position of the main resistance hump (i.e. similar handicap as speed-independent MMs).

Note that for the speed-dependent MMs, speed is usually the most dominant variable, so that care must be exercised to ensure that the other variables are visible. In general, speed-dependent MMs seem to be better for the integrated power prediction approaches.

There is a hybrid approach (Swift et al. 1973; Radojčić 1985). Essentially, with this method the speed-independent equations (with common independent variables for each Froude number) are developed first. A second regression analysis is then performed with the regression coefficients cross-faired against speed (or Froude number). Thus accurate speed-independent equations are obtained for discrete Froude numbers through the first step, and the second step provides a speed-dependent equation. Of course, either of these equations may be used independently to estimate resistance.

An important and delicate part of this method is the development of the "best subsets" from the initial, for all speeds the same, equation (see discussion on "the best equation" in Sect. 2.2.1). The usual statistical metrics (coefficient of determination, t-test or F-test, standard deviation, significance test for each variable etc.) were found to be insufficient, and in fact were sometimes even misleading. Therefore, a trial and error technique is used to define the best subsets for the whole speed range (see Radojčić 1985). Some variables, judged to be less significant, were rejected deliberately, although a stepwise method was used throughout the analysis. It should be pointed out however, that with this technique, several very good, although dissimilar, models may be derived.

3.3 Systematic Series Applicable to Conventional High-Speed Craft

A systematic or methodical series consists of ship models that are based on a given parent hull, which at the time of testing is usually representative of a state-of-the-art hull form. The principal parameters of these models are obtained by varying their particular dimensions, which results in a systematic change of, for instance, length-beam ratio, dead rise angle, block coefficient etc. There are several systematic series of conventional ships, however only the HSC series are of interest here. Note that hull form, test procedures, or the way results are presented often evolve over time. Consequently, some series may no longer be of interest and may even be outdated.

This section presents a review of the available systematic series applicable to the HSC, upon which the MMs discussed here are based. Monohulls are sorted in chronological order and presented in Table 3.1. Similar information, together with parent body planes, is given in Blount and McGrath (2009).

The catamaran Series 89' is considerably different and is placed at the end of Table 3.1. Series 89' has an additional peculiarity in that it is the only one for which resistance and self-propulsion tests were done. Hence its results are used for extracting MMs for both the resistance and for the power predictions.

Series 62 and DSDS

Series 62 and DSDS (Delft Systematic Deadrise Series) require additional clarification. Namely, Series 62 consists of 5 models with deadrise angle $\beta = 12.5°$ It is tested in DTMB and the results were published in 1963 (Clement and Blount 1963). A sequence of follow-up experiments were carried out in DUT to investigate the influence of dead rise angle. These experiments were performed in several phases over a long period of time. Each phase consisted of models—subseries—where several parameters were systematically changed, except for the dead rise angle (β) which was kept constant.

Table 3.1 Systematic series of HSC for which calm-water resistance and dynamic trim was modeled (for the catamaran Series' 89 power too)

Name Reference	Main characteristics	Range of main parameters	Remarks
Series 50 Davidson and Suarez (1941) original data Morabito (2013) data converted to contemporary format	Hard chine planing hulls. Warped bottom with negative keel angle and deep forefoot. See Fig. 8.27 Consists of 20 models.	$Fn_\nabla = 1.0–6.0$ $L_P/\nabla^{1/3} = 5.3–8.9$ $L_P/B_{PX} = 2.2–8.5$ $LCG/L_P = 0.15–0.43$ $C_B = 0.407$ Interrelated parameters are: $\frac{B_{PX}}{T} = C_B \cdot \frac{(L_P/\nabla^{1/3})^3}{(L_P/B_{PX})^2}$ $\beta_M \approx \tan^{-1}\left(\frac{1.35}{B_{PX}/T}\right)$	Obsolete hull form for planing regimes, nowadays applicable for the semi-displacement regimes. Hull form enables placement of less inclined propeller shafts.
Series 62 (see DSDS) Clement and Blount (1963) $\beta = 12.5°$	Hard chine, narrow-stern, planing hull form. Constant deadrise over the after 50% of hull. See Fig. 8.29 Consists of 5 models with $\beta = 12.5°$ $L_P = 1.2–2.4$ m	$Fn_\nabla = 0.75–6$ $A_P/\nabla^{2/3} = 4.0–8.5$ $L_P/B_{PX} = 2–7$ $LCG/L_P = 0.368–0.488$ $\beta = 12.5°$ $(CA_P\text{-}LCG)\cdot 100/L_P = 0\text{-}12\%$ $CA_P/L_P = 0.488$; $B_{PT}/B_{PX} = 0.64$ (for wider models a bit different) $L_P/\nabla^{1/3} \approx 1.1\,(A_P/\nabla^{2/3} \cdot L_P/B_{PX})^{1/2}$	Often analyzed together with DSDS. Extensively investigated series (seakeeping, shallow water, appendages etc.). Used for bench-marking.
Series 65-B (65-A is not of interest here) Holling and Hubble (1974) Hadler et al. (1974)	Hard chine, wide-stern, planing hull form. Small warp over after portion. See Fig. 8.23 Three groups of models, in total 9.	$Fn_\nabla = 1.0–4.0$ $\beta_T = 16.3$, 22.5 and 30.4° $A_P/\nabla^{2/3} = 5.5–8.5$, for lower Fn_∇ from 4 $L_P/B_{PX} = 2.35–9.38$ $LCG/L_P = 0.38$ $CA_P/L_P = 0.397$; $B_{PT}/B_{PX} = 0.99$	Fixed-trim method of testing. Results converted to free-to-trim. Initially developed for hydrofoils (65-A of aeroplane and 65-B of canard type).

(continued)

Table 3.1 (continued)

Name Reference	Main characteristics	Range of main parameters	Remarks
NPL series Bailey (1976)	Important, high-speed, transom-stern, round bilge form for displacement and semi-displacement regimes. See Fig. 8.16 Consists of 22 models of L = 2.54 m	$Fn_\nabla = 0.6–3.0$ $Fn_L = 0.3–1.2$ $L/B = 3.33–7.50$ $L/\nabla^{1/3} = 4.5–8.3$ $B/T = 1.75–10.77$ LCB = 6.4%L aft amidship $C_B = 0.397$ $A_T/A_X = 0.52$ $\beta_T = 12°$	Static wetted surface used for all speeds. Given maneuvering, seakeeping, stability underway, propulsion coefficients, influence of transom wedge and spray rails. Investigated influence of LCB position. Some loll instability at high speeds.
VTT series Lahtiharju et al. (1991)	High-speed, transom-stern, small-bilge radius, semi-displacement form. Modified NPL. See Fig. 8.14 Consists of 5 models. One model hard chine (converted from round bilge).	$Fn_\nabla = 0.6–3.8$, hard chine to 5.0 $L/\nabla^{1/3} = 4.86–6.59$ $L/B = 2.75–6.25$ $B/T = 4.39–6.90$ $C_B = 0.397–0.60$ $C_X = 0.57–0.87$ LCB =0.436L B_T/B =0.815 A_T/A_X =0.52 Tests carried also for 130% and 85% of design displacement.	Static wetted surface used for all speeds. Hull form suitable for shallow draught and waterjets. Seakeeping tests.
TUNS series Delgado-Saldivar (1993)	Hard chine, wide-stern, planing hull form. Prismatic hull form (constant deadrise and beam over the after 50% of hull). See Fig. 8.25 Consists of 9 models of 1–3.5 kg	$Fn_\nabla = 0.6–4.0$ $L_P/\nabla^{1/3} = 3.9–6.9$ $A_P/\nabla^{2/3} = 5.2–11.6$ $L_P/B_{PX} = 2.5–3.5$ $\beta_{Bpx} = 12–24$ $LCG/L_P = 0.27–0.38$	Experiments performed in a small university basin only 27 m long. Disadvantage: small models. Match the contemporary hulls.

(continued)

Table 3.1 (continued)

Name Reference	Main characteristics	Range of main parameters	Remarks
SKLAD series Gamulin (1996)	High-speed, transom-stern, round bilge, semi-displacement form. Built-in stern knuckle and spray rail; forward sections deep V. See Fig. 8.18 Consists of 27 models, length 3–6m.	$Fn_\nabla = 1.0–3.0$ $L/\nabla^{1/3} = 4.5–9.7$ $L/B = 4–8$ $B/T = 3–5$ $C_B = 0.35$, 0.45 and 0.55 LCB = (8.8–9.2)% L_{DWL} aft amidship $\beta_T = 12°$	Stern-knuckle (chine) 20%L, spray rail 40%L. Good for seakeeping. The afterbody bottom flat and hooked (space for propellers). Given propulsion coefficients, C_A and appendage resistance.
NTUA series Grigoropoulos and Loukakis (1999) resistance Grigoropoulos and Damala (2001) dynamic trim	Double-chine (70%L), wide-stern, semi-displacement form with bottom warp. See Fig. 8.21 Consists of 6 models, length above 2 m.	$Fn_L = 0.5–0.9$ $L/B = 4.3–7.5$ $L/\nabla^{1/3} = 6.2–10.0$ $B/T = 3.2–6.2$ $C_B = 0.34–0.54$ LCB = (12.4–14.6)%L aft amidship $\beta_T = 10°$	Hull form similar to that in Savitsky et al. (1972). May be advantageous for Fn_L>0.8. Focus on resistance, seakeeping, and hull simplicity. Intended for large (above 20 m) high-speed ferries. Wide transom for waterjets.
USCG series Kowalyshyn and Metcalf (2006)	Hard chine, wide-stern, planing hull form. Slight bottom warp and deadrise variation. See Fig. 8.25 Consists of 4 models of 135–220 kg.	$Fn_\nabla = 0.6–6.0$ $L_P/\nabla^{1/3} = 5.2–6.1$ $A_P/\nabla^{2/3} = 4.8–9.1$ $L_P/B_{PX} = 3.3–4.6$ β_{Bpx} =1 8–21 $LCG/L_P = 0.37–0.41$	Representative length of wetted surface is L_{KEEL}. Reliable results as models are large. Fn_∇ range up to 6. Contemporary hull form.

(continued)

Table 3.1 (continued)

Name Reference	Main characteristics	Range of main parameters	Remarks
DSDS (based on Series 62) Keuning and Geritsma (1982) $\beta = 25°$ (prismatic hulls) Keuning et al. (1993) $\beta = 30°$ (prismatic hulls) Keuning and Hillege (2017a, b) $\beta = 19°$ (prismatic hulls) $\beta = 25°$ (variable deadrise with warp and rocker)	Hard chine, narrow-stern, planing hull form. $L_P = 1–1.5$ m a) Constant deadrise over the after 50% of hull. See Fig. 8.29. Consists of 13 models (4+5+4 with β=19, 25, and 30 deg. respect.). b) Variable deadrise over the after 50% of hull. Consists of 6 models.	a) $Fn_\nabla = 0.75–3.0$ (for some models to $Fn_\nabla=4$) β =19, 25 and 30° $A_P/\nabla^{2/3} = 4.0–8.5$ $L_P/B_{PX} = 2–7$ $(CA_P\text{-}LCG)\cdot 100/L_P = 0\text{-}12\%$ $CA_P/L_P = 0.488$; $B_{PT}/B_{PX} = 0.64$ (for wider models a bit different) b) $\beta = 25°$ $\varepsilon = 10$ and 20° $\gamma = -1.69$ to $-4.93°$ $L_P/B_{PX} = 4.1–7.0$ Other parameters are same as for the prismatic hulls.	Parent models of DSDS subseries with β=19, 25, and 30° are developed from the Series 62 parent 4667-1. Subseries with β=19° was tested in 1996 but released in 2017. Entire database (with the Series 62) consist of 24 models. Subseries with warp and rocker has β_M=25° and β_T=5 or 15° and negative keel angle.
NSS De Luca and Pensa (2017)	Hard chine, wide-stern, planing hull form. Slight bottom warp and deadrise variation. See Fig. 8.31 Consists of 5 models of 40–122 kg.	$Fn_L = 0.5–1.6$ $Fn_\nabla = 1.1–4.2$ $L/\nabla^{1/3} = 4.83–7.49$ $L_P/B_C = 3.45–6.25$ $LCG/L_P = 0.33–0.38$ $\beta_T = 14.3°$ $\beta_M = 22.3°$ $A_T/A_X = 0.94$	Contemporary hull form. Suggested to be used with the interceptors. NSS notation: L_{WLD}, B_C and L correspond to L_{WL} dynamic, B_{PX} and L_{WL} static.
Series '89 Müller-Graf (1999) (And several other papers published between 1989-2000)	Systematic series of hard chine catamarans. See Fig. 10.9 Standard configurat. consist of 12 models of 3.6–5.1 m. Resistance and self-propulsion tests.	$Fn_L = 0.8–1.4$ $L_{WL}/B_{WLDH} = 7.55, 9.55, 11.55, 13.55$ $\beta_M = 38, 27, 16°$ $\delta_W = 0, 4, 8, 12°$ ($L_{wedge} = 2.8\%L_{WL}$) $L_P/(\nabla/2)^{1/3} = 6.1–9.6$ $LCB = (0.38–0.42)L_{WL}$ $B_T/L_{WL} = 0.167$ $B_{TR}/B_{XDH} = 0.92$ $T_{TR}/T_H = 0.54$ – for fully wetted prop. $T_{TR}/T_H = 0.68$ – for surface piercing $T_{TR}/T_H = 0.95$ – for waterjets	Different hull forms tested (symmetric, semi-symmetric, and asymmetric hard chine and round bilge), afterbodies and propulsion configurations (fully wetted propellers (FP and CPP), SPP and WJ). Tests and results validated for trial conditions.

The parent model of each subseries was based on the Series 62 parent DTMB model 4667-1. The results of the subseries with $\beta = 25$ and 30° were published in Keuning and Geritsma (1982) and Keuning et al. (1993), respectively. The cluster of subseries datasets for $\beta = 12.5$ and 25° (10 models), or 12.5, 25, and 30° (14 models) was named, "Series 62", "PHF", or recently "DSDS", depending on the user/author. Obviously, none is fully correct. Present authors used the name "Series 62". Then in 1996 the subseries with $\beta = 19°$ was tested (additional 4 models), but these results were released twenty years later in Keuning and Hillege (2017a, b), actually through the DUT website. The same publications included results for 6 variable dead rise models with negative keel angle (i.e. with twist[1] and rocker) also tested in two phases. To summarize, the whole series, released over the period of more than 50 years, consists now of 24 models, all based upon the Series 62, and *now* merit enough to be called DSDS.

3.4 Mathematical Modeling of Resistance and Dynamic Trim for High-Speed Craft

This section addresses the MMs for resistance prediction (and dynamic trim where available) of high-speed craft (HSC's MMs). Similar subjects were addressed earlier, for instance by van Oossanen (1980), Almeter (1993), and others, but these are now outdated and merit an update. Some resistance prediction MMs for high-speed round bilge hull forms were re-evaluated recently, see Sahoo et al. (2011). In the following text 18 MMs are discussed and 10 are recommended as they have not yet been superseded. These MMs are summarized in Tables 8.1 and 8.14 which are given later in Part II, Chap. 8. Some of them form the basis of current computerized resistance prediction software packages.

3.4.1 BK and MBK (Yegorov et al. 1978)

A relatively unusual resistance prediction MM, based on the Soviet hard chine BK and MBK series, was developed in the 1970s. This MM had two main differences from all following MMs. Namely:

- From the outset, both BK and MBK series were based on obsolete hull forms when conceived and model tested in the 1960s and 1970s. BK and MBK series were intended to be used for large semi-planing patrol craft and small planing

[1]Note: *Bottom twist* is equivalent to *bottom warp*. The term *twist* is used by the authors from DUT; for the remainder of this text the term *warp* is used.

craft, respectively. Their parent forms were actually similar to the Series 50. BK consisted of 16 2.1–3.9 m and MBK of 13 1.6–2.4 m long models with length-beam ratio (L_P/B_M), ranging between 3.75–7.00 and 2.50–3.75, respectively. Tested speed range corresponded to $Fn_\nabla = 1.0$–4.5. The principal reason for obsolescence is the low transom deadrise and high after body warp as $\beta_{TR} = 0$ and 5° and $\beta_M = 12$–18 and 7–18° respectively for BK and MBK series. Incomplete calm water resistance results were published in the late 1960s through to the late 1970s, ending with the book by Yegorov et al. (1978).

- A relatively unusual MM for calm water resistance prediction was proposed, using a Taylor series expansion, as a f(C_Δ, LCG, β_M, L_P/B_M). Therefore, the resistance of a target hull, with a form within the range of either MK or MBK series, had to be expressed via a Taylor's series expansion for each Froude number. Based on the abovementioned, a technique for resistance prediction and form optimization of low dead rise hard chine hull forms is given in Almeter (1988).

3.4.2 Mercier and Savitsky—Transom-Stern, Semi-displacement (Mercier and Savitsky 1973)

The first HSC MM that enabled evaluation of total resistance of transom-stern craft in the non-planing regime ($Fn_\nabla = 1$–2) built by application of regression analysis was developed by Mercier and Savitsky (1973). Their MM is based on seven transom-stern hull series (see Sect. 8.2.1 and Fig. 8.12) and focused on the lower speed range of HSC. For higher planing speeds, the Savitsky (1964) method was already in use. Selection of four hull form and loading parameters as input variables (slenderness ratio ($L_{WL}/\nabla^{1/3}$), static beam load coefficient (∇/B_X^3), half-angle of entrance of waterline at bow ($i_{e.}$), and transom area ratio (A_T/A_X)), was based on the authors' experience, and no statistical correlation analysis was performed. The speed-independent approach was employed as wave-making-resistance-theories for speed-dependent MMs were judged not to be applicable for high-speed transom-stern craft. The speed range covered the main resistance hump, as well as several local humps that occur at lower speeds (in terms $R_W = f(Fn_L)$)—which is generally difficult to model with a single equation.

3.4.3 Transom-Stern, Round Bilge, Semi-displacement (Jin et al. 1980)

Jin et al. (1980) followed the Mercier and Savitsky (1973) approach and replaced Series 62 (hard chine hulls), Series 64 (slender forms), and SSPA (flattened-off after body), with the Chinese built round bottom hulls, forming the database of 87 hull forms. The objective was speed-independent modeling of residuary resistance of high-speed round bilge displacement hulls, for Fn_L range 0.4–1.0. After examining the influence of various parameters, Mercier and Savitsky's input variables $L_{WL}/\nabla^{1/3}$, ∇/B_X^3, i_e, and A_T/A_X that modeled $(R_T/\Delta)_{100,000}$, were replaced by C_∇, C_P, A_T/A_X, LCB, and i_e and modeled C_R. Note that in both cases the most significant parameter is the ratio of length and displacement, whether expressed as $L_{WL}/\nabla^{1/3}$ or $C_\nabla = \nabla/(0.1L)^3$. Thus, practically only ∇/B_X^3 is replaced with C_P and LCB. It is reported, but not quantified, that accuracy was a bit better than that of Mercier and Savitsky. Two speed ranges were identified which resulted in a suggestion to adjust the hull-form parameters for each of them.

3.4.4 62 and 65—Hard Chine, Semi-planing and Planing (Radojčić 1985)

Calm-water resistance (R/Δ) and dynamic trim (τ) of stepless planing hulls for a wide speed range corresponding to Fn_∇ = 1.0–3.5 (to Fn_∇ = 4 for τ) can be estimated from the speed-independent and speed-dependent mathematical models given in Radojčić (1985). The speed-independent MM is a derivative of the previous work (Radojčić 1984), with an altered number of cases, regression equations etc. This MM (Radojčić 1985) is based on the systematic Series 62 with deadrise angles (β) 12.5° and 25°, Series 65-B, and a single DL-62-A model (see Sect. 8.3.1 and Fig. 8.23). Actually, the data for β = 25° (Keuning and Geritsma 1982) were transformed and added to the existing database of Hubble (1974), hence all models were represented on a unique basis regarding model tow force (horizontal through CG), water density and viscosity, ATTC-1947 friction line, correlation allowance etc.

The MM is a function of four hull-form and loading parameters representative for the planing hulls: $A_P/\nabla^{2/3}$, LCG/L_P, L_P/B_{PA} and β. A "dummy" variable was introduced to separately predict R/Δ and τ using a single equation, while still distinguishing between the two different hull form types, i.e. those resembling Series 62 and Series 65-B hull form respectively. This approach avoided the introduction of a fifth input variable.

Further refinement of this MM involved the introduction of a 4-dimensional ellipsoid (actually a single non-linear equation, see Radojčić 1991) which defined the bounds of applicability of the MM, and hence replaced the use of typical linear bounds, normally expressed through 60 equations, see Fig. 8.24. Note that the size of the 4-D ellipsoid is defined by the user, and the use of a smaller ellipsoid results in increased accuracy. The motivation for this novel approach was: (a) the bounds of applicability are often strange and not necessarily logical, and (b) the accuracy of the MM is usually reduced at the outer limits of the database. On the other hand, optimization routines require precise applicability bounds.

3.4.5 VTT—Transom-Stern, Semi-displacement (Lahtiharju et al. 1991)

Lahtiharju et al. (1991) developed a 24-term, speed-dependent resistance prediction equation using a regression analysis based on the round bilge NPL, SSPA, and VTT series (see Sect. 8.2.2 and Fig. 8.14). A separate, much simpler, 6-term MM was also developed for hard chine craft. Both equations are applied for $(R_T/\Delta)_{100,000}$ for speed range Fn_∇ = 1.8–3.2, while the Mercier and Savitsky MM is recommended for lower speeds and Savitsky's empirical method for higher speeds. The database contained five new systematically developed and experimentally tested models; see Table 3.1. Note that in this work relatively unusual draught based parameters were used as input variables. A MM for dynamic trim predictions, however, was not given. Altogether, the Lahtiharju et al. (1991) work is important not only because MMs for resistance evaluation were derived, but also because resistance and sea-keeping tests were performed for new high-speed, round bilge systematic VTT series.

3.4.6 PHF—Series 62 (Keuning et al. 1993)

For predictions of resistance (R/Δ), dynamic trim (τ), and rise of center of gravity ($RCG/\nabla^{1/3}$), for the entire Series 62 (termed now PHF—Planing Hull Form), i.e. β = 12.5°, 25°, and 30°, are given in Keuning et al. (1993). See discussion on Series 62 and DSDS in Sect. 3.3. This paper also presented experimental results for the Series 62 with β = 30°, thus augmenting the data with β = 12.5° and 25° which was given in Clement and Blount (1963) and Keuning and Geritsma (1982), respectively. Input variables were $A_P/\nabla^{2/3}$, LCG, and L_P/B_{PX} and separate equations were given for each, β = 12.5°, 25°, and 30°. These 12-term MMs were both deadrise-independent and speed-independent, which required interpolation for intermediate β values between the measured points of β = 12.5°, 25°, and 30°, not to mention 10 discrete Froude numbers ranging between Fn_∇ = 0.75 and 3.0. Moreover, R/Δ for displacements of 5 m^3 and 50 m^3 was given, which required further interpolations. MMs for evaluation of length of wetted area and wetted area however were not given. Additional 7-term polynomial equations for evaluation of the effect of warped bottom, i.e. evaluation of the differences in resistance (Δ R/Δ), dynamic trim (Δ τ), and sinkage (Δ $RCG/\nabla^{1/3}$), compared to parent model of β = 25°, were also developed using two new input variables (centerline inclination angle (γ) and warp angle[2] (ε)). The PHF MM is recently superseded by DSDS MM; see Sect. 3.4.16.

[2]Usual measure for longitudinal variation of deadrise is *warp rate*, which is change of deadrise over longitudinal length equivalent to chine beam, and is expressed in *degrees per beam* (see Savitsky 2012; Blount 2014). DUT use *warp angle*, see list of Symbols.

3.4.7 NPL (Radojčić et al. 1997)

Three groups of MMs based on high-speed, round bilge, NPL series and a broad speed range of $Fn_\nabla = 0.8–3.0$ were derived in Radojčić et al. (1997). These included speed-independent and speed-dependent models for resistance ($(R/\Delta)_{100,000}$) and dynamic trim (τ), and one for wetted surface coefficient ($(S) = S/\nabla^{2/3}$); see Sect. 8.2.3. Note that these MMs are based on the NPL hull-form series exclusively (Fig. 8.16), as opposed to the previous works (Mercier and Savitsky 1973; Lahtiharju et al. 1991) which combined the NPL series with the data of other series. Furthermore, since $(R/\Delta)_{100,000}$, τ, and (S), were chosen as dependent variables, data for model-size wetted area and $R_R/\Delta = f\ (L/\nabla^{1/3}, Fn_\nabla, L/B)$ (see Bailey 1976) were transferred to a new format $(R/\Delta)_{100,000}, \tau, (S) = f(L/B, L/\nabla^{1/3}, B/T)$. Through the first step, 12 27-term speed-independent equations were formed for $(R/\Delta)_{100,000}$ and τ (one for each Fn_∇ ranging from 0.8 to 3.0) and one for (S) as S does not depend on speed. In the next step, the speed-dependent models for $(R/\Delta)_{100,000}$ and τ were developed through cross-fairing of resistance and dynamic trim regression coefficients against Fn_∇ (as previously done in Radojčić 1985). Satisfactory results were obtained when Fn_∇ was raised to the eight power, resulting in 126- and 134-terms MMs for $(R/\Delta)_{100,000}$ and τ respectively. Speed-independent models are valid for Fn_∇ range between 0.8 and 3.0, while speed-dependent ones were valid for $Fn_\nabla = 1.0–3.0$, due to instability between $Fn_\nabla = 0.8–1.0$.

3.4.8 SKLAD (Radojčić et al. 1999)

A mathematical representation of both calm water resistance and dynamic trim for the systematic round bilge, transom-stern, semi-displacement SKLAD series (Gamulin 1996) for a speed range of $Fn_\nabla = 1.0–3.0$ was presented in Radojčić et al. (1999); see Sect. 8.2.4. The dependent variables are the residuary resistant coefficient (C_R), wetted surface coefficient ($S/\nabla^{2/3}$), length-displacement ratio ($L/\nabla^{1/3}$), and dynamic trim (τ), while the independent variables are L/B, B/T, and C_B. All initial polynomial equations started out with 100 terms, while the final MMs for C_R, $S/\nabla^{2/3}$, $L/\nabla^{1/3}$, and τ had 21, 29, 20, and 18 terms respectively; one for each of 8 discrete Fn_∇ ranging from 1 to 3. Consequently, all derived equations were speed-independent. At the time, considerable effort was made to develop speed-dependent MMs, but without success.

3.4.9 Round Bilge and Hard Chine (Robinson 1999)

The results of regression analysis (WUMTIA regression per Molland et al. 2011) of 30 and 66 models of round bilge and chine hull forms, respectively, tested at the

Wolfson Unit over some 30 years, is presented in Robinson (1999). The hull forms are not of systematic form, but do broadly fit into the round bilge or hard chine hull-form category, and thus this database is classified as random hulls. Relying on the fact that a relationship between C-Factor (actually C^2) and relative speed and specific total resistance ($v/L^{1/2} \cdot \Delta/R_T$) does exist, a relatively unusual and obsolete method, set up on the so called C-Factor, was applied ($C = 30.1266 \cdot v/(L)^{1/4} \cdot (\Delta/2P_E)^{1/2}$). These results were valid for $Fn_\nabla = 0.5–2.75$ and the initially trim-optimized, but not specified, conditions (i.e. LCG position was shifted or wedge/trim-tabs were used). The dependent variable was the C-Factor, and the independent variables for chine and round bilge 7-term and 10-term speed-independent equations were $L/\nabla^{1/3}$ and L/B, and $L/\nabla^{1/3}$, L/B, and S/L^2, respectively. Simple 3-term MMs for wetted area were also derived for chine craft (for 6 discrete values of Fn_∇) and displacement craft (for static condition only), and depend on Δ, L, and B. According to Molland et al. (2011) the abovementioned MMs overestimate power by some 3–4%.

3.4.10 Transom-Stern, Round Bilge (Grubišić and Begović 2000)

Resistance data of 12 fast round bilge systematic series with 186 models were analyzed in Grubišić and Begović (2000). In addition to the aforementioned round bilge series, a fast twin-screw displacement ship series and a semi-planing series given in Compton (1986) were added. Using regression analysis, a 12-term speed-independent MM was defined with 7 independent variables. These are: slenderness ratio $L/\nabla^{1/3}$, beam draught ratio B/T, prismatic coefficient C_P, maximum area coefficient C_X, transom area coefficient A_T/A_X, Taylor wetted surface coefficient C_S, and longitudinal center of buoyancy LCB/L_{WL}. The dependent variable is residuary resistance coefficient C_R. This MM is valid for the speed range that corresponds to $Fn_L = 0.3–1.2$ and is essentially similar to Mercier and Savitsky (1973) and Fung (1991). It is a random hull form type MM as is based on similar round bilge hulls and is usable in the concept design phase.

3.4.11 NTUA (Radojčić et al. 2001)

In Radojčić et al. (2001) raw model test data of the NTUA Series was used for the development of the MMs for resistance (C_R) and dynamic trim (τ), hence regression analysis was applied for both, fairing of the raw data and actual model extraction. Modeling of the NPL and SKLAD series did not require this since the data were already faired. The speed-dependent approach for modeling of C_R and τ was chosen with L/B, $L/\nabla^{1/3}$, B/T, and Fn_L being the independent variables, see Sect. 8.2.5.

Note that two of the independent variables defined the third one, so there was no need to use all three; however all three were incorporated (even though that was wrong from the statistical point of view), because (a) this simplified the mathematical models and, (b) it prevented the use of the MM for cases that are very dissimilar to the NTUA series. This is de facto a new modeling approach applied for resistance and trim predictions.

3.4.12 Displacement, Semi-displacement, and Planing Hull Forms (Bertram and Mesbahi 2004)

ANN as an extraction tool was used for derivation of simple equations for resistance and dynamic trim evaluation (see Bertram and Mesbahi 2004). The MMs derived are based upon an extremely wide hull form spectrum, ranging from the fast displacement ships to hard chine planing hulls. This means that the secondary hull form parameters are neglected. Moreover, oversimplified formulas and diagrams underlie the dataset, which inherently resulted in neglecting some important parameters (e.g. L/B, which is usually one of the primary parameters). A nonlinear relationship between the input variables and an output is derived by a simple ANN with a single hidden layer. For evaluation of total resistance coefficient $C_{T\nabla} = R_T/(\rho/2 \cdot v^2 \cdot \nabla^{2/3}) = f(Fn_L, C_\nabla)$ three groups of equations were derived, covering speed (Fn_L), and slenderness ratios ($C_\nabla = \nabla/L^3$), of 0.2 to 1.2 and 0.016 to 0.007, respectively. Three equations for evaluation of appendage resistance R_{APP}/R_T % = $f(Fn_L)$ were also derived, for configurations with 2, 3, and 4 propellers, and two for dynamic trim $\tau = f(\nabla^{2/3}/B \cdot T)$, for semi-displacement and planing hull forms. Resistance, appendage resistance, and dynamic trim equations have on average 20, 10, and 15 terms, respectively. To summarize, an overly sophisticated extraction tool was used for an oversimplified approach (where $R_T = f(Fn_L, C_\nabla)$).

It should be noted however, that this is probably the first application of ANN to HSC monohulls. ANN modeling of catamaran resistance was described in Couser et al. (2004) and Mason et al. (2005), also mentioned in Sect. 2.2.2, but this topic is not discussed further because an adequate MM for resistance prediction has not been released (see Sect. 1.4.3.2).

3.4.13 USCG and TUNS—Hard Chine, Wide Transom, Planing (Radojčić et al. 2014a)

A mathematical representation of resistance for wide-transom planing hull forms based on the USCG and TUNS Series (see Kowalyshyn and Metcalf 2006; Delgado-Saldivar 1993, respectively) for the speed range corresponding to $Fn_\nabla = 0.6–3.5$ is presented in Radojčić et al. (2014a). Regression analysis and

Artificial Neural Network (ANN) techniques are used to establish, respectively, "Simple" and "Complex" mathematical models, see Sect. 8.3.2. For the Simple model, the dependent variable was $(R/\Delta)_{100,000}$, while Fn_∇ and $L/\nabla^{1/3}$ were chosen as the independent variables (being the two dominant high-speed parameters). For the Complex model, additional independent variables were L/B, LCG/L, and β. Both types of MMs are obviously speed-dependent. Being a 2-parameter formulation, the Simple model is intended for use during the concept design phases where the reduced quality of resistance predictions is acceptable. The Complex MM is intended for use for various performance predictions during all design phases. Relatively simple MMs for wetted surface ($S/\nabla^{2/3}$) and its length (L_K/L) were also developed, without β and even without $L/\nabla^{1/3}$, L/B, and β as dependent variables, respectively. Dynamic trim however, could not be modeled. The principal disadvantage of all the derived models is that the TUNS Series consists of very small models, which to some extent introduces unreliability (see Moore and Hawkins 1969; Morabito and Snodgrass 2012; Tanaka et al. 1991 regarding the usefulness of small models).

The simple model for $(R/\Delta)_{100,000}$ was developed in two phases: (a) Data having same $L/\nabla^{1/3}$ were grouped (regardless of the other hull form and loading parameters), then a trend line $R/\Delta = f(Fn_\nabla)$ was produced for each group, and (b) A second regression analysis is then performed with the regression coefficients cross-faired against the slenderness ratio. The Complex speed-dependent version, derived with an ANN routine, for $(R/\Delta)_{100,000}$ and $S/\nabla^{2/3}$ have 116 and 23 equation terms, respectively. The MM for L_K/L has 16 terms only and was derived with regression analysis.

Comparing the methods used (tools for MM development) ANN showed to be a very good modeling tool for resistance predictions. Regression analysis required more time and higher levels of skill, at least for complex relations with many polynomial terms; it seems to be more convenient for simpler relations/equations.

3.4.14 Series 50 (Radojčić et al. 2014b)

Further assessments of contemporary ANN and conventional regression analysis for modeling of resistance were investigated in Radojčić et al. (2014b). Mathematical representations for predicting resistance (this time R_R/Δ), dynamic trim (baseline trim τ_{BL}), and wetted length (L_M/L_P), of the EMB Series 50 are given in the same reference. The Series 50 database consisted of a re-analyzed full data set, as discussed in Morabito (Morabito 2013), because the original model-test data (Davidson and Suarez 1941) were prepared before contemporary planing hull coefficients were introduced. Models derived by regression analysis were somewhat "stiffer" than ANN models. This is not only due to the smaller number of terms, but also to the different format, as regression used polynomials while ANN used a much more complex nonlinear function. The "double hump" phenomenon for dynamic trim, between Fn_∇ 2.0 and 3.0, often connected with dynamic instability, was also noted.

$L_P/\nabla^{1/3}$, L_P/B_{PX}, LCG/L_P, and Fn_∇ were used as the independent variables throughout the work and hence the MMs were speed-dependent. The dependent variables were R_R/Δ, τ_{BL}, and L_M/L_P. Three mathematical models, finalists amongst the several hundred tested, were developed for each of the dependent variables. The final MMs for resistance and trim evaluation were 65-term and 91-term equations derived by regression and ANN, respectively. MMs for predicting L_M/L_P were much simpler, so there was no need to use ANN as the regression technique sufficed (see Sect. 8.3.3).

A detailed quantification of the differences between ANN and regression methods for developing stable and accurate MMs is provided. ANN has been demonstrated as a viable technique for fitting complex data sets accurately—as required for modeling HSC resistance. Modeling dynamic trim is even more challenging and ANN-derived-models successfully replicated the double hump—something that was not achieved with the regression analysis approach.

3.4.15 Series 62 (Radojčić et al. 2017a, b)

After the transition period—from regression to ANN (previous two references)—the authors were encouraged to continue application of ANN for similar problems. Follow resistance and trim modeling of the well-known planing hull Series 62 (Radojčić et al. 2017a, b). When this work was performed Series 62 (or DSDS, see discussion in Sect. 3.3) consisted of three groups of experiments, each with a deadrise angle of $\beta = 12.5$, 25, and 30°, conducted across three decades; see Table 3.1 and Clement and Blount 1963, Keuning and Geritsma 1982 and Keuning et al. 1993, respectively. The first two groups of models ($\beta = 12.5$ and 25) were also used in Radojčić (1985), while all three in Keuning et al. (1993), where derived equations were both dead rise-independent and speed-independent.

Consequently, the original experimental data given in the abovementioned references was rearranged and used for formation of a revised database (Radojčić et al. 2017a) which consisted of original and 'virtual' measurements. The virtual measurements were introduced to cover zones which were not sufficiently covered by the experimental data, ensuring the polynomial fits were fair and continuous between data points. Then, stable speed- and deadrise-dependent MMs for R/Δ and τ predictions for Fn_∇ and β range 1–4 and 12.5–30° respectively, were extracted. The independent variables were β, L_P/B_{PX}, $A_P/\nabla^{2/3}$, LCG/L_P, and Fn_∇, and the dependent ones were R/Δ and τ (see Sect. 8.3.4).

It should be noted that a novel *multiple output* ANN technique was applied, in contrast to *the single output* technique (where resistance data was used only for modeling of R/Δ, and separately, trim data was only used for modeling of trim (τ), hence the structure of ANN models are inherently dissimilar, yielding entirely different equations for R/Δ and τ). With the *multiple output* ANN, all available R/Δ and τ data are used simultaneously, producing slightly different equations for R/Δ and τ. This implied that τ data influenced the model for R/Δ and vice versa, which makes sense in the high-speed regime where R/Δ and τ curves mirror each other. The multiple-output ANN technique resulted in an equation with 233 terms defining both R/Δ and τ values, with only 8 terms differing between them (i.e. 217 terms are common, 225 terms define each, R/Δ and τ). Independently derived, single-output MMs for R/Δ and τ include 190 and 160 terms respectively. The multiple output approach is a novel ANN application for this kind of problems. Single output models, however, have been found to have a bit better accuracy, which is to be expected since they track independently either R/Δ or τ.

3.4.16 DSDS (Keuning and Hillege 2017a, b)

MMs for DSDS (Delft Systematic Deadrise Series) consist of two different sets of MMs developed by the same authors a month apart - Keuning and Hillege (2017a, b). Each MM is for resistance and dynamic trim prediction. Both are derived by regression analysis and are similar to MMs for PHF given in Keuning et al. (1993), with the exception that the new MMs are based on the results of the entire DSDS subseries, and hence include $\beta = 19°$ as well. Specifically, MMs (separate equations for each deadrise angle) for evaluation of R_T/Δ, $\tau = f(A_P/\nabla^{2/3}, L_P/B_{PX}, LCG)$, are given in the first paper. They are both deadrise- and speed-independent, as in Keuning et al. (1993) 24 years earlier. A more advanced approach is given in the second paper with a set of reformatted equations, i.e. R_R/Δ, τ, $L_M/\nabla^{1/3}$, $S/\nabla^{2/3} = f(\beta, A_P/\nabla^{2/3}, L_P/B_{PX}, LCG)$. Since β is among the input parameters, these equations are now deadrise-dependent, but are still speed-independent. These polynomial equations have 17 and 8 terms for evaluation of R_R/Δ & τ and $L_M/\nabla^{1/3}$ & $S/\nabla^{2/3}$ respectively.

Two sets of equations for warped hulls follow the logic and format of prismatic hull equations, and hence depend on the same input variables as the prismatic hulls, in addition to warp and buttock angles (ε and γ, respectively). Note that the PHF equations for warped hulls (Keuning et al. 1993) did not depend on L_P/B_{PX}. The new MM now does, owing to the extension of the warped hull subseries (presently it consists of six models having L_P/B_{PX} range between 4.1 and 7.0). The warped hull equations are essentially evaluating corrections dR_R/Δ, $d\tau$, $dL_M/\nabla^{1/3}$, and $dS/\nabla^{2/3}$ relative to a prismatic hull with $\beta = 25°$ It is assumed that the same correction may be applied to other deadrises.

3.4.17 NSS (De Luca and Pensa 2017)

Model test results in calm water for the NSS (Naples Systematic Series) have been presented in a recent paper by De Luca and Pensa (2017) for dynamic trim (τ), total and residuary resistance coefficients (C_T and C_R), wetted surface (S), and waterline length (L_{WL}); see Table 3.1. NSS is envisaged to be used with the interceptors, but interceptor effect is yet to be published. In the referenced work, for each of the five NSS models and $LCG/L_P = 0.38$, regression analysis was used to derive a 20-term polynomial equation for evaluation of resistance (actually C_R), running wetted surface, and waterline length as a function of $L_P/\nabla^{1/3}$ and Fn_L. These MMs are speed-dependent, but L_P/B_{PX} and LCG independent. The equations actually evaluate C_R, S, and L_{WL} within the tested $L_P/\nabla^{1/3}$ range, but do not interpolate between the two LCG positions and the various L_P/B_{PX} values. This is similar to the "Simple" 2-parameter MM of Radojčić et al. (2014a) except that each of the five polynomial equations are valid for a particular NSS model, rather than for the whole series.

3.4.18 NSS (Radojčić and Kalajdžić 2018)

Radojčić and Kalajdžić (2018)[3] fill in the gaps in the abovementioned simplified 2-parameter NSS MM and present MMs of resistance (actually $(R_T/\Delta)_{100,000}$), dynamic trim (τ), wetted area ($S/\nabla^{2/3}$), and length of wetted area (L_{WL}/L_P), as functions of L_P/B_{PX}, $L_P/\nabla^{1/3}$, LCG/L_P, and Fn_∇ (see Sect. 8.3.5). An Artificial Neural Network (ANN) method with multiple outputs is used to develop the enhanced mathematical models enabling simultaneous use of all the available R/Δ and τ data on one side, and $S/\nabla^{2/3}$ and L_{WL}/L_P on another. Moreover, the ANN structures for both datasets, R/Δ & τ, and $S/\nabla^{2/3}$ & L_{WL}/L_P, are identical, thereby further simplifying the programming. Two equations with 122 terms define each, R/Δ & τ and $S/\nabla^{2/3}$ & L_{WL}/L_P, with even 113 terms in common for R/Δ & τ and for $S/\nabla^{2/3}$ & L_{WL}/L_P. This demonstrates the relationship between dynamic trim and resistance, and wetted surface and its length, for planing craft. Once the influence of the interceptors is revealed, new MMs can be developed with, for example depth of interceptor, as a possible fifth independent variable.

[3] This is an upgrated and corrected version of work published under the same title at the High Speed Marine Vehicles Conference (HSMV 2017) in Napleas 2017.

3.5 Additional Comments on Modeling Resistance and Dynamic Trim for High-Speed Craft

3.5.1 MMs for Series 62/DSDS

Additional elaboration is required for the MMs for Series 62/DSDS (see also discussion on Series 62 and DSDS in Sect. 3.3) as there are four of them. The authors' opinion follows:

- MM 62 and 65 (Radojčić 1985) is the oldest. Nevertheless, in some cases it still might be useful since it contains some of Series 65-B characteristics. It should be used well within its own boundaries of applicability.
- PHF MM (Keuning et al. 1993) was not practical even when released (speed- and deadrise-independent, for 5 and 50 m^3 only). It is superseded by MMs released in Keuning and Hillege (2017a, b).
- Two new versions of DSDS MMs for prismatic (constant deadrise) hulls have been released recently (Keuning and Hillege 2017a, b; regression coefficients are available on DUT's website). The second version is more advanced than the first, even though it is still speed-independent. Both MMs are based on the subseries with $\beta = 19°$, which should be an advantage. However, since these experimental results seem to be incorrect, these MMs may be invalid too, at least for $\beta = 19°$ This conclusion was reached when (a) $\beta = 19°$ results were compared with predictions of other MMs, and (b) $\beta = 19°$ data was correlated with the data of other "siblings" (DSDS subseries with $\beta = 12.5, 25$, and $30°$). MMs for warped hulls, however, may be useful.
- Series 62 MM (Radojčić et al. 2017b), is by far the most advanced as is a $f(A_P/\nabla^{2/3}, L_P/B_{PX}, LCG/L_P, \beta, Fn_V)$. It seems to be the most accurate and reliable, especially in the intermediate range (between the data points). Lack of $\beta = 19°$ data in its foundation is substituted with the so called "virtual measurements" (see Radojčić et al. 2017a). Therefore, for the time being, this MM is recommended.

It is expected, however, that once a reliable database is available, a new and probably final Series 62/DSDS MM will be developed, and will probably include the characteristics for the DSDS subseries for warped hull forms. In any case, contemporary MMs should be both deadrise- and speed- dependent.

3.5.2 Stepped Hulls

Speeds corresponding to Fn_∇ of up to 8 are relevant for the small planing craft. Stepped hull forms would be advantageous for these speeds (not only from the resistance, but also from the dynamic stability viewpoint). For these hull types some experimental data exist (for instance, Taunton et al. 2010; De Marco et al. 2017), although adequate MM for resistance and dynamic trim evaluations are missing (except some tries based on the Savitsky method).

3.6 Mathematical Model Use

MM produce incorrect results for two main reasons:

1. MM is not good enough, as
   (a) It does not satisfactorily represent the experimental results it is based on, and/or
   (b) There is an unexpected behavior between the original data points.
2. MM is used incorrectly (discussed also in Sect. 1.5), as
   (a) Boundaries of applicability are violated, and/or
   (b) MM is not applicable for the target hull form.
   (Note that this is same as when inadequate prototype or wrong experimental results are used for the target hull's resistance evaluation).

The MM-developer is responsible for the errors of the first type, while MM-user is responsible for those of the second type. This should be taken into consideration when the MM's quality is judged. In fact, the only truly correct metric of MM's quality is a comparison of predicted versus measured value (errors ad 1.a above). Evaluation of the predicted model values that lie between the measured data points (ad 1.b) is not trivial and usually is performed by the MM developer.

The cause of the errors in 2.a is obvious, since the boundaries of MM applicability must be obeyed. However, the errors in 2.b require further discussion. Namely, just because the input parameters of a target hull are within the applicability boundaries of a MM, it does not mean that the MM is really usable for the particular target hull (discussed also in Sect. 2.1.1). That is, the target hull's secondary characteristics (i.e. hull form) may be dissimilar from those upon which the MM is based. For instance, the input parameters of a target semi-displacement yacht (L/B, $L/\nabla^{1/3}$ etc.) may satisfy the applicability boundaries of a MM for trawlers, but the hull form of these two vessel types are considerably different, which obviously disqualifies this MM.

Nevertheless, there are cases when the available MM is developed for the same or similar vessel types as the target hull, but the secondary parameters of the respective hulls differ (e.g. MM is based on the narrow stern planing hull forms, while the target hull is of a wide transom form). In these cases the MM may be applied, but the target hull's input parameters required by the MM should be modified. This modification is called "mapping of input parameters" and consists of replacing the input parameters with the suitably adjusted *effective* values. For instance, the required input parameter could be L_P/B_{PA} and/or β_M, while the effective ones could be L_P/B_{PX} and/or β_{Bpx}, respectively. Consequently, the effective input parameters are not necessarily the same as the ones suggested for the MM. Mapping is not a straightforward procedure because an understanding of the issue is necessary and hence, in most cases is either ignored or is wrongly employed. No wonder then that mapping is amongst the principal causes of errors (see Blount and Fox 1976; Savitsky and Brown 1976; Savitsky 2012, etc.). Note that mapping is much more important (influential) for dynamic trim prediction than for resistance prediction (see Radojčić et al. 2017b).

It follows that the MM's users should be aware of the hull forms and development methods used for derivation of the MM. Moreover, without adequate interpretation of the input parameters and prediction results, statistically based MMs can lead to erroneous results. MacPherson (2003) summarized typical problems with numerical model performance predictions and proposed practical solutions. Similarly, regardless of how resistance and propulsion are estimated (i.e. by model experiments, analytical method, CFD etc.), common powering prediction errors are given in Almeter (2008). Over fifty possible sources of errors are discussed with a focus on the design. It is concluded that the most accurate prediction method is not always the most expensive one.

A MM's predictive accuracy is of paramount importance for the user; see Sect. 2.1. However, it is not really possible to determine the accuracy of the MM's predictions even when the developer took all possible precautions and checked the relevant statistics, assessed the discrepancies between measured and evaluated values, and validated the behavior of the MM between the data points. A "correlation procedure" which could be used to improve a MM's accuracy is given in Van Hees (2017). This approach essentially consists of tuning the MM to hull forms similar to the subject hull, but with known and available performance results. The same paper clarifies the correlation[4] procedures applying the Holtrop and Mennen method (Holtrop and Mennen 1982). Note that a similar approach, i.e. correlation procedure, could be used for improving the mapping.

The predictions of the four MMs *based* on the Series 62 are compared with the Series 62 (or DSDS) measurements in Figs. 3.1 and 3.2. A comparison of predicted results with the test results of *similar* hull forms is more demanding; see Figs. 3.3 and 3.4. Note that a user can be sure of the results only by using as many predictive methods as possible. That is, in everyday design application the user has to rely entirely on the MM's predictions as the actual value is not known.

[4]"Correlation is the process of comparing experimental and numerical results in ship hydrodynamics with the aim to improve prediction accuracy" (Van Hees 2017).

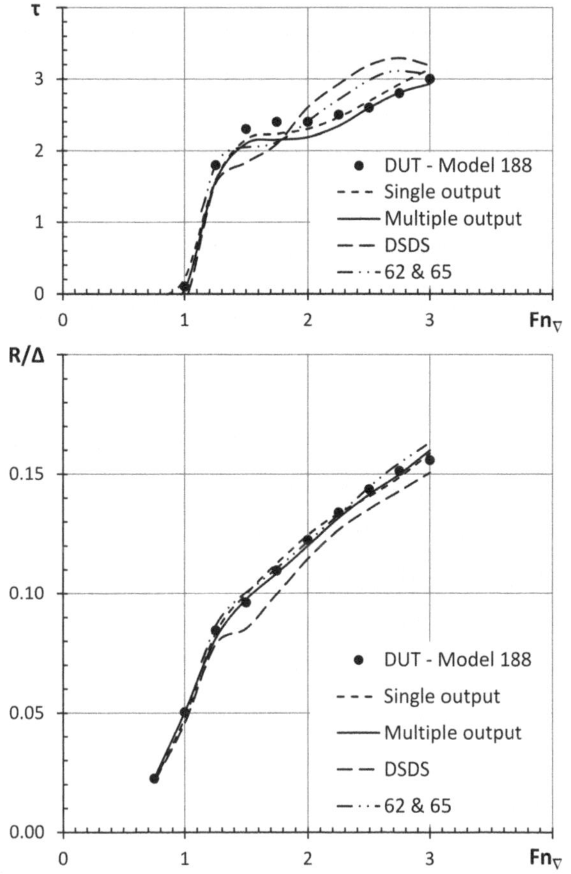

Fig. 3.1 Comparison of four MMs with the Series 62/DSDS measurements—$\beta = 25°$, $A_P/\nabla^{2/3} = 7$, %LCG = 4%, $L_P/B_{PX} = 4.1$ (Single- and Multiple-output—Radojčić et al. 2017b, DSDS—Keuning and Hillege 2017a, 62 and 65—Radojčić 1985)

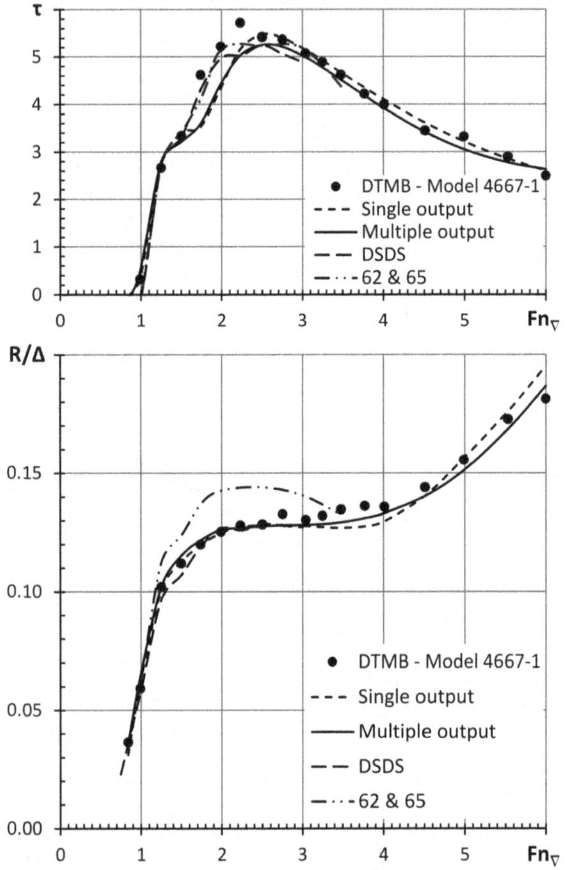

Fig. 3.2 Comparison of four MMs with the Series 62/DSDS measurements—$\beta = 12.5°$, $A_P/\nabla^{2/3} = 5.5$, $\%LCG = 4\%$, $L_P/B_{PX} = 4.1$ (Single- and Multiple-output—Radojčić et al. 2017b, DSDS—Keuning and Hillege 2017a, 62 and 65—Radojčić 1985). *Note* In the $\tau = f(Fn_\nabla)$ graph, three data points for $Fn_\nabla = 1.75$, 2.00 and 2.25 are outliers, i.e. they are measurement errors, according to Radojčić et al. (2017b). This was detected by ANN based MMs, while regression based MMs obviously tracked erroneous measurements

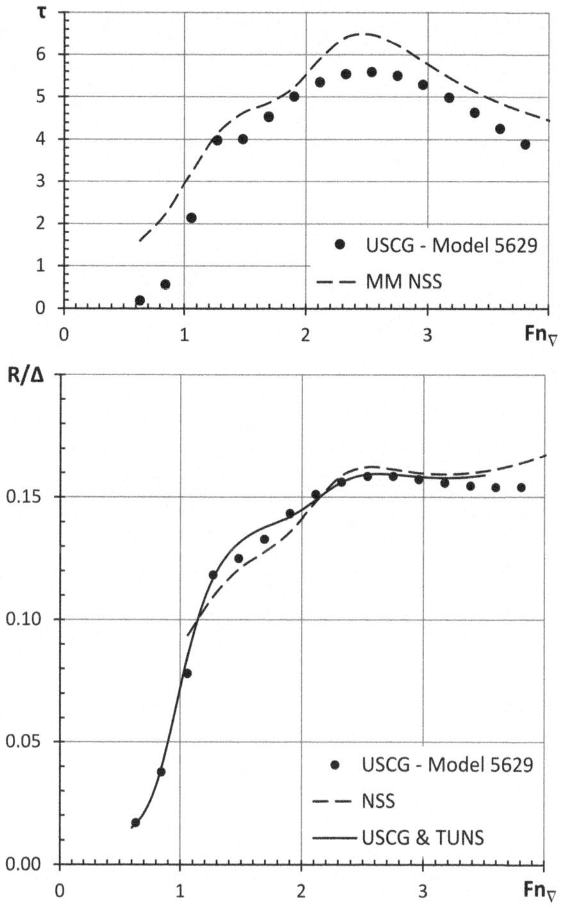

Fig. 3.3 Comparison of two hard chine, wide-transom MMs with the USCG (Model 5629) measurements—$\beta = 23^{\circ}$, $L_P/\nabla^{1/3} = 5.15$, $LCG/L_P = 0.373$, $L_P/B_{PX} = 4.09$ (NSS—Radojčić and Kalajdžić 2018, USCG and TUNS—Radojčić et al. 2014a)

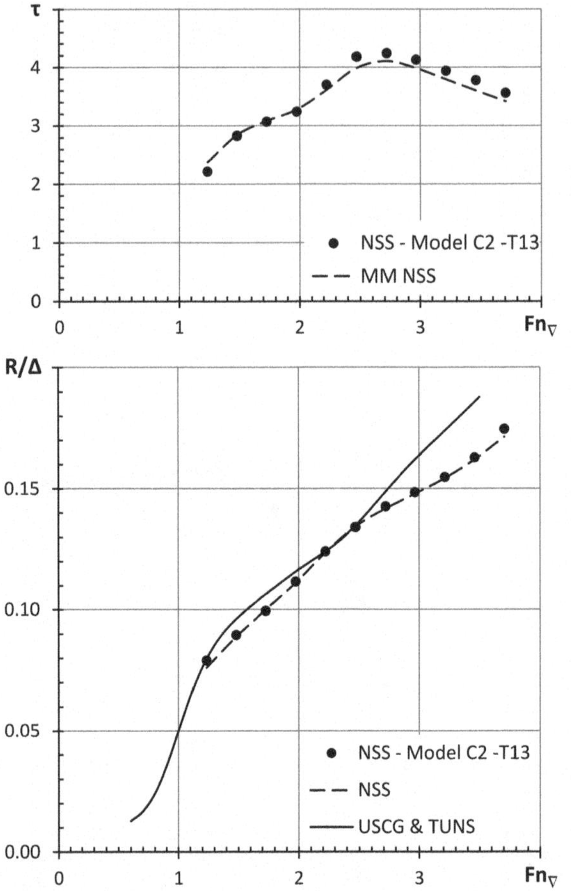

Fig. 3.4 Comparison of two hard chine, wide-transom MMs with the NSS (Model C2-T13) measurements—$\beta = 22.3°$, $L_P/\nabla^{1/3} = 5.95$, $LCG/L_P = 0.38$, $L_P/B_{PX} = 3.89$ (NSS—Radojčić and Kalajdžić 2018, USCG and TUNS—Radojčić et al. 2014a)

3.7 Recommended Mathematical Models for Resistance and Dynamic Trim Prediction

The following MMs for resistance and dynamic trim prediction, out of the 18 that have been discussed here, are recommended.

3.7.1 MMs for Semi-displacement Hull Forms

Transom-stern Mercier and Savitsky (MM of random hull form class) and NPL, VTT, SKLAD, and NTUA (methodical series class MMs). The length Froude number (Fn_L) of up to 1.1 is of interest for all of them. Applicability range of these MMs is shown in Fig. 3.5.

3.7.2 MMs for Semi-planing or Planing Hull Forms

62 & 65 and USCG & TUNS (each MM comprises two methodical series, and are therefore a kind of random hull form class), and 50, 62, and NSS (methodical series

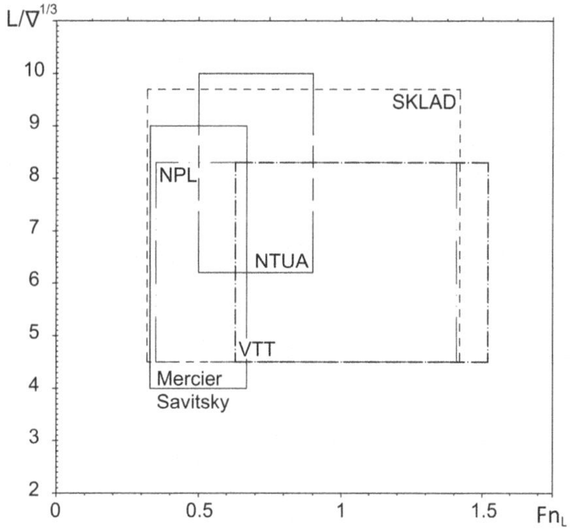

Fig. 3.5 Applicability range of MMs for semi-displacement hull forms (Mercier and Savitsky—Mercier and Savitsky 1973, VTT—Lahtiharju et al. 1991, NPL—Radojčić et al. 1997, SKLAD—Radojčić et al. 1999, NTUA—Radojčić et al. 2001)

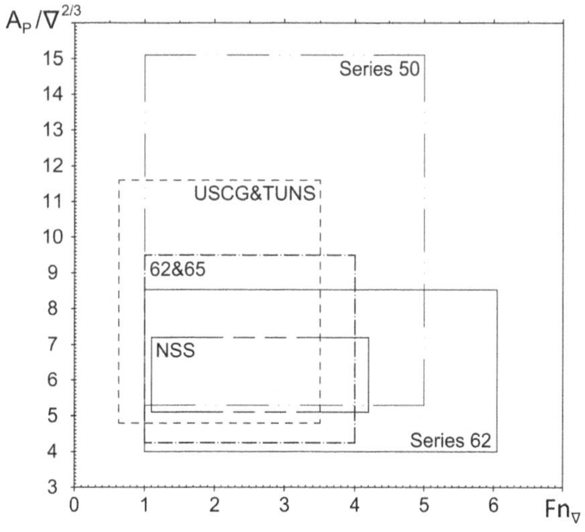

Fig. 3.6 Applicability range of MMs for semi-planing and planing hull forms (62 and 65—Radojčić 1985, USCG and TUNS—Radojčić et al. 2014a, Series 50—Radojčić et al. 2014b, Series 62—Radojčić et al. 2017b, NSS—Radojčić and Kalajdžić 2018)

class MMs). For this group volumetric Froude number (Fn_∇) of up to 5.5 or so is appropriate. Applicability range of these MMs is shown in Fig. 3.6.

For quick reference the main characteristics of the recommended MM for semi-displacement, semi-planing and planing hull forms are presented in the Tables 8.1 and 8.14 respectively (Sects. 8.2 and 8.3). Note that none of the MMs for semi-displacement hull forms depend on LCG, while LCG is among the primary input variables for all hard chine forms.

Note for Figs. 3.5 and 3.6: Dimensionless speed and loading parameters are interrelated, i.e. $Fn_\nabla = Fn_L \cdot (L/\nabla^{1/3})^{1/2}$ and $L_P/\nabla^{1/3} \approx 1.1 \cdot [A_P/\nabla^{2/3} \cdot L_P/B_{PX}]^{1/2}$ (assuming $L/L_P \approx 0.98 \approx 1$ and $A_P \approx 0.83 \cdot L_P \cdot B_{PX}$, which stems from the Series 62 with average value for $B_{PX}/B_{PA} \approx 1.22$); see Blount (2014).

References

Almeter JM (1988) Resistance prediction and optimization of law deadrise, hard chine, stepless planing hulls. SNAME STAR Symposium

Almeter JM (1993) Resistance prediction of planing hulls: state of the art. Marine Technol 30(4)

Alemeter JM (2008) Avoiding common errors in high-speed craft powering predictions. In: 6th international conference on high performance marine vehicles, Naples

Bailey D (1976) The NPL high speed round bilge displacement hull series. In: Maritime Technology Monograph No. 4. RINA

Bertram V, Mesbahi E (2004) Estimating resistance and power of fast monohulls employing artificial neural nets. Int. Conf High Performance Marine Vehicles (HIPER), Rome

Blount DL (2014) Performance by design. ISBN 0-978-9890837-1-3

Blount DL, Fox DL (1976) Small craft power prediction. Marine Technol 13(1)

Blount DL, McGrath JA (2009) Resistance characteristics of semi-displacement mega yacht hull forms. RINA Trans, vol 151, Part B2, Int. J. Small Craft Technol, July–Dec

Clement PE, Blount DL (1963) Resistance tests of a systematic series of planing hull forms. SNAME Trans 71

Compton RH (1986) The resistance of a systematic series of semi-planing transom stern hulls. Marine Technol 23(4)

Couser P, Mason A, Mason G, Smith CR, Konsky BR von (2004) Artificial neural network for hull resistance prediction. In: 3rd international conference on computer and IT applications in the maritime industries (COMPIT' 04), Siguenza

Davidson KSM, Suarez A (1941) Tests of twenty related models of V-bottom motor boats—U.S. E.M.B. Series 50. Report No. 170, Experimental Towing Tank, Stevens Institute of Technology, Hoboken, NJ

Delgado-Saldivar G (1993) Experimental investigation of a new series of planing hulls. M.Sc Thesis. Technical University of Nova Scotia, Halifax, Nova Scotia

De Luca F, Pensa C (2017) The Naples warped hard chine hulls systematic series. Ocean Engineering, 139

De Marco A, Mancini S, Miranda S, Scognamiglio R (2017) Experimental and numerical hydrodynamic analysis of a stepped planing hull, Appl Ocean Res, 64

Doust DJ (1960) Statistical analysis of resistance data for trawlers. In: Fishing boats of the world: 2 fishing. News (Books) Ltd., London

Farlie-Clarke AC (1975) Regression analysis of ship data. Int Shipbuilding Progress 22(251)

Fung SC (1991) Resistance and powering prediction for transom stern hull forms during early stage ship design. SNAME Trans 99

Fung SC, Leibman L (1993) Statistically-based speed-dependent powering predictions for high-speed transom stern hull forms. Chesapeake Section of SNAME

Gamulin A (1996) A semidisplacement series of ships. Int Shipbuilding Progress 43(43)

Grigoropoulos GJ, Damala DP (2001) The effect of trim on the resistance of high-speed craft. In: 2nd international EURO conference on high-performance marine vehicles, HIPER' 01, Hamburg

Grigoropoulos GJ, Loukakis TA (1999) Resistance of double–chine large high–speed craft. In: Aeronautique ATMA, vol. 99. Paris

Grubišić I, Begović E (2000) resistance prediction of the fast round-bilge hulls at the concept design level. In: Proceedings of the 9th international congress of the international Association of Mediterranean, IMAM, Ischia

Hadler JB, Hubble EN, Holling HD (1974) Resistance characteristics of a systematic series of planing hull forms—Series 65. Chesapeake Section of SNAME

Holling HD, Hubble EN (1974) Model resistance data of a series 65 hull forms applicable to hydrofoils and planing craft. NSRDC Report 4121

Holtrop J, Mennen GGJ (1982) An approximate power prediction method. Int Shipbuilding Progress. 29(335)

Hubble EN (1974) Resistance of hard-chine stepless planing craft with systematic variation of hull form, longitudinal centre of gravity and loading. DTNSRDC R&D Report 4307

Jin P, Su B, Tan Z (1980, September) A parametric study on high-speed round bilge displacement hulls. High-Speed Surface Craft

Keuning JA, Geritsma J (1982) Resistance tests of a series of planing hull forms with 25 degrees deadrise angle. Int Shipbuilding Progress 29(337)

Keuning JA, Hillege L (2017a) The results of delft systematic deadrise series. In: Proceedings of 14th international conference on fast sea transportation (FAST 2017), Nantes

Keuning JA, Hillege L (2017b) Influence of rocker and twist and the results of the delft systematic deadrise series. In: High speed marine vehicles conference on (HSMV 2017), Naples

Keuning JA, Gerritsma J, Terwisga PF (1993) Resistance tests of a series planing hull forma with 30° deadrise angle, and a calculation model based on this and similar systematic series. Int Shipbuilding Progress 40(424)

Kowalyshyn DH, Metcalf B (2006) A USCG systematic series of high speed planing hulls. SNAME Trans 114

Lahtiharju E, Karppinen T, Hellevaara M, Aitta T (1991) Resistance and seakeeping characteristics of fast transom stern hulls with systematically varied form. SNAME Trans 99

MacPherson DM (2003) Comments on reliable prediction accuracy. A HydroComp Technical Report 103

Mason A, Couser P, Mason G, Smith CR, Konsky BR von (2005) Optimisation of vessel resistance using genetic algorithms and artificial neural networks. In: 4th international conference on computer and IT applications in the maritime industries (COMPIT' 05), Hamburg

Mercier JA, Savitsky D (1973) Resistance of transom-stern craft in the pre-planing regime. Davidson Laboratory Report 1667

Molland AF, Turnock SR, Hudson DA (2011) Ship resistance and propulsion—practical estimation of ship propulsive power. Cambridge University Press, ISBN 978-0-521-76052-2

Moore WL, Hawkins F (1969) Planing boat scale effects on trim and drag (TMB Series 56). NSRDC Technical Note No. 128, Washington

Morabito MG (2013) Re-analysis of series 50 tests of V-bottom motor boats. SNAME Trans 121

Morabito M, Snodgrass J (2012) The use of small model testing and full scale trials in the design of motor yacht. In: SNAME's 3rd Chesapeake Power Boat Symposium, Annapolis

Müller-Graf B (1999) Widerstand und hydrodynamische Eigenschaftender schnellen Knickspant-Katamarane der VWS Serie'89 (Resistance and hydrodynamic characteristics of the VWS fast hard chine catamaran series' 89), 20th Symposium Yachtenwurf und Yachtbau, Hamburg

Radojčić D (1984) A statistical method for calculation of resistance of the stepless planing hulls. Int Shipbuilding Progress 31(364)

Radojčić D (1985) An approximate method for calculation of resistance and trim of the planing hulls. University of Southampton, Ship Science Report No. 23. Paper presented on SNAME Symposium on Powerboats

Radojčić D (1991) An engineering approach to predicting the hydrodynamic performance of planing craft using computer techniques. RINA Trans 133

Radojčić D, Kalajdžić M (2018) Resistance and trim modeling of naples hard chine systematic series. RINA Trans Int J Small Craft Technol https://doi.org/10.3940/rina.ijsct.2018.b1.211)

Radojčić D, Rodić T, Kostić N (1997) Resistance and trim predictions for the npl high speed round bilge displacement hull series. RINA Conf On Power, Performance and Operability of Small Craft, Southampton

Radojčić D, Prinčevac M, Rodić T (1999) Resistance and trim predictions for the SKLAD semidisplacement hull series. Oceanic Eng Int 3(1)

Radojčić D, Grigoropoulos GJ, Rodić T, Kuvelić T, Damala DP (2001) The resistance and trim of semi-displacement, double-chine, transom-stern hull Series. In: Proceedings of 6th international conference on fast sea transportation (FAST 2001), Southampton

Radojčić D, Zgradić A, Kalajdžić M, Simić A (2014a) Resistance prediction for hard chine hulls in the pre-planing regime. Polish Maritime Res 21,2(82). Gdansk

Radojčić D, Morabito M, Simić A, Zgradić A (2014b) Modeling with regression analysis and artificial neural networks the resistance and trim of series 50 experiments with V-bottom motor boats. J Ship Prod Design 30(4)

Radojčić DV, Zgradić AB, Kalajdžić MD, Simić AP (2017a) Resistance and trim modeling of systematic planing hull series 62 (With 12.5, 25 and 30 Degrees Deadrise Angles) using artificial neural networks, part 1: the database. J Ship Prod Design 33(3)

Radojčić DV, Kalajdžić MD, Zgradić AB, Simić AP (2017b) Resistance and trim modeling of systematic planing hull series 62 (With 12.5, 25 and 30 Degrees Deadrise Angles) using artificial neural networks, part 2: mathematical models. J Ship Prod Design 33(4)

Robinson JL (1999) Performance prediction of chine and round bilge hull forms. RINA Int. Conf. on Hydrodynamics of High Speed Craft, London

Sabit AS (1971) Regression analysis of the resistance results of the BSRA series. Int Shipbuilding Progress. 18(197)

Sahoo P, Peng H, Won J, Sangarasigamany D (2011) Re-evaluation of resistance prediction for high-speed round bilge hull forms. In: Proceedings of 11th international conference on fast sea transportation (FAST 2011), Honolulu

Savitsky D (1964) Hydrodynamic design of planing hulls. Marine Technol 1(1)

Savitsky D (2012) The effect of bottom warp on the performance of planing hulls. In: SNAME's 3rd Chesapeake Power Boat Symposium, Annapolis

Savitsky D, Brown PW (1976) Procedure for hydrodynamic evaluation of planing hulls in smooth and rough water. Marine Technol 13(4)

Savitsky D, Roper JK, Benen L (1972) Hydrodynamic development of a high speed planing hull for rough water. In: 9th symposium naval hydrodynamics, ONR, Paris

Swift PM, Nowacki H, Fischer JP (1973) Estimation of great lakes bulk carrier resistance based on model test data regression. Marine Technol 10(4)

Tanaka H, Nakato M, Nakatake K, Ueda T, Araki S (1991) Cooperative resistance tests with geosim models of a high-speed semi-displacement craft. J SNAJ 169

Taunton DJ, Hudson DA, Shenoi RA (2010) Characteristics of a series of high speed hard chine planing hulls—part 1: performance in calm water. RINA Trans 152. Part B2 Int J Small Craft Technol

Van Hees MT (2017) Statistical and theoretical prediction methods. In: Encyclopedia of maritime and offshore engineering Wiley

Van Oortmerssen G (1971) A power prediction method and its application to small ships. Int Shipbuilding Progress 18

Van Oossanen P (1980) Resistance prediction of small high-speed displacement vessels: state of the art. Int Shipbuilding Progress 27(313)

Yegorov IT, Bunkov MM, Sadovnikov YM (1978) Propulsive performance and seaworthiness of planing vessels. Sudostroenie, Leningrad (in Russian)

Chapter 4
Propeller's Open-Water Efficiency Modeling

4.1 An Overview of Modeling Propeller's Hydrodynamic Characteristics

Modeling propeller's open water hydrodynamic characteristics is in many respects different from modeling resistance, although the same tools and methods are used. Two main differences should be emphasized:

1. Dependent variables that should be modeled simultaneously are thrust coefficient (K_T) and torque coefficient (K_Q). By definition, these coefficients are interrelated (linked) through the expression for the open water efficiency—$\eta_o = (K_T/K_Q) \cdot (J/2\pi)$. There is nothing similar for modeling resistance as there is no explicit relationship between resistance and any other dependent variable. Resistance and trim, for instance, are correlated, but are not linked.
2. While the dependent variables are always K_T and K_Q, the independent ones are some or all of the following: advance coefficient ($J = v_a/nD$), pitch ratio (P/D), area ratio (A_E/A_O or A_D/A_O), number of blades (z), and cavitation number (σ or $\sigma_{0.7R}$). This pre-determination makes modeling easier, since there is no need to search for optimum independent variables best suited for a particular propeller series.

Note that although $K_T - K_Q - \eta_O$ interdependence is a fact, some MMs improperly ignore this, often resulting in a chaotic η_O curve. Namely, modeling a propeller's open water hydrodynamic characteristics is, in mathematical terms, *a multiple objective (or multicriteria) optimization with constraints. Multiple objective* because a MM should be simultaneously obtained for both K_T and K_Q, and *with constraints* because K_T and K_Q are linked through an equality constraint $\eta_o = (K_T/K_Q) \cdot (J/2\pi)$.

Input variables depend on the size of the propeller series that is modeled. From the complexity viewpoint, MMs may be classified as follows (similarly done in Radojčić 1988):

D. Radojčić et al., *Power Prediction Modeling of Conventional High-Speed Craft*,
https://doi.org/10.1007/978-3-030-30607-6_4

Rank-1—A single propeller may be modeled satisfactorily by a third degree polynomial—K_T, K_Q = f(J).
Rank-2—A small series where P/D is varied (A_E/A_O = const., z = const.) polynomial eq. is K_T, K_Q = f(J, P/D).
Rank-3—A series with only z = const., polynomial eq. is K_T, K_Q = f(J, P/D, A_E/A_O).
Rank-4—A series where all four variables are varied, polynomial eq. is K_T, K_Q = f(J, P/D, A_E/A_O, z).
When cavitation becomes influential the cavitation number σ or $\sigma_{0.7R}$ should also be one of the independent variables (and the database comes from the cavitation tunnel experiments):
Rank-5—Extension to Rank-3 above as cavitation number is varied, so eq. is K_T, K_Q = f(J, P/D, A_E/A_O, σ).
Rank-6—A series where all five variables are varied, polynomial eq. is K_T, K_Q = f(J, P/D, A_E/A_O, z, σ).

Two facts should be noted here:

- All lower level MMs, belonging to Ranks-1 to 5, may be regarded as a special case of a Rank-6 MM, and
- Given that HSC propellers inherently cavitate, to a smaller or greater extent, MMs of interest in this work belong to Rank 5 or 6. However, due to a variety of reasons, models belonging to Rank 5 were developed only recently and to the best of authors' knowledge Rank 6 models do not yet exist.[1]

The modeling methods discussed above imply that some mathematical expression adequately represents the dataset. A different approach is given in Bukarica (2014) where the entire dataset is a part of a MM, while the modeling of multi-dimensional surfaces (i.e. K_T and K_Q) is done with the *interpolating spline functions*. Although the multidimensional surface obtained through the application of splines is smooth, this approach does not seem to be of practical use, hence will not be elaborated further.

4.1.1 MARIN Propeller Series

The first Rank-2 polynomials for 4- and 5-bladed Wageningen B-series propellers were presented in Van Lammeren et al. (1969). After elimination of the negligible terms these polynomials had 10 terms for both K_T and K_Q. The same paper presented Rank-3 polynomials for 4- and 5-bladed B-series propellers with 18–26

[1]With the exception of cases where ANN was applied directly for obtaining K_T and K_Q (e.g. Neocleous and Schizas (2002) where commercial propellers presented in Denny et al. (1988) were modeled), and not for MM development (which then can be used by other users who do not have knowledge about ANN whatsoever).

terms. Finally, the entire B-series was presented with a Rank-4 MM with 39 and 47 polynomial terms for K_T and K_Q respectively, in Oosterveld and van Oossanen (1975). Ducted propeller K_a-series in various nozzles are presented with Rank-2 polynomials and have around 10 terms for each K_T and K_Q.

Polynomial equations of Rank-4 for B-series, as well as Rank-2 for various pairs of propeller + nozzle, have been finalized and have been in worldwide use for the last four decades. Traditional charts practically do not exist anymore. MARIN's book, published on the 60th anniversary of the Institute (Kuiper 1992), is the crossover between chart-based and computer-based presentation of the above-mentioned Wageningen propeller series. Nowadays the purpose of charts, if provided, is only to prove the modeling was done correctly.

Note that the B-series is by far the largest series, consisting of 120 propellers that were tested over a period of almost 40 years. Experiments were conducted in different facilities, under different conditions and for different Reynolds numbers, so that regression analysis, amongst other faired experimental results and formed unified database too. Derived polynomials are valid for $Rn = 2 \times 10^6$; corrections for other Reynolds numbers are given with the additional polynomials ΔK_T and ΔK_Q. Note that the traditional charts did not even state the Rn for which they were valid.

The B-series, with inner blade sections of airfoil type and outer sections of segmental type (see Sect. 9.2.1 and Fig. 9.10) is a fixed-pitch general purpose propeller series, and is currently used extensively for both design and benchmarking purposes (see Carlton 2012). B-series polynomials are usable for speeds of up to 20–30 kn or so, for z = 2–7, P/D = 0.5–1.4, and A_E/A_O = 0.3–1.05. Note however that caution should be exercised with the boundaries of applicability (see Radojčić 1985).

Common open-water $K_T - K_Q - J$ charts are practically just graphical presentation of B-series polynomials. However, there are other representations or formats (Troost $B_P - \delta$ and $B_u - \delta$ for instance) that enable easier propeller optimization, and are important mostly if charts, rather than MMs, are used. All parameters necessary for other formats may be easily recalculated from the open water $K_T - K_Q - J$ values. Consequently, there are several propeller optimization tools based on the B-series polynomials, often developed with chart presentation in mind (see for instance Radojčić 1985; Yosifov et al. 1986; Loukakis and Gelegeris 1989; Shen and Marchal 1993 etc.). Application of an ANN for selection of a maximum efficiency ship propeller is given in Matulja et al. (2010), for instance.

Propeller characteristics used for ship maneuvering (stopping, astern sailing etc.) are usually presented with four quadrant diagrams where $C_T^* - C_Q^* - \beta$ are used instead of $K_T - K_Q - J$ parameters. C_T^* and C_Q^* are, respectively, thrust and torque indexes, while β is a hydrodynamic pitch angle at 0.7R. The effect of P/D, A_E/A_O, and z in the 4-quadrant presentation is analyzed traditionally with a Fourier analysis (e.g. Van Lammeren et al. 1969), but the ANN technique was recently also used (see Roddy et al. 2006). The B-series and ducted K_a-series propellers in nozzles, both being trademarks of MARIN, are succeeded by new 4- and 5-bladed C- and ducted D-Series of Controllable Pitch Propellers (CPP), as described in

Dang et al. (2013) where 2-quadrant diagrams in $C_T^* - C_Q^* - \beta$ format are given (important for propeller's off-design conditions).

Note that a direct relationship between open water K_T, K_Q, and J, and respectively, 4-quadrant C_T^*, C_Q^*, and β parameters does exist. Conventional open water $K_T - K_Q - J$ presentation, corresponding to the first quadrant, is in fact just a special case of the 4-quadrant presentation. Being simpler and more practical for the subject in hand (i.e. power prediction), traditional $K_T - K_Q - J$ coefficients are used in all further discussions here.

When RPM is required $K_T - K_Q - J$ can be easily preformatted to K_T/J^2 or K_Q/J^3, while the known input values are R, D, and v, or P, D, and v, respectively. However, if D is required and RPM is known, transformation K_T/J^4 or K_Q/J^5, respectively would be adequate. The thrust-loading format η_O, $J = f(K_T/J^2)$ is frequently used by the HSC community.

Summarizing, it may be concluded that the propeller series, compared to ship systematic series, are more resistant to the passage of time (Van Hees 2017). The reason for this is the fact that propellers' hydrodynamic characteristics can be modeled with just a few parameters, which is not the case with the ship hull hydrodynamic characteristics.

4.2 Mathematical Modeling of K_T, K_Q, and η_O of High-Speed Propellers

While MMs for HSC resistance prediction were presented in a chronological order, with each more or less more sophisticated than the previous one, MMs for propellers, ranked as explained above, are presented from the application viewpoint, i.e. going from (for the HSC) lower to higher vessel speeds.

4.2.1 AEW and KCA Propeller Series

HSC propellers should be more resistant to cavitation than the B-series. Flat-faced, segmental section propellers are inherently more resistant to the inception of cavitation than those with the airfoil sections. They also have good open-water characteristics and are easier to manufacture and repair. Amongst the first systematic tests of these propellers were those reported by Gawn (1953) and Gawn and Burrill (1957). Parameter space of the series is roughly P/D = 0.6–2.0, $A_E/A_O = 0.5–1.1$, and $\sigma = 0.5–6.3$, but the range of MMs discussed here vary from one MM to another. Both series consist of similar three bladed propellers (see Figs. 9.11 and 9.12), but are tested in different experimental facilities, and under different conditions. Gawn (1953) AEW data are valid for open-water conditions, and Gawn and Burrill (1957) KCA data is for atmospheric and cavitating conditions. Note that the results of open-water experiments and cavitation experiments at atmospheric conditions are not necessarily the same, as is often assumed, but any further discussion on this topic is beyond the scope of this work.

Based on the aforementioned experimental data, four MM were developed. The first model is based on the AEW data (for non-cavitating conditions only), and the other three on the KCA data. Note that up to 10% back cavitation might be regarded as non-cavitating, as the breakdown point is not reached yet. The first three MMs were developed using multiple regression techniques, while the fourth one is derived by ANN. These MMs are compared in Radojčić et al. (2009).

Model 1—(Blount and Hubble 1981)

Consists of separate equations for the non-cavitating (open-water) and for the cavitating regime (see Sect. 9.2.2). Open water equations are actually the B series polynomials with 39 and 47 terms for K_T and K_Q respectively (i.e. Rank-4 MM), but new regression coefficients were evaluated. The cavitating regime however, is represented with a relatively simple set of equations (see Blount and Fox 1978, also discussed in Sect. 4.3). Note that from the mathematical viewpoint, use of the same polynomial terms to model different test data, i.e. AEW and B-series propellers, and hence extending model's validity to four bladed propellers, is not correct, although it may be considered practical.

Model 2—(Kozhukharov 1986)

Consists of a pair of single equations for both regimes, with 121 and 116 polynomial terms for K_T and K_Q respectively. Hence this is a Rank-5 MM, although a relatively simple function transformation was used for the dependent variables.

Model 3—(Radojčić 1988)

Consists of a pair of equations for the non-cavitating, and an additional pair for the cavitating regime (see Sect. 9.2.3). The first set, belonging to Rank-3 MMs, has 16 and 17 polynomial terms for K_T and K_Q. The second set has 20 and 18 terms for evaluation of ΔK_T and ΔK_Q reductions for cavitating conditions. These are f(A_D/A_O, P/D, $\sigma_{0.7R}$, K_T) so that for the cavitating regime there are in total 36 and 35 terms respectively.

Model 4—(Koushan 2007)

Separate equations are derived for the non-cavitating and cavitating conditions, with 34 equation terms for each K_T and K_Q for the non-cavitating regime, and 89 and 107 terms respectively for the cavitating regime (see Sect. 9.2.4). Linear and hyperbolic tangent functions were used for the activation function for the non-cavitating and the cavitating regime, respectively. In a broader sense the first set of equations would belong to the Rank-3 MM and the second to the Rank-5 MM.

Discussion of Models 1 to 4—(Radojčić et al. 2009)

For all four abovementioned MMs additional narrower, and more conservative, boundaries of applicability are suggested in Radojčić et al. (2009). Also, special attention was paid to the η_o representation for all four models, since modeling errors in independently evaluated K_T and K_Q are well illustrated by η_o (see Radojčić 1988). Model 1 seems to have overlooked this in the cavitating regime only. Model 2 and Model 4 produce inconsistent values of η_o even for non-cavitating regimes. Models 1 and 3 however, do have disadvantages around the K_T and K_Q cavitation breakdown points, as the transition zone (multidimensional plane) should be faired smoothly into the open-water data (another multidimensional plane).

Model 1 and Model 3 are valid for the non-cavitating regime all the way through to the inception of cavitation (intersection of open-water and transition zone, i.e. K_T and K_Q breakdown points). Model 4 however, consists of separate equations for the cavitating and non-cavitating regimes for the whole J-range. Note that from the physical viewpoint this is not correct. Model 2 is based on the single equations for both cavitating and non-cavitating regimes. In principle this is correct, but is difficult for modeling because the character of K_T and K_Q curves are completely different for the non-cavitating and cavitating conditions.

Finally, it was concluded that the advantage of Model 1 is its simplicity and validity for heavily-cavitating propellers. Model 3 is probably the best for non-cavitating conditions, while Model 4 is advantageous for the transition zone. Model 2 appears to have no advantages compared to the others. Model 1 is applicable for 3- and 4-bladed propellers, while all other models are valid for 3-bladed propellers only (as is the case for both AEW and KCA propeller series).

4.2.2 Newton-Rader Propeller Series

Newton-Rader three bladed propeller series (Newton and Rader 1961) covered P/D, A_E/A_O, and σ range from about 1–2, 0.5–1.1, and 0.25–2.5, respectively. This hollow-faced blade section series (see Fig. 9.14) is intended for very high speeds of 50 or so knots, while the flat-faced segmental section propellers are intended for speeds of up to approximately 40 kn. Note however, that this kind of propellers generally require custom-design and manufacture, in contrast to the B-series and flat-faced segmental section propellers.

There are two MMs that represent the Newton-Rader propeller series. The first one (Kozhukarov and Zlatev 1983) was developed using the multiple regression technique, while the second one (Koushan 2005) used ANN. Both MMs are very similar to Model 2 and Model 4 respectively, for the segmental section propeller series. The regression Rank 5 model has 101 polynomial terms for both K_T and K_Q, while ANN derived models for the non-cavitating and cavitating conditions have, respectively, 34 and 101 equation terms for each K_T and K_Q (see Sect. 9.2.5). Note however, that the authors could not obtain reasonable results using K_T and K_Q polynomials given in Kozhukarov and Zlatev (1983). This was also reported by Diadola and Johnson (1993).

Taking into account: (a) the abovementioned similarity between the MMs (Kozhukarov and Zlatev 1983 versus Kozhukharov 1986, and Koushan 2005 versus Koushan 2007); and (b) negligence of comparison of recently developed ANN models with the existing regression model (Koushan 2005, and Kozhukarov and Zlatev 1983), it might be expected that most of the conclusions derived in Radojčić et al. (2009) for AEW and KCA propeller series are also valid for the Newton-Rader MMs.

4.2.3 Swedish SSPA Ma and Russian SK Series

Both the Swedish SSPA Ma 3- and 5-bladed series ($A_E/A_O = 0.75$–1.2, P/D = 1–1.45, $\sigma = 0.25$–atm., see Lindgren 1961), and Russian 3-bladed SK series ($A_E/A_O = 0.65$–1.1, P/D = 1–1.8, $\sigma = 0.3$–atm., see Mavludov et al. 1982) are similar to the Newton-Rader series.

For the SSPA Ma 3-075 propeller series for open water conditions the polynomial equations for evaluation of K_T and $K_Q = f\,(J, \sigma, P/D)$, with 22 and 18 terms respectively, are given in Blount and Bjarne (1989). Note that K_T and $K_Q = f\,(\sigma)$, which is unusual.

Concerning the SK series, the first author initiated a project (Milićević 1998) with the purpose to regress the K_T and K_Q data for cavitating conditions, as done in Radojčić (1988), but without success. Namely, K_T, K_Q, and η_o curves, as published in Mavludov et al. (1982), were found not to be consistent for the SK series (K_T and K_Q curves are linked through $\eta_o = (K_T/K_Q) \cdot (J/2\pi)$ as already discussed in Sect. 4.1). Only a Rank 3 MM for open water conditions with 21 and 26 polynomial terms for K_T and K_Q, respectively, could be obtained.

4.2.4 SPP Series

Surface Piercing Propellers (SPP) are also intended for very high speeds, above 40 or so knots (see Allison 1978 and Kruppa 1990). SPP's modus operandi at high speeds is: (a) reduced cavitation danger due to ventilation on the suction side (the pressure side remains fully wetted), and (b) avoidance of appendage resistance (the appendages are above the water level). Additional side benefits include a vertical force, but only when the SPPs are properly installed. SPPs have few disadvantages; torque absorption depends on propeller immersion which, depending on the SPP type, may be influenced by shaft inclination. Consequently, amongst the SPP's input variables is immersion ratio (h/D), or shaft inclination (Ψ).

A MM for the SPP series based on five 4-bladed Rolla Series ($A_E/A_O = 0.8$, P/D = 0.9–1.6) is given In Radojčić and Matić (1997). Tests were carried out in a free-surface cavitation tunnel under $\sigma = 0.2$, 0.5, and atmospheric pressure, while the immersion ratio (h/D) was 30, 47.4, and 58% corresponding to shaft inclination of 4, 8, and 12°, respectively (see Rose and Kruppa 1991 and Rose et al. 1993). Consequently, the SPP MM consists of three *cavitation independent* pairs, each with 13 to 20 polynomial terms of K_T, $K_Q = f$ (h/D, P/D, J), one for each of the tested cavitation numbers σ. Special attention was paid to selecting an optimum K_T and K_Q pair to give the best representation of $\eta_o = (K_T/K_Q) \cdot (J/2\pi)$. This however meant that, from the statistical point of view, the finally selected individual K_T and K_Q curves were not the best fit.

Experimental results of another 4 and 5 bladed systematic SPP series with P/D ratio between 0.8 and 1.4 are represented by regression equations for K_T' and K_Q' as a f(J_ψ, P/D), see Ferrando et al. (2007). That is, the main parameters are slightly modified, so that the advance, thrust and torque coefficients are revised to $J_\Psi = v_A \cdot \cos\Psi/nD$, $K_T' = T/(\rho n^2 D^2 A_O')$, and $K_Q' = Q/(\rho n^2 D^3 A_O')$ respectively, where Ψ is longitudinal shaft inclination and A_O' is immersed propeller area. As a result of this unique *SPP nondimensionalization*, different K_T and K_Q curves for various shaft inclinations respectively, collapse into single curves K_T' and K_Q', but only for advance coefficients J_ψ above the critical values J_{CR}. This fact enabled formulation of the relatively simple equations K_T', $K_Q' = f(J_\psi$, P/D), valid for $J_\psi > J_{CR}$, which actually correspond to the partially vented regime. Note that the fully vented regime (i.e. regime below J_{CR}) is not modeled.

4.3 Loading Criteria for High-Speed Propellers

The design objectives for high-speed propellers include maximal efficiency, minimal cavitation, and minimal propeller diameter. Some guidelines for allowable propeller loading are required to achieve these objectives, particularly when cavitation becomes significant. The 10% back cavitation curve (stemming from the segmental section Gawn-Burrill KCA propeller series) is usually assumed to be an erosion-free criterion for high-speed propellers. Note that the loss of thrust due to excessive cavitation (i.e. the thrust breakdown point) usually occurs between 10 and 20% back cavitation.

Further development of loading limits requires a new propeller format based on the local cavitation number ($\sigma_{0.7R}$, using resultant water velocity at 0.7 radius—$v_{0.7R}$), thrust and torque load coefficients τ_C and Q_C, respectively. These are:

$$\sigma_{0.7\,R} = (p_A + p_H - p_V)/\left(\rho/2 \cdot v_{0.7\,R}^2\right)$$
$$\tau_C = T/\left(\rho/2 \cdot A_P \cdot v_{0.7\,R}^2\right)$$
$$Q_C = Q/\left(\rho/2 \cdot D \cdot A_P \cdot v_{0.7\,R}^2\right)$$

For each blade section, three groups of simple equations (in the log-log coordinate system these are straight lines) approximate the data scatter τ_C & Q_C versus $\sigma_{0.7R}$:

1. τ_C & Q_C showing inception of 10% back cavitation,
2. τ_C & Q_C indicating maximal K_T & K_Q values respectively, in the transition (partly cavitating) zone, and
3. τ_C & Q_C in the fully cavitating zone.

Essentially, the B-, KCA-, Newton-Rader-, and super-cavitating propeller series (surface piercing propeller (SPP) series was added later) characterize the various blade sections, each represented with a distinct set of equations. These equations actually outline a high-speed propeller loading criteria, i.e. MMs ready for computerization. The guidelines (obtained under experimental conditions) suggest the upper limits of τ_C and Q_C for transition and fully cavitating regimes, with about 90% confidence. For the full-scale conditions, it is recommended to reduce the limits to 80% of maximal allowable τ_C & Q_C value. The viability and usefulness of these simple loading criteria has been proven through experience since their initial release in 1978 (Blount and Fox 1978). The complete set of equations, including those for SPPs, is given in Blount's book (2014). See also the recommendations given in Bjarne (1993).

Note that these design criteria are by no means sufficient metrics for sizing an optimal propeller. That is, loading limits can easily identify an optimal, but inappropriate propeller with, for instance, a large pitch ratio (e.g. P/D $\approx$ 2). The off-design conditions must also be checked, resulting in overall optimal propeller characteristics (typically with P/D between 1.1 and 1.4, see Blount and Fox 1978) at the cost of slight efficiency loss at one (usually maximal) speed. Note that these guidelines should be used with care since the abovementioned criteria are functions only of propeller characteristics, while other influential parameters are omitted (e.g. wake).

4.4 Recommended Mathematical Models for High-Speed Propellers

The following MMs for HSC propellers are recommended:

1. B series (3–6 blades, with inner blade sections of aerofoil type, outer of segmental type, for non-cavitating regime and speeds of up to 20–30 kn),
2. AEW/KCA series (3- and 4-bladed, segmental flat-faced sections throughout, non-cavitating and cavitating regimes, for speeds 25–40+ kn), and
3. Newton-Rader series (3-bladed, hollow-faced blade sections, cavitating regime, for speeds 40–50+ kn).

For speeds above 30, 40, and 50 knots, the waterjets, surface piercing (SPP), and submerged fully cavitating propellers respectively, should be considered (in-depth discussion is given in Blount 2014).

Commercial off-the-stock propellers usually belong to the first two groups, while the hollow-faced blade section propellers are usually custom-made. In some cases flat-faced propellers are convenient for cupping (bending of propeller's trailing edge, corresponding to pitch increase), which bring flat-faced sections closer to hollow-faced sections, hence increase their ability in partly- and fully-cavitating regimes (see Blount and Hubble 1981 and MacPherson 1997, for instance).

Propellers of the second group are of principal interest for the type of HSC discussed here. Recommended MMs for segmental flat-faced section propellers are:

- 4-bladed, non-cavitating regime—Blount and Hubble (1981)
- 3-bladed, non-cavitating regime—Radojčić (1988)
- 3-bladed, transition region—Koushan (2007)
- 3-bladed, cavitating regime—Blount and Hubble (1981), Radojčić (1988), and Koushan (2007), but use all three with caution.

For quick reference the recommended MMs for B, AEW, KCA, and the Newton-Rader series are presented in Table 9.1 (Sect. 9.2). Their boundaries of applicability are shown in Fig. 4.1. Other MMs mentioned earlier are incomplete (with respect to the range of applicability), do not seem to have any advantage over the recommended ones, or are not considered to be reliable enough. Note that equations discussed in Sect. 4.3 are partly incorporated in the MM for the cavitating regime of Blount and Hubble (1981), shown in Table 9.1.

Note: For the KCA series, DAR (Developed Blade Area) is used. All other series use EAR (Expanded Area Ratio). In Fig. 4.1, KCA series' DAR is recalculated to EAR according to O'Brien (1969): $EAR \approx 0.34 \cdot DAR \cdot [2.75 + DAR/z]$.

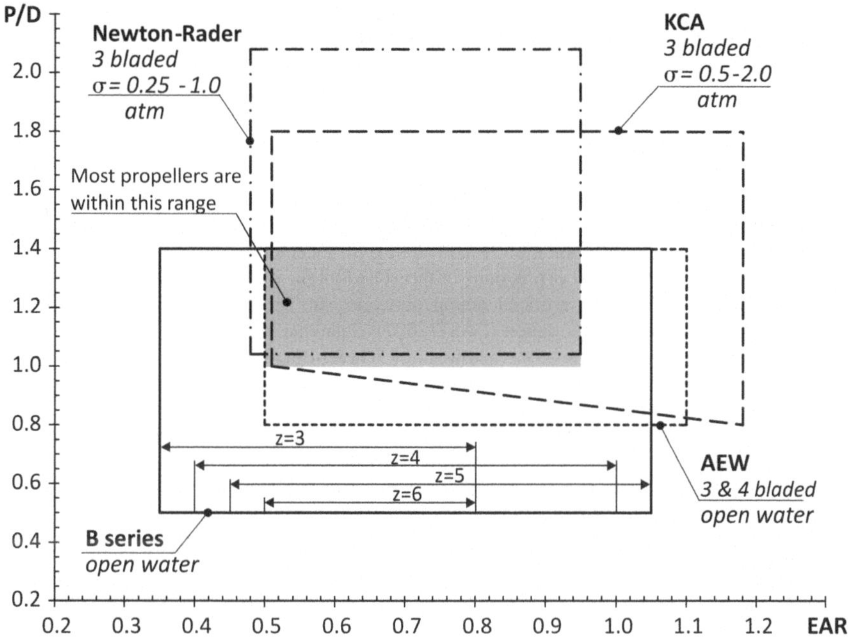

Fig. 4.1 Applicability range of MMs for B-, AEW-, KCA-, and Newton-Rader series (Oosterveld and van Oossanen 1975; Blount and Hubble 1981; Radojčić 1988; Koushan 2007, 2005, respectively)

References

Allison J (1978) Propellers for high performance craft. Marine Technol 15(4)

Bjarne E (1993) Completely submerged propellers for high speed craft. In: Proceedings of 2nd international conference on fast sea transportation (FAST'93), Yokohama

Blount DL (2014) Performance by design. ISBN 0-978-9890837-1-3

Blount DL, Bjarne E (1989) Design and selection of propulsors for high speed craft. In: 7th lips propeller symposium, Nordwijk-on-Sea

Blount DL, Fox DL (1978) Design considerations for propellers in cavitating environment. Marine Technol 15(2)

Blount DL, Hubble EN (1981) Sizing segmental section commercially available propellers for small craft. In: Propellers'81 symposium, SNAME, Virginia Beach

Bukarica M (2014) Mathematical modeling of propeller series. RINA Trans 156, Part B2, Int J Small Craft Technol Jul–Dec

Carlton JC (2012) Marine propellers and propulsion. 3rd edn. Butterworth-Heinemann. ISBN 9780080971230

Dang J, van den Boom HJJ, Ligtelijn JT (2013) The Wageningen C- and D-series propellers. In: Proceedings of 12th international conference on fast sea transportation (FAST 2013), Amsterdam

Denny SB, Puckette LT, Hubble EN, Smith, SK, Najarian RF (1988) A new usable propeller series. SNAME, Hampton Road Section

Diadola JC, Johnson MF (1993) Software user's manual for propeller selection and optimization. Program (PSOP). SNAME Technical and Research Bulletin No. 7-7
Ferrando M, Crotti S, Viviani M (2007) Performance of a family of surface piercing propellers. In: 2nd international conference on marine research and transportation (ICMRT), Ischia
Gawn RWL (1953) Effect of pitch and blade width on propeller performance. INA Trans 95
Gawn RWL, Burrill LC (1957) Effect of cavitation on the performance of a series of 16 in. Model propellers. INA Trans 99
Koushan K (2005) Mathematical expressions of thrust and torque of Newton-Rader propeller series for high speed crafts using artificial neural networks. In: Proceedings of 8th international conference on fast sea transportation (FAST 2005), St. Petersburg
Koushan K (2007) Mathematical expressions of thrust and torque of Gawn-Burrill propeller series for high speed crafts using artificial neural networks. In: Proceedings of 9th international conference on fast sea transportation (FAST 2007), Shanghai
Kozhukarov PG (1986) Regression analysis of Gawn-Burrill series for application in computer-aided high-speed propeller design. Proceedings. In: 5th international conference on high-speed surface craft, Southampton
Kozhukarov PG, Zlatev ZZ (1983) Cavitating propeller characteristics and their use in propeller design. In: High speed surface craft conference, London
Kruppa C (1990) Propulsion systems for high speed marine vehicles. In: Second conference on high speed marine craft, Kristiansand
Kuiper G (1992) The Wageningen propeller series. MARIN Publication 92-001 (ISBN 90-900 7247-0)
Lindgren H (1961) Model tests with a family of three and five bladed propellers. SSPA Publication No. 47
Loukakis TA, Gelegeris GJ (1989) A new form of optimization diagrams for preliminary propeller design. RINA Trans., Part B, vol 131
MacPherson DM (1997) Small propeller cup: a proposed geometry standard and a new performance model. In: SNAME propellers/shafting symposium, Virginia Beach
Matulja D, Dejhalla R, Bukovac O (2010) Application of an artificial neural network to the selection of a maximum efficiency ship screw propeller. J Ship Prod Des 26(3)
Mavludov MA, Roussetsky AA, Sadovnikov YM, Fisher EA (1982) Propellers for high speed ships. Sudostroenie, Leningrad (in Russian)
Milićević M (1998) Mathematical modeling of supercavitating SK series. Diploma degree thesis, Faculty of Mechanical Engineering, Department of Naval Architecture, University of Belgrade (in Serbian)
Neocleous CC, Schizas CN (2002) Artificial neural networks in estimating marine propeller cavitation. In: Proceedings of the international joint conference on neural networks, vol 2
Newton RN, Rader HP (1961) Performance data of propellers for high-speed craft. RINA Trans 103(2)
O'Brien TP (1969) The design of marine screw propellers. Hutchinson and Co. Publishers Ltd., London
Oosterveld MWC, van Oossanen P (1975) Further computer-analyzed data of the Wageningen B-screw series. Int Shipbuild Prog 22(251)
Radojčić D (1985) Optimal preliminary propeller design using nonlinear constrained mathematical programming technique. University of Southampton, Ship Science Report No. 21
Radojčić D (1988) Mathematical model of segmental section propeller series for open-water and cavitating conditions applicable in CAD. In: Propellers'88 symposium, SNAME, Virginia Beach
Radojčić D, Matić D (1997) Regression analysis of surface piercing propeller series. In: High speed marine vehicles conference (HSMV 1997), Sorrento
Radojčić D, Simić A, Kalajdžić M (2009) Fifty years of the Gawn-Burrill KCA propeller series. RINA Trans 151, Part B2, Int J Small Craft Technol July–Dec
Roddy RF, Hess DE, Faller W (2006) Neural network predictions of the 4-Quadrant Wageningen propeller series. NSWCCD-50-TR-2006/004, DTMB Carderock Division, Bethesda

Rose J, Kruppa C (1991) Surface piercing propellers, methodical series model test results. In: Proceedings of 1st international conference on fast sea transportation (FAST'91), Trondheim

Rose J, Kruppa C, Koushan K (1993) Surface piercing propellers, propeller/hull inteaction. In: Proceedings of 2nd international conference on fast sea transportation (FAST'93), Yokohama

Shen Y, Marchal LJ (1993) Expressions of the $B_P - \delta$ diagrams in polynomial for marine propeller series. RINA W10 (1993) paper issued for written discussion

Van Hees MT (2017) Statistical and theoretical prediction methods. Encyclopedia of Maritime and Offshore Engineering, Wiley

Van Lammeren WPA, van Manen JD, Oosterveld MWC (1969) The Wageningen B-screw series. SNAME Trans, vol 77

Yosifov K, Zlatev Z, Staneva A (1986) Optimum characteristics equations for the 'K-J' propeller design charts, Based on the Wageningen B-screw series. Int Shipbuild Prog 33(382)

Chapter 5
Additional Resistance Components and Propulsive Coefficients

5.1 Evaluation of In-service Power Performance

Additional components necessary for the evaluation of a HSC's in-service power performance are briefly discussed here (see also Sect. 1.3). These components consist of those that:

1. Increase the resistance from bare hull total resistance in deep and calm water to in-service total resistance (i.e. from R_T to R_T*), and
2. Account for the hull-propeller interaction (i.e. propulsive coefficients).

Note that there are different ways how R_T* is split into its components; see for instance, Müller-Graf (1981, 1997a) and Savitsky (1981) for round bilge semi-displacement and hard chine planing hulls, respectively. For better insights of both resistance components and hull-propeller interaction, see ITTC (2011) "Recommended Procedures and Guidelines for High Speed Marine Vehicles". Practical and up-to-date recommendations for the evaluation of various additional components, as well as step-by-step HSC powering procedures, are discussed in the Blount book (2014).

D. Radojčić et al., *Power Prediction Modeling of Conventional High-Speed Craft*,
https://doi.org/10.1007/978-3-030-30607-6_5

5.2 Resistance Components—Calm and Deep Water

5.2.1 Appendages

In general, the approach for appendage resistance evaluation of semi-displacement (round bilge) and planing (hard chine) vessels is similar, except that the appendage configuration is slightly different (rudder type may differ, skeg may be applied to semi-displacement round bilge but not planing hulls, etc.).

The usual design practice is to calculate the appendage resistance (shaft, strut, rudder, skeg etc.), as:

1. A percentage of bare hull resistance, or from a simple expression for all appendages (see Müller-Graf 1981; Blount and Fox 1976, respectively).
2. From expressions which calculate a component of each appendage separately (most frequently used are those published in Hoerner 1965 and combined into a complete method by Hadler 1966 (Sect. 10.1.1); see also Lasky 1980). For rudder resistance see for instance Molland and Turnock (2007) and Gregory and Dobay (1973). When the appendage configuration resembles that of an outboard or sterndrive propulsion (i.e. surface piercing drive systems); see Scherer and Patil (2011).
3. Data from model experiments (in which case the scale of the model, i.e. Reynolds number, may be important). Test results of typical full-scale appendages are given in Gregory and Beach (1979).

A procedure for predicting HSC turning characteristics is given in Denny and Hubble (1991) and Lewandowski (1993); see also Bowles (2012).

Spray Rails, Wedges, Flaps, Interceptors

Wedges, Flaps, Interceptors and Spray Rails (or spray strips) may be treated as appendages with negative resistance (i.e. if well-designed, overall resistance reduction can be achieved).

Spray Rails

Spray rails (i.e. longitudinal steps) when fitted on the bottom (near and below calm water surface), reduce the wetted area and are beneficial at higher planing speeds (see Clement 1964; Grigoropoulos and Loukakis 1995). Somewhat different spray rails (i.e. spray separators or deflectors, acting as side bow knuckles) fitted above or near the waterline reduce spray of semi-displacement hulls (particularly for $Fn_L > 0.7$) but also increase dynamic trim. The *spray rail system* for semi-displacement vessels (spray rails in combination with wedge) that reduce resistance by 6–10% in the speed range $Fn_L = 0.5–0.9$ and increase dynamic stability is proposed by Müller-Graf (1991).

Stern Wedges and Flaps

Stern wedges (buttocks' aft hook), flaps, and interceptors are beneficial in the resistance-hump region (i.e. above $Fn_L \approx 0.5$) since they increase hydrodynamic lift and reduce the resistance and dynamic trim. Flap and wedge design references are Brown (1971), Savitsky and Brown (1976) and Chen et al. (1993), and Millward (1976), for planing and round bilge hulls, respectively. *Integrated wedge-flap* combination for fast displacement and semi-displacement vessels with reported power savings of up to 12% are investigated in Cusanelli and Karafiath (1997).

Interceptors

Interceptors are nowadays used as an alternative to wedges and flaps, see for instance De Luca and Pensa (2012), where it is concluded that: (a) Interceptors are beneficial for $Fn_\nabla = 1.8{-}2.4$; and (b) Savings of $15\%R_T$ are feasible. Flaps and interceptors are movable/controllable devices and are often used as ride control appliances to improve seakeeping performance.

Dynamic Instability

Note that planing-craft-specific longitudinal instability (i.e. pitch-heave instability), often called porpoising, may be reduced or prevented by trim reduction. Caution is necessary however, as transverse dynamic stability may be degraded at low trim due to generation of a low dynamic pressure on the bow. Risk of dynamic instability for round bilge and hard chine craft should be considered for Fn_L larger than 0.75 and 0.95, respectively (Blount and McGrath 2009). Recommendations for avoidance of dynamic instability are given in Blount and Codega (1992) and Müller-Graf (1997c). A method to evaluate coupled roll-yaw-sway dynamic stability of planing craft is given in Lewandowski (1997). A useful overview of planing hull transverse dynamic stability is given in Ruscelli et al. (2012). The prediction of porpoising inception for planing craft is elaborated in Celano (1998). These methods should be integrated in the power prediction routines.

5.2.2 *Air and Wind Resistance*

Still-air drag (resistance) may be evaluated from:

1. A simple expression as a function of above-water projected frontal area, superstructure shape (depends on aerodynamics or superstructure streamlining), and aerodynamic drag coefficient (which has to be estimated), see Sect. 10.1.2, or
2. A percentage of total bare hull resistance (e.g. $C_{AA} = 2{-}3\%R_T$ to even $9\%R_T$ at speeds of 30+ knots, see Müller-Graf (1981) and Blount and Bartee (1997), respectively). Air resistance is also discussed in Blount (2014).

Wind resistance however, may be a much larger additional component, which depends on wind velocity and wind direction relative to the vessel direction (peak values are for relative headwind of around 30°). Note that other underwater components also increase due to wind-induced waves.

A noteworthy reference on this topic is Fossati et al. (2013). It suggests that the aerodynamic forces should not be accounted for just as additional resistance components, but that they should be included in the equations of equilibrium (discussed later in Sect. 6.1), since they impact both the resistance and the dynamic trim. Non-dimensional coefficients for evaluation of air and wind resistance of high speed motor yachts with standard superstructure profile are provided in the same reference. Mega yacht aerodynamics is treated in Fossati et al. (2014).

5.2.3 Correlation Allowance and Margins

Correlation allowance (C_A) should account for the differences between the ideal towing tank conditions (hence also the predictions of the MMs), and full-scale real conditions. For the slower HSC, i.e. the displacement and semi-displacement vessels, C_A depends on the extrapolation method used (ATTC-1947, ITTC-1957, ITTC-1978), and on whether the resistance components of lesser magnitude are accounted for separately (e.g. added resistance due to course keeping, spray wetted area etc.). Usually, within a correlation allowance a small allowance is accounted for, resulting in C_A being on the conservative side. For better insights see Müller-Graf (1997b). For vessels less than 80 m a correlation allowance of 0.4×10^{-3} approximately corresponds to a power increase of 10% (Blount and Bjarne 1989). Note that correlation was also discussed in Sect. 3.6.

For higher HSC speeds, actually for vessels that operate through a relatively wide speed range (semi-planing and planing), the use of excess *design margin* is recommended (see Blount 2014), since a single C_A value cannot be correct over a wide speed rage (e.g. hump and top speed). That is, the standard correlation allowance procedure, derived from the conventional model-to-ship extrapolation methods, may result in insufficient margin for the hump-speed region. The solution is to use excess design margin, which need not necessarily be uniform across the entire speed range. The design margins necessary to enable a vessel to accelerate and overcome overloading due to increase of in-service resistance is discussed in Blount and Bartee (1997). In-service margin is analogous to the power prediction factor, or load factor, usually denoted as $(1 + x)$, for conventional ships.

5.3 Resistance in a Seaway

Performance of HSC in a seaway is an extremely important topic, but is beyond the scope of this work, because it requires special attention. However, for power predictions, only added resistance in a seaway is of interest. Nevertheless, MMs for powering predictions should also account for the HSC's motion (pitch, heave, accelerations etc.) which are normally a part of the basic design criteria. The simplest approach is to restrict (reduce) speed in a rough sea, although better results—i.e. enhanced HSC performance in a seaway—would be achieved if some primary and/or secondary hull characteristics are altered.

Classical references for added resistance in a seaway are Savitsky and Brown (1976), Fridsma (1971), Hoggard (1979), Hoggard and Jones (1980), etc. It should be noted, however, that the predictions of these MMs often disagree. A very useful overview of various aspects that concern seakeeping of hard chine planing craft is given in Savitsky and Koelbel (1993) and Blount (2014). HSC seakeeping and various safety, comfort, structures, machinery, etc. considerations are discussed in Faltinsen (2005). Habitability and human factors criteria, for instance, are nowadays in the focus of HSC community; see Schleicher (2008), amongst others.

5.4 Resistance in Shallow Water

Shallow water effects are noticeable whenever $h/L_{OA} < 0.80$ or $Fn_h > 0.6–0.7$. The maximum effect is in the critical region ($0.7 < Fn_h < 1.2$), where wave-making resistance (R_W) increases dramatically. Shallow water effects subside in supercritical region at $Fn_h > 1.2$, where the resistance may be lower than in deep water. Therefore, a vessel's speed in the critical region may be substantially lower than expected, and/or the power demand is substantially higher. However, when operating in the super-critical region the reverse occurs and required power to achieve a specific speed may be less than in deep water, due to reduced resistance and somewhat enhanced propulsive efficiency. Consequently, larger and faster modern commercial vessels and mega yachts are likely to experience shallow water effects and in some cases may be sailing in, what is hydrodynamically considered to be shallow or littoral waters, throughout their life.

Note that the design logic for vessels intended for littoral and shallow water operation is fundamentally different from the usual deep-water logic. Namely, HSC which must operate in all water depths (shallow and deep) must be able to operate at speeds above $Fn_L \approx 0.7$ (see Hofman and Radojčić 1997 and Radojčić and Bowles 2010). At this speed, wave wash—an important aspect for the littoral environment—is inherently significant. Moreover, wave wash decay is lower in shallow water than in deep water. So, for shallow water operation it is not only the propulsor type and size that matters, it is also desirable to have a low wave-wash hull form. This calls for careful selection of speed and waterline length, although reduction of the HSC's weight is the only measure that can effectively lower both resistance and wave-wash (see Hofman and Kozarski 2000).

Shallow water resistance can be predicted within engineering accuracy, see Radojčić and Bowles (2010) and Sect. 10.1.4. The recommended approach is based on an evaluation of the ratio of shallow- to deep-water wave resistance (R_{Wh}/R_{Wd}), where the final results rely primarily on deep water data, and the possible inaccuracies (of evaluated R_{Wh}/R_{Wd}) do not influence shallow water total resistance (R_{Th}) to a great extent (MM is given in Sect. 10.1.4). This method enables an evaluation of the resistance increment in the critical region, by utilizing deep water data only, while decrements in the supercritical regime are neglected, so that the shallow water resistance predictions are on the safe side. Note however, that shallow water propulsive efficiency requires further research, i.e. water depth effects on propulsive factors are not investigated and are not sufficiently accurate (see Sect. 10.2). Lyakhovitsky (2007), Hofman and Radojčić (1997) and Radojčić and Bowles (2010) provide more details on shallow water effects. See Millward and Sproston (1988), and Toro (1969) and Morabito (2013) for shallow water experiments for semi-displacement and planing hull forms respectively. Self-propulsion testing of a single HSC model is discussed in Friedhoff et al. (2007).

5.5 Propulsive Coefficients

The total propulsive efficiency (η_P) consists of propeller open water efficiency (η_O), hull efficiency (η_H), relative rotative efficiency (η_R), and shaft (including a gearbox) efficiency (η_S), i.e. $\eta_P = \eta_H \cdot \eta_R \cdot \eta_S \cdot \eta_O$; see for instance Molland et al. (2011). Hull efficiency can be further fragmented and expressed as $\eta_H = (1-t)/(1-w)$, where "t" and "w" are thrust deduction fraction and wake fraction, respectively. Similarly, $\eta_R = \eta_B/\eta_O$, where η_B is propeller efficiency behind the vessel. Consequently, for power prediction it is necessary to evaluate overall propulsive efficiency η_P, which consists of propulsive coefficients $\eta_H = f(t, w)$, $\eta_R = f(\eta_O, \eta_B)$, η_S, and η_O.

Note that the hull-propeller interaction elaborated above is correct for conventional ships (displacement vessels). Hull-propeller interaction for HSC is much more complex (hence the need for an integrated approach) because:

1. Water inflow to the propeller is usually not axial (as is for displacement vessels), but is oblique (i.e. $v_a\cos(\psi + \tau)$), due to shaft inclination relative to hull (ψ) and dynamic trim (τ); see Fig. 6.1, and
2. Cavitation effects.

Moreover, correct "behind" condition testing under cavitation conditions is still not feasible, so that an alternative approach is required. Two separate types of experiments are usually performed—one under atmospheric (in the conventional towing tank), and one under depressurised conditions (in the cavitation tunnel/channel); see ITTC (1984). As a consequence propulsive coefficients for HSC, particularly for the cavitation regime, are not as reliable as those for conventional ships.

MMs for prediction of open water efficiency are already discussed in Sect. 4.2, while other propulsive coefficients for displacement, semi-displacement, and planing craft are given in Blount and Bjarne (1989). See Blount (1997) if the HSC's propellers are recessed into tunnels, and hence a reduction of shaft inclination is enabled. See Katayama et al. (2012) for outboard and sterndrive configurations. HSC propulsive coefficients for various propeller configurations are also given in Blount (2014). Influence of cavitation on the propeller-hull interaction is elaborated in Rutgersson (1982).

Regression analysis is employed for mathematical modeling of propulsive coefficients for high-speed round bilge hulls tested in the speed range of $Fn_L = 0.4-1.05$ and conventional installations where power is delivered to the propeller along an inclined shaft; see Bailey (1982). Independent variables are standard high-speed round bilge hull and propeller parameters. MMs are simple with 4–8 terms only. For each propulsive coefficient two equations are derived for $C_B < 0.45$ and for $C_B = 0.45-0.51$, matching faster and slower hulls respectively (see Sect. 10.2).

When published data is used, attention should be paid to the proper interpretation of propulsive coefficients, and particularly to the thrust deduction factor (t). Shaft inclination for instance, influences propeller characteristics more than hull-propeller interaction, and yet this influence is often expressed solely by the propulsive coefficients. Therefore, the specific conditions upon which the propulsive coefficients are based should be clarified (e.g. whether the t value is valid for the horizontal resistance component or inclined ($\psi + \tau$) thrust vector, or whether the appendage resistance is accounted for, etc.); see Molland et al. (2011) and Blount (2014).

5.6 Recommended References for Evaluation of Additional Resistance Components and Propulsive Coefficients

This section provides an overview of references linked to the:

- Additional resistance components (related to bare hull resistance in calm and deep water), and
- Propulsive coefficients (other than open water efficiency).

Suggested references are presented in Table 5.1. See also Sects. 10.1 and 10.2.

Table 5.1 Recommended references for evaluation of additional resistance components and propulsive coefficients

	Reference	
Additional resistence components	**Calm and deep water**	
	Appendages	
	Struts, shafts etc.	Hadler (1966), Lasky (1980)
	Rudders*	Gregory and Dobay (1973)
	Sterndrives, outboards	Scherer and Patil (2011)
	Model experiments	Gregory and Beach (1979)
	Flaps and wedges**	Brown (1971) – planing hulls Millward (1976) – semi-displacement
	Interceptors**	De Luca and Pensa (2012)
	(*) Maneuverability and turning characteristics: - Denny and Hubble (1991), Lewandowski (1993)	
	(**) Reduce dynamic trim and may cause dynamic instability: - Transverse (dynamic trim too low) – Blount and Codega (1992), Müller-Graf (1997c) - Longitudinal, porposing (dynamic trim too high) – Celano (1998)	
	Air resistance	Blount (2014), Fossati et al. (2013)
	Correlation allowance and margins	Blount (2014)
	Resistance in a seaway	Savitsky and Brown (1976) Hoggard (1979) Hoggard and Jones (1980) Blount (2014)
	Resistance in shallow water	Radojčić and Bowles (2010) – MM Millward and Sproston (1988) – tests semi-displacement hulls Toro (1969) – tests planing hulls Friedhoff et al. (2007) – self-propulsion tests, planing hulls
Propulsive COEF.		Blount and Bjarne (1989) – HSC, all kind Bailey (1982) – Round bilge Blount (1997) – Propeller in tunnel Katayama et al. (2012) – Outboards, sterndrives

References

Bailey D (1982) A statistical analysis of propulsion data obtained from models of high speed round bilge hulls. In: RINA symposium on small fast warships and security. Vessels, London
Blount DL (1997) Design of propeller tunnels for high speed craft. In: Proceedings of the 4th international conference on fast sea transportation (FAST '97). Sydney
Blount DL (2014) Performance by design. ISBN 0-978-9890837-1-3
Blount DL, Bartee RJ (1997) Design of propulsion systems for high-speed craft. Mar Technol 34(4)
Blount DL, Bjarne E (1989) Design and selection of propulsors for high speed craft. In: 7th Lips propeller symposium. Nordwijk-on-Sea
Blount DL, Codega LT (1992) Dynamic stability of planing boats. Mar Technol 29
Blount DL, Fox DL (1976) Small craft power prediction. Mar Technol 13(1)
Blount DL, McGrath JA (2009) Resistance characteristics of semi-displacement mega yacht hull forms. RINA Trans, vol 151, Part B2, Int. J. Small Craft Technol, July–Dec
Bowles J (2012) Turning characteristics and capabilities of high speed Monohulls. In: SNAME's 3rd Chesapeake power boat symposium. Annapolis
Brown PW (1971) An experimental and theoretical study of planing surfaces with trim flaps. Davidson Laboratory Report 1463. Stevens Institute of Technology
Celano T (1998) The prediction of porpoising inception for modern planing craft. USNA Trident Report No. 254. Annapolis
Chen CS, Hsueh TJ, Fwu J (1993) the systematic test of wedge on flat plate planing surface. In: Proceedings of 2nd international conference on fast sea transportation (FAST '93). Yokohama
Clement EP (1964) Reduction of planing boat resistance by deflection of the whisker spray. DTMB Report 1920
Cusanelli D, Karafiath G (1997) Integrated wedge-flap for enhanced powering performance. In: Proceedings of 4th international conference on fast sea transportation (FAST '97). Sydney
De Luca F, Pensa C (2012) Experimental investigation on conventional and unconventional interceptors. RINA Trans, vol. 153, Part B2, Int. J. Small Craft Technol, July–Dec
Denny SB, Hubble EN (1991) Predicting of craft turning characteristics. Mar Technol 28(1)
Faltinsen OM (2005) Hydrodynamics of high-speed marine vehicles. Cambridge University Press. ISBN-13 978-0-521-84568-7
Fossati F, Muggiasca S, Bertorello C (2013) Aerodynamics of high speed small craft. In: Proceedings of 12th international conference on fast sea transportation (FAST 2013). Amsterdam
Fossati F, Robustelli F, Belloli M, Bertorello C, Dellepiane S (2014) Experimental assessment of mega-yacht aerodynamic performance and characteristics. RINA Trans, vol. 156, Part B2. London. https://doi.org/10.3940/rina.ijsct.2014.b2.157)
Fridsma G (1971) A systematic study of the rough water performance of planing boats (Irregular Waves—Part II). Davidson Laboratory Report 1495
Friedhoff B, Henn R, Jiang T, Stuntz N (2007) Investigation of planing craft in shallow water. In: Proceedings of 9th international conference on fast sea transportation (FAST 2007). Shanghai
Gregory D, Beach T (1979) Resistance measurements of typical planing boat appendages. DTNSRDC Report SPD-0911-01
Gregory DL, Dobay GF (1973) The performance of high-speed rudders in a cavitating environment. SNAME Spring Meeting, Florida
Grigoropoulos GJ, Loukakis TA (1995) Effect of spray rails on the resistance of planing hulls. In: Proceedings of 3rd International conferences on fast sea transportation (FAST '95). Lubeck-Travemunde
Hadler JB (1966) The prediction of power performance of planing craft. SNAME Trans, vol 74
Hoerner SF (1965) Fluid dynamic drag. Midland Park, NJ (Book published by the author)
Hofman M, Kozarski V (2000) Shallow water resistance charts for preliminary vessel design. Int Shipbuilding Prog 47(449)
Hofman M, Radojčić D (1997) Resistance and propulsion of fast ships in shallow water. University of Belgrade, Faculty of Mechanical Engineering, Belgrade (in Serbian). ISBN 86-7083-297-6

Hoggard MM (1979) Examining added drag of planing craft operating in the seaway. Hampton Road Section of SNAME

Hoggard MM, Jones MP (1980) Examining pitch, heave and accelerations of planing craft operating in a seaway. In: High speed surface craft conference. Brighton

ITTC (1984) Proceedings of the 17th international towing tank conference, high-speed propulsion, vol 1. Goteborg

ITTC (2011) Recommended procedures and guidelines—high speed marine vehicles (Section 7.5-02-05)—Resistance test, Section 7.5-02-05-01; Propulsion test, Section 7.5-02-05-02

Katayama T, Nishihara Y, Sato T (2012) A study on the characteristics of self-propulsion factors of planing craft with outboard engine. In: SNAME's 3rd Chesapeake power boat symposium. Annapolis

Lasky MP (1980) An investigation of appendage drag. DTNSRDC Report SPD-458-01

Lewandowski E (1993) Manueverability of high-speed power boats. In: 5th Power boat symposium, SNAME Southeast Section

Lewandowski E (1997) Transverse dynamic stability of planing craft. Mar Technol 34(2)

Lyakhovitsky A (2007) Shallow water and supercritical ships. Backbone Publishing Company, Hoboken, NJ

Millward A (1976) Effect of wedges on the performance characteristics of two planing hulls. J Ship Res 20(4)

Millward A, Sproston J (1988) The prediction of resistance of fast displacement hulls in shallow water. RINA Maritime Technology Monograph No. 9, London

Molland AF, Turnock SR (2007) Marine rudders and control surfaces—principles, data, design and applications. Elsevier. ISBN 978-0-75-066944-3

Molland AF, Turnock SR, Hudson DA (2011) Ship resistance and propulsion—practical estimation of ship propulsive power. Cambridge University Press. ISBN 978-0-521-76052-2

Morabito MG (2013) Planing in shallow water at critical speed. J Ship Res 57(2)

Müller-Graf B (1981) Semi-displacement round bilge vessels. In: Status of hydrodynamic technology as related to model tests of high speed marine vehicles (Section 3.2). DTNSRDC Report 81/026

Müller-Graf B (1991) The effect of an advanced spray rail system on resistance and development of spray on semi-displacement round bilge hulls. In: Proceedings of 1st international conference on fast sea transportation (FAST '91). Trondheim

Müller-Graf B (1997a) Part I: Resistance components of high speed small craft. In: 25th WEGEMT School, Small Craft Technology, NTUA, Athens. ISBN I 900 453 053

Müller-Graf B (1997b) Part III: Factors affecting the reliability and accuracy of the resistance prediction. In: 25th WEGEMT School, Small Craft Technology, NTUA, Athens. ISBN I 900 453 053

Müller-Graf B (1997c) Dynamic stability of high speed small craft. In: 25th WEGEMT School, Small Craft Technology, NTUA, Athens. ISBN I 900 453 053

Radojčić D, Bowles J (2010) On high speed monohulls in shallow water. In: SNAME's 2nd Chesapeake power boat symposium. Annapolis

Ruscelli D, Gualeni P, Viviani M (2012) An overview of planing monohulls transverse dynamic stability and possible implications with static intact stability rules. RINA Trans, vol. 154, Part B2. London. https://doi.org/10.3940/rina.ijsct.2012.b2.134)

Rutgersson O (1982) High speed propeller performance—influence of cavitation on the propeller-hull interaction. Ph.D. thesis, Chalmers University of Technology, Goteborg. ISBN 91-7032-072-1

Savitsky D (1981) Planing hulls. In: Status of hydrodynamic technology as related to model tests of high speed marine vehicles (Section 3.3). DTNSRDC Report 81/026

Savitsky D, Brown PW (1976) Procedure for hydrodynamic evaluation of planing hulls in smooth and rough water. Mar Technol 13(4)

Savitsky D, Koelbel JG (1993) Seakeeping of hard chine planing hulls. SNAME's Technical and Research Panel SC-1 (Power Craft), Report R-42

Scherer JO, Patil SKR (2011) Hydrodynamics of surface piercing outboard and sterndrive propulsion systems. In: Proceedings of 11th international conference on fast sea transportation (FAST 2011). Honolulu

Schleicher DM (2008) Regarding small craft seakeeping. In: SNAME's 1st Chesapeake power boat symposium. Annapolis

Toro A (1969) Shallow-water performance of a planing boat. Michigan University (AD-A016 682), Department of Naval Architecture and Marine Engineering. Report No. 019

Chapter 6
Power Prediction

6.1 Power and Performance Predictions for High-Speed Craft

The performance prediction models are typically composed of modules (subroutines) for the calculation of bare hull resistance, appendage resistance, propeller characteristics, etc., which are derived individually and may be used independently of each other. For conventional ships the Holtrop and Mennen method (Holtrop and Mennen 1982, since then slightly corrected) applicable to a wide variety of ship types is still in use despite its age, and is typically included in multiple software packages. Adequate MMs for HSC do not really exist, probably due to the unique nature of HSC, operating in displacement, semi-displacement, and often planing regimes, as discussed in Sect. 1.2. Specifically, with increasing speed HSC change both the displacement and trim, which is not the case with conventional (displacement) ships. Therefore, in order to model the HSC's operating conditions and power requirement, equations of equilibrium must be formed.

Equations of Equilibrium

Note that the equations of equilibrium (including interpretation of propeller forces), are closely connected with propulsive coefficients, and in particular with propulsive efficiency η_P and hence also power prediction. That is, in general

$$T_h = T_a \cdot \cos(\psi + \tau) - N \cdot \sin(\psi + \tau) \quad \text{and} \quad L = T_a \cdot \sin(\psi + \tau) + N \cdot \cos(\psi + \tau),$$

where T_h, L, T_a, and N are the propeller's horizontal force, lift (vertical force), axial force, and normal force, respectively; see Fig. 6.1. Basically, three different approaches of accounting for propeller forces may be considered:

1. Shaft inclination ($\psi + \tau$) is neglected. Simplification of this kind is permissible for $\psi < 6°{-}8°$ (Blount and Bjarne 1989), i.e. for the displacement vessels when

D. Radojčić et al., *Power Prediction Modeling of Conventional High-Speed Craft*,
https://doi.org/10.1007/978-3-030-30607-6_6

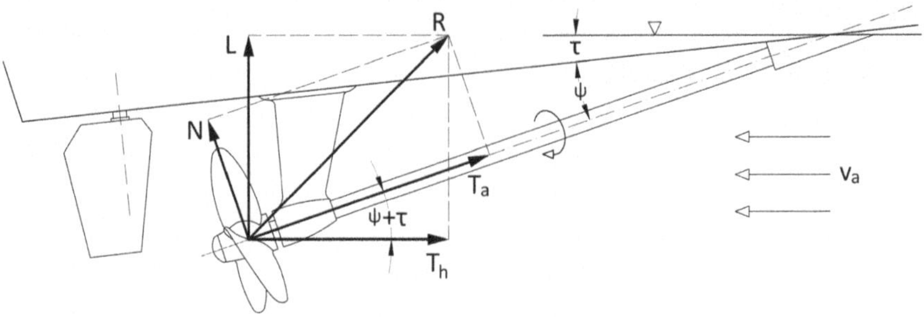

Fig. 6.1 Propeller forces on inclined shaft

$\psi + \tau$ is small. Nevertheless, this is often, albeit wrongly, done even when $\psi > 6°–8°$. This simplification also misinterprets propeller-hull interaction with altered propeller characteristics due to oblique inflow.

2. Only the propeller's horizontal forces are accounted for, through

$$T_h = T_a \cdot \cos(\psi + \tau) - N \cdot \sin(\psi + \tau),$$

while vertical force L (lift) is neglected. This is a more realistic approach than the previous one, and most power prediction routines use it.

3. Both the propeller's horizontal T_h and vertical force L are accounted for through the equilibrium equations. This means that the propeller's vertical force L reduces both the displacement and the trim, which further impacts the resistance and significantly complicates power predictions. In addition, the ratio of T_a/N for non-cavitating and cavitating conditions is not the same, i.e. N decreases as cavitation increases (see ITTC 1984), so $(T_a/N)_{CAV} > (T_a/N)_{NON\text{-}CAV}$, which additionally complicates the analysis.

This discussion resembles the one presented in ITTC (1984), where model testing, high-speed propulsion, inclined shaft propeller forces, cavitation effects on propulsive efficiency etc., were elaborated. Accordingly, model experiments where all influences are taken into account (corresponding to Approach 3) are difficult to execute; see discussion in Sect. 5.5. Simpler approaches for model testing, and consequently for mathematical modeling, corresponding to methods 1 and 2 are usually practiced, and are often combined with some correlation factors.

Finally, it may be concluded that the interaction between the resistance and the propulsion modules, expressed through the equations of equilibrium and propulsive coefficients, in essence differentiates the powering prediction concepts among themselves.

6.2 Classics[1]

Papers in this category are written before the wide use of computers in everyday engineering practice. They include graphs and tables, and often instructions/routines for hand-held calculators. Nevertheless, in the authors' opinion these papers have "enduring quality" (i.e. value) and contain information necessary for understanding HSC hydrodynamics. They are therefore highly recommended as background reading, even for those who rely on commercial software in their design work.

Hadler (1966)

Equations of equilibrium are examined in Hadler (1966). Propeller forces on an inclined shaft were superimposed on the bare hull planing hydrodynamics (in essence following Savitsky's method for resistance evaluation). Appendage lift and drag were examined too (skeg, rudders, shafts, struts etc.), since it was necessary to balance all relevant forces in the equilibrium equations. The methodology for performance analysis was established by examining the interaction of propeller and planing surface forces. Amongst the conclusions was that in "optimizing the design of high-performance planing hull, the whole hydrodynamic system must be considered".

Hadler and Hubble (1971)

Hadler's approach was soon upgraded (Hadler and Hubble 1971) and prismatic hulls and Troost open water propeller series were replaced with the Series 62 (12.5° deadrise) and Gawn-Burrill KCA propeller series respectively. Single, twin, and quadruple-screw configurations were investigated for wide speed and hull size ranges, giving optimum diameters versus RPMs, optimum propulsive coefficients etc. The charts presented are especially useful to planing hull designers for power prediction decisions typical in the preliminary design phases.

Blount and Fox (1976)

The technical data available at the time has been further organized in Blount and Fox (1976), resulting in a methodology for hard chine planing craft power prediction. This power prediction method accounts for bare hull resistance, various appendage configurations, propeller characteristics under cavitating and non-cavitating conditions, and resistance augmentation due to rough water. For bare hull resistance evaluation, Savitsky's method for prismatic hulls was modified with the so called *multiplying factor M*. The *effective beam and deadrise* were also investigated for non-prismatic hulls. Propeller selection procedure, beside the

[1]The definition for the *classic literature*, according to Dictionary.com, is: "an author or literary work of the first rank, especially one of demonstrably enduring quality". Here, word *classics* should be replaced with *classic papers, references* or *classic power prediction approaches*.

Gawn-Burrill series, allowed the use of other propeller series (Newton-Rader for instance). For rough water performance the Fridsma (1971) method was applied.

Savitsky and Brown (1976)[2]

Studies of planing hull hydrodynamics, conducted by the recognized Davidson Laboratory, are presented in Savitsky and Brown (1976). MMs for resistance prediction for the pre-planing and full planing regimes were pulled from Mercier and Savitsky (1973) and Savitsky (1964) respectively, with the addition of bottom warp discussion. An in-depth study of the effects of bottom warp are given four decades later in Savitsky (2012) and in Begović and Bertorello (2012). Seakeeping prediction is according to the Fridsma (1971) method, and the effectiveness of trim control flaps according to Brown (1971).

Müller-Graf (1997a, b, c, d)

In spite of being presented well within the "computer era", Müller-Graf's WEGEMT Lectures on Small Craft Technology (Müller-Graf 1997a, b, c, d) are certainly amongst the unavoidable classics. Namely, in Part I are various HSC resistance components and subcomponents under trial conditions and procedures for their determination. Part II treats published systematical resistance data and deals with power predictions of different HSC types. Part III discusses various uncertainties of prediction methods and suggests appropriate safety margins. Different types of dynamic instabilities and measures to improve them are discussed in Müller-Graf (1997d).

6.3 Modernism[3]

MMs presented in this section are developed essentially with computer application in mind. Moreover, the aim is often not only to predict the performance of a HSC with given principal characteristics, but also to enable the designer to choose the optimal characteristics of the hull and of the propulsor during the preliminary design stages. In general this might be done through the:

1. Systematic variation of input variables (i.e. through a parametric study, nowadays using a simple spread-sheet program)—for which the user's interaction is necessary to some extent.

---

[2]Note that Savitsky and Brown (1976) treats hull hydrodynamics only, and is an extension to prismatic hull method approach. That is, propulsion is not discussed, and hence, strictly speaking, this work doesn't belong in the "Power prediction" category.

[3]The definition for *modernism*, according to Merriam-Webster.com, is: "a style of art, architecture, literature, etc., that uses ideas and methods which are very different from those used in the past". Here, the word *modernism* should be replaced with *modern power prediction approaches*.

2. Application of general nonlinear constrained optimization techniques. This, however, requires strict definition of MM's applicability bounds, and hence these bounds practically became a part of the MM itself. Consequently, the complete MM consists of an objective function for performance prediction and the constraints—which are usually nonlinear equations of equality and inequality type. Note that the comprehensive MM often must be presented in a form which can be efficiently handled by the chosen optimization technique (see Radojčić 1985 for instance, where the optimization method used was *Sequential Unconstrained Minimization Technique—SUMT*).

Four MMs, or approaches, with the aim to minimize power, rather than to minimize the resistance and maximize efficiency of propellers, are mentioned here.

Hubble (1980)

This paper gives a procedure for prediction of power and vertical accelerations for planing hulls, with propellers on inclined shafts. Speed versus wave heights is developed based on the power limits of the propulsion system, and endurance limits of the crew due to vertical acceleration. Power optimization is done through a systematic variation of input variables, i.e. it is a parametric study. To evaluate power, several independent state-of-the-art modules were used: for prediction of bare hull resistance (Hubble 1974); appendage resistance and propulsive coefficients (Blount and Fox 1976); propeller characteristics (Oosterveld and van Oossanen 1975; Blount and Fox 1978); habitability limits (Hoggard and Jones 1980) etc. A much broader planing craft feasibility model, which in addition to power prediction, incorporates weight groups, structures, engines, loads etc., is explained in Hubble (1978). Both procedures, however, are based on an approach that is nowadays somewhat obsolete, since computer power is used mainly for the interpolations etc. This is rather similar to what would have been done "by hand" in the pre-computer era.

Calkins (1983)

This routine is developed for recreational powerboats and is based on the design spiral concept, where powering "absorbs" just a few modules (a phrase *technological areas* was used). That is, the abovementioned Hubble (1980, 1978) MMs consisted of 7 and 10 modules respectively, while this procedure consists of 10 modules. However, since the module's scope is not identical, direct comparison does not make much sense. This is actually an interactive CAD synthesis program, which inherently relies much more on computers and graphics than previous instances. In summary, although powerboat modules and sub-modules are simple equations, the logic of this program essentially corresponds to present-day CAD methods.

Radojčić (1991)

The primary objective of this paper is to present a new methodology for planing craft power prediction. A nonlinear constrained optimization technique is used, and design optimization is done on the whole system (min P_D), rather than on any of the components (min R_T and max η_D). Power prediction modules (subroutines) are those for: bare hull resistance (Radojčić 1985), appendage resistance (Savitsky and Brown 1976), rough water performance (Hoggard 1979; Hoggard and Jones 1980), KCA propeller performance for open-water and cavitating environment (Radojčić 1988b), and propeller performance on inclined shaft (Radojčić 1988a, based on Gutche's quasi-steady method[4]—Gutsche 1964; Taniguchi et al. 1967 for respectively, non-cavitating and cavitating propellers). The equilibrium equations (essentially similar to those in Hadler 1966) articulate the interaction between the planing craft's resistance and propulsion forces. This, to a certain extent reduces the importance of propulsive coefficients as Approach 3 explained in Sect. 6.1 is used (i.e. both propeller's horizontal T_h and vertical force L are taken into account through the equilibrium equations). Since several hull and propeller variables have to be examined simultaneously, it is necessary to apply one of the optimization routines. For this purpose the least efficient method is chosen: the Monte Carlo technique (particularly, *multistage Monte Carlo integer optimization technique*; see Conley 1981), mainly because it is simple, robust and easy to program. The goal is to choose the best nonfixed parameters of the hull and propulsors from the performance point of view (min P_D) and then to conduct a tradeoff study, which is important in the early design process.

An illustrative example of Approach 3 and interdependence between the resistance and propulsion forces is depicted in Fig. 6.2 (from Radojčić 1991). Namely, the resistance components R_{BH} and R_{AP}, as well as the dynamic trim τ, change with variation of a normal propeller force N which depend on RPM, hence is a f(D, P/D and $\psi + \tau$). In this particular case, RPM is varied between hypothetical 400 and 1600, approximately corresponding to the shaft inclination variation ψ of 15° to 8° respectively. For each RPM an optimal propeller is chosen (see Fig. 6.2) satisfying the following restrictions: propeller clearance of 15%D, allowed up to 10% back cavitation and various bounds of applicability of all applied MMs (modules). Note that if the above-mentioned interdependence would not be accounted for, displacement Δ, dynamic trim τ and bare hull resistance R_{BH} would all have constant values, i.e. 73 t, 4.5° and 92.5 kN respectively, depicted with dash lines in Fig. 6.2. Moreover, it follows that RPM_{opt} from the viewpoint of inclined propeller efficiency $\eta_{\psi+\tau}$, total resistance R_T and delivered power P_D are 700, 1100 and 800 respectively. In this case, for all three above optimums differences in predicted Pd are essentially negligible, though differences in the propeller diameter D are relatively large (around 20%D).

Use of general optimization routines, as explained above, is not so common; improvements may be implemented in the following three areas:

[4]This method allows recalculation of the commonly available data for axial water inflow to those that correspond to oblique inflow conditions (i.e. propeller on inclined shaft). In essence, water inflow velocity is assumed to be $v_{OBLOUE} = v_{AXIAL} \cdot \cos(\psi + \tau)$ and the propeller's inclined shaft unsteady (oscillatory) forces are averaged across one revolution.

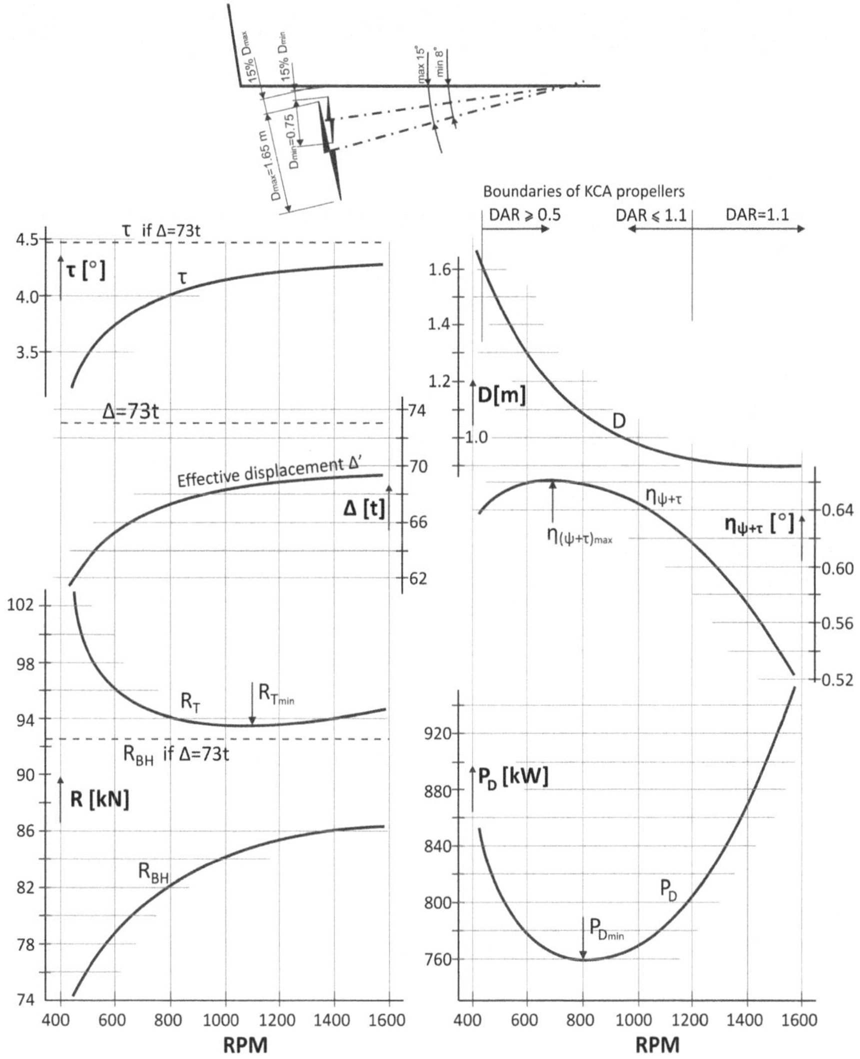

Fig. 6.2 An example of interaction between resistance and propulsion forces (planing hull form, $\Delta = 73$ t, $L_P = 25$ m, $P_B = 3 \times 1000$ kW, $Fn_\nabla = 2.5$ = const. corresponds to speed of 31 kn)

1. Existing modules (individual MMs) within the synthesis routine can be replaced with better ones, or new modules could be added. This would also enhance the power prediction routine. Modules are important because they shape the prediction. Individual modules are in fact the focus of present work and have therefore been discussed in the previous sections.
2. The interaction between the modules, expressed by the equations of equilibrium and hull-propeller interaction in non-cavitating and cavitating conditions, are the

next area of possible enhancements. Also, introduction of correlation factors seems to be inevitable.

3. The optimization technique itself is a third important part of the powering prediction routines, because it provides the tradeoff information and enables the choice of optimal characteristics of the hull and of the propulsor. A short and informative article about optimization is given in Collete (2014).

   Although discussion about optimization techniques is beyond the scope of present work, two recent references will be mentioned:
   - Mohamad Ayob et al. (2011) showing the application of a single-objective minimization technique for calm-water resistance and multiobjective minimization of total resistance, steady turning diameter and vertical impact acceleration.
   - Knight et al. (2014) using a multiobjective optimization technique to minimize resistance and vertical acceleration.

   Incidentally, in both cases the Savitsky (1964) method is used for calm-water planing hull resistance evaluation.

Müller-Graf et al. (2003a, b)

Series '89 is different from everything discussed up to here, not only because it deals with high-speed, hard chine, planing catamaran hulls, but also because extensive resistance and self-propulsion tests were carried out (see also Sect. 10.4). An enormous quantity of data was gathered through resistance and self-propulsion tests carried out with relatively large models. An immense stock of data is represented by regression derived MMs for resistance, trim, propulsion coefficients, and delivered power. The tests and results were validated for trial conditions and thus are believed to be very reliable. Prior to this modeling, Zips (1995) presented a regression analysis for evaluation of the Series '89 residuary resistance. Zips' mathematical model is simpler and also less accurate than the MMs presented in Müller-Graf et al. (2003a).

Through the application of regression analysis, three groups of mathematical models for the reliable evaluation of power have been derived (MM is given in Sect. 10.4):

- Speed-independent model for the residuary resistance-to-weight ratio $\varepsilon_R = R_R/\Delta \cdot g = f(\beta_M, \delta_W, L_{WL}/B_{XDH})$,
- Deadrise-independent models for the evaluation of dynamic trim (denoted here θ), propulsive coefficients, and specific power ratio (i.e. θ, η_D, η_O, η_H, w, t and $\varepsilon_B = P_B/\Delta \cdot g \cdot v$), all as $f(Fn_{\nabla/2}, L_{WL}/B_{XDH})$,
- Models for wetted surface coefficient ($WSC = S/(\nabla/2)^{2/3}$), and length-to-displacement ratio ($L_{WL}/(\nabla/2)^{1/3}$), as a $f(\beta_M, L_{WL}/B_{XDH})$,

where β_M—angle of deadrise amidship, δ_W-angle of transom wedge, L_{WL}/B_{XDH}—ratio of waterline length and maximum beam of demihull.

The power prediction method is reliable because the propulsive coefficients for Series '89 were obtained from the self-propulsion tests under trial conditions (i.e. the model propellers are working at a loading which is equal to that of the full-scale propellers). Therefore, none of the less accurate speed-independent trial allowances for power and RPM have to be applied to the power obtained by these regression models. The method also enables the optimization of hull-form parameters and prediction of power in early design stage for catamarans with length-to-beam ratios of $L_{WL}/B_{XDH} = 7.55–13.55$ and midship deadrise values of $\beta_M = 16°–38°$ with an optimal wedge inclination of $\delta_W = 8°$. Residuary resistance and hull-propeller interaction coefficients (i.e. propulsive coefficients) versus speed are given for a large speed range from the hump speed to planing speeds (up to $Fn_{\nabla/2} = 4$ or $Fn_L \approx 1.5$). Power may be evaluated by: (a) conventional method ($P_{BTR} = R_{TTR} \cdot v/(\eta_D \cdot \eta_S)$), or (b) short method ($P_{BTR} = \varepsilon_{BTR} \cdot \Delta \cdot g \cdot v$).

Note that the regression models for propulsive coefficients may be used relatively successfully for other catamaran forms, as propulsive coefficients are influenced more by length-to-beam ratio than by section shape. So, power predictions for round bilge hull form catamarans are possible if accurate resistance characteristics are available. The Southampton round bilge catamaran series[5] is by far the best known, see Molland et al. (1996) and Molland and Lee (1997), for instance, or Molland et al. (2011); resistance prediction for this series was also mentioned in the Sects. 2.2.2 and 3.4.12.

References

Begović E, Bertorello C (2012) Resistance assessment of warped hull form. Ocean Eng 56

Blount DL, Bjarne E (1989) Design and selection of propulsors for high speed craft. In: 7th Lips propeller symposium. Nordwijk-on-Sea

Blount DL, Fox DL (1976) Small craft power prediction. Mar Technol 13(1)

Blount DL, Fox DL (1978) Design considerations for propellers in cavitating environment. Mar Technol 15(2)

Brown PW (1971) An experimental and theoretical study of planing surfaces with trim flaps. Davidson Laboratory Report 1463. Stevens Institute of Technology

Calkins DE (1983) An interactive computer aided design synthesis program for recreational powerboats. SNAME Trans, vol 91

Collete M (2014) Effective optimization. Mar Technol (July)

Conley W (1981) Optimization, a simplified approach. Petrocelli Books, Inc., New York

Couser PR, Molland AF, Armstrong NA, Utama IKAP (1997) Calm water powering predictions for high speed catamarans. In: Proceedings of 4th international conference on fast sea transportation (FAST '97). Sydney

Fridsma G (1971) A systematic study of the rough water performance of planing boats (Irregular Waves—Part II). Davidson Laboratory Report 1495

[5]Couser et al. (1997) paper entitled "Calm water powering predictions for high speed catamarans", actually treats various resistance components, extrapolation methods for catamarans etc., not powering as designated here.

Gutsche F (1964) Untersuchung von Schiffsscrauben in schrager Austromung. Schiffbauforschung 3, 3/4, Rostock

Hadler JB (1966) The prediction of power performance of planing craft. SNAME Trans, vol 74

Hadler JB, Hubble EN (1971) Prediction of the power performance of the series 62 planing hull forms. SNAME Trans, vol 79

Hoggard MM (1979) Examining added drag of planing craft operating in the seaway. Hampton Road Section of SNAME

Hoggard MM, Jones MP (1980) examining pitch, heave and accelerations of planing craft operating in a seaway. In: High speed surface craft conference. Brighton

Holtrop J, Mennen GGJ (1982) An approximate power prediction method. Int Shipbuilding Prog 29(335)

Hubble EN (1974) Resistance of hard-chine stepless planing craft with systematic variation of hull form, longitudinal centre of gravity and loading. DTNSRDC R&D Report 4307

Hubble EN (1978) Planing craft feasibility model, user's manual. Report DTNSRDC/SPD-0840-01

Hubble EN (1980) Performance prediction of planing craft in a seaway. Report DTNSRDC/SPD-0840-02

ITTC (1984) Proceedings of the 17th international towing tank conference, high-speed propulsion, vol 1. Goteborg

Knight JT, Zahradka FT, Singer DJ, Collette MD (2014) Multiobjective particle swarm optimization of a planing craft with uncertainty. J Ship Prod Des 30(4)

Mercier JA, Savitsky D (1973) Resistance of transom-stern craft in the pre-planing regime. Davidson Laboratory Report 1667

Mohamad Ayob AF, Ray T, Smith WF (2011) Beyond hydrodynamic design optimization of planing craft. J Ship Prod Des 27(1)

Molland AF, Lee AR (1997) An investigation into the effect of prismatic coefficient on catamaran resistance. RINA Trans, vol 139

Molland AF, Wellicome JF, Couser PR (1996) Resistance experiments on a systematic series of high speed displacement catamaran forms: variation of length-displacement ratio and breadth-draught ratio. RINA Trans, vol 138

Molland AF, Turnock SR, Hudson DA (2011) Ship resistance and propulsion—practical estimation of ship propulsive power. Cambridge University Press. ISBN 978-0-521-76052-2

Müller-Graf B (1997a) Part I: Resistance components of high speed small craft. In: 25th WEGEMT School, Small Craft Technology, NTUA, Athens. ISBN I 900 453 053

Müller-Graf B (1997b) Part II: Powering performance prediction of high speed small craft. In: 25th WEGEMT School, Small Craft Technology, NTUA, Athens. ISBN I 900 453 053

Müller-Graf B (1997c) Part III: Factors affecting the reliability and accuracy of the resistance prediction. In: 25th WEGEMT School, Small Craft Technology, NTUA, Athens. ISBN I 900 453 053

Müller-Graf B (1997d) Dynamic stability of high speed small craft. In: 25th WEGEMT School, Small Craft Technology, NTUA, Athens. ISBN I 900 453 053

Müller-Graf B, Radojčić D, Simić A (2003a) Resistance and propulsion characteristics of the VWS hard chine catamaran hull Series '89. SNAME Trans, vol 110

Müller-Graf B, Radojčić D, Simić A (2003b) Discussion of Paper 1: Resistance and propulsion characteristics of the VWS hard chine catamaran hull Series '89. Mar Technol 40(4)

Oosterveld MWC, van Oossanen P (1975) Further computer-analyzed data of the Wageningen B-screw series. Int Shipbuilding Prog 22(251)

Radojčić D (1985) Optimal preliminary propeller design using nonlinear constrained mathematical programming technique. Ship Science Report No. 21. University of Southampton

Radojčić D (1988a) Evaluation of propeller performance in oblique flow. In: 8th symposium on theory and practice of shipbuilding, in memoriam of Prof. Sorta, Zagreb (in Serbian)

Radojčić D (1988b) Mathematical model of segmental section propeller series for open-water and cavitating conditions applicable in CAD. In: Propellers'88 symposium, SNAME, Virginia Beach

Radojčić D (1991) An engineering approach to predicting the hydrodynamic performance of planing craft using computer techniques. RINA Trans, vol 133

Savitsky D (1964) Hydrodynamic design of planing hulls. Mar Technol 1(1)

Savitsky D (2012) The effect of bottom warp on the performance of planing hulls. In: SNAME's 3rd Chesapeake power boat symposium. Annapolis

Savitsky D, Brown PW (1976) Procedure for hydrodynamic evaluation of planing hulls in smooth and rough water. Mar Technol 13(4)

Taniguchi K, Tanibayashi H, Chiba N (1967) Investigation into the propeller cavitation in oblique flow. Mitsubishi Technical Bulletin No. 143

Zips JM (1995) Numerical resistance prediction based on the results of the VWS hard chine catamaran hull Series '89. In: Proceedings of 3rd international conference on fast sea transportation (FAST '95). Lübeck-Travemünde

Chapter 7
Concluding Remarks

7.1 Evolutionary Process

As stated in the Introduction, *MM development is an evolutionary process*. Namely, new MMs are expected to be better than the previous ones. In general this pattern is followed and the results of any new procedures are compared with the previous ones. For example, the authors used an ANN technique first in Radojčić et al. (2014a, b), but cautiously and together with already proven regression analysis. Since encouraging results were obtained, modeling was executed solely with ANN in Radojčić et al. (2017b). In that paper single- and multiple-output ANN routines were compared. Subsequently, only a multiple output ANN was used in Radojčić and Kalajdžić (2018) for derivation of MMs for R/Δ & τ and $S/\nabla^{2/3}$ & L_{WL}/L_P. This historical perspective is illustrated in Fig. 7.1, showing a timeline of some notable MMs along with their main attributes. In the authors' opinion the MMs shown are important milestones that have left a footprint on MM development methodologies.

7.2 Conclusions

This work is an overview that outlines the authors' views. It is believed to be of value for MM developers and the HSC community. In addition, key references important for HSC hydrodynamics, and in particular for resistance, propulsion, and power prediction, are provided.

MMs for resistance prediction are the core of present work. The MMs presented here address conventional HSC, which may be classified as those belonging to the semi-displacement type (NPL, VTT, SKLAD, NTUA) and planing type (Series 50, 62, 65-B, TUNS, USCG, NSS). For the first group, the length Froude number (Fn_L) up to 1.1 is of interest. For the second group the volumetric Froude number (Fn_∇) up to 5.5 or so is appropriate. Adequate resistance and trim MMs for small HSC

D. Radojčić et al., *Power Prediction Modeling of Conventional High-Speed Craft*,
https://doi.org/10.1007/978-3-030-30607-6_7

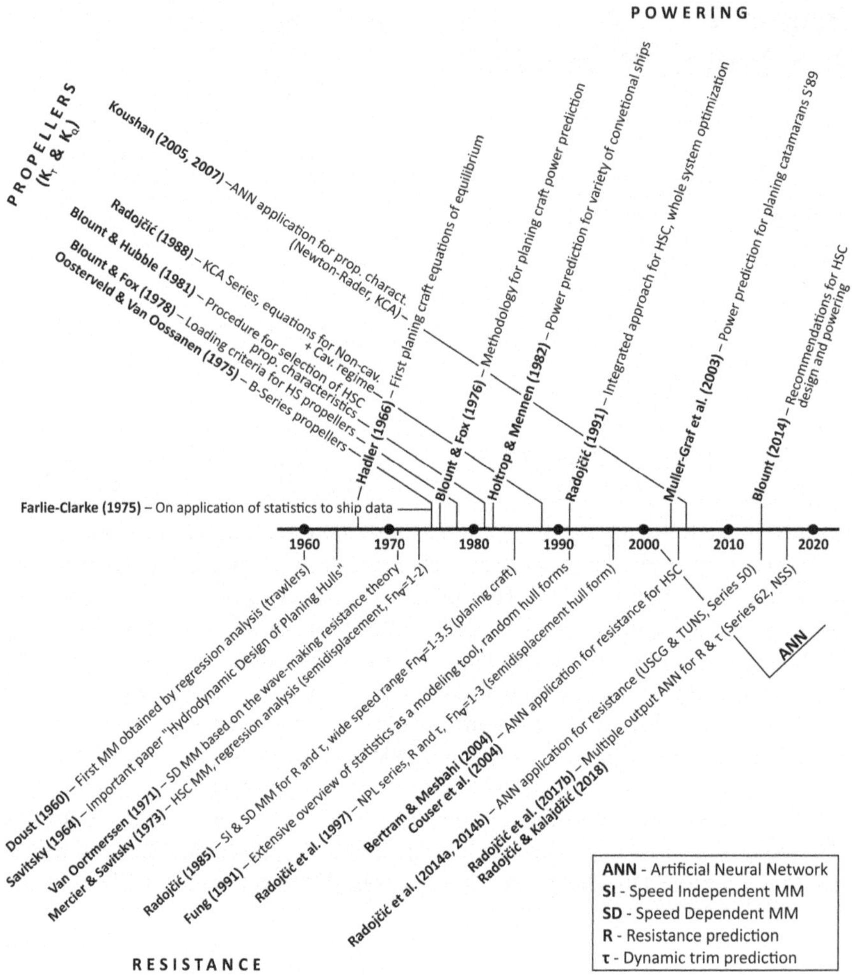

Fig. 7.1 Timeline of some notable MMs with their principal characteristics

(boats) for speeds corresponding to Fn_∇ of up to 8—for which the stepped hull form would be advantageous—is missing.

In order to achieve better accuracy, contemporary MMs are complex, and include approximately 10 times as many equation terms compared to those created a few decades ago. Note however, that the number of base parameters (input variables) did not change much. The complexity of the MMs is not an issue nowadays, given the available computer power. Furthermore, the unlimited computer power enables new techniques for model extraction, and ANN is gradually replacing regression methods.

Note also the novelty of the iterative procedure adopted in Radojčić et al. (2017a, b) used to shape an incomplete database, eventually resulting in a MM for predicting resistance and dynamic trim of Series 62. To some extent this reverses the conventional sequential set of steps depicted in Fig. 2.1.

Development of MMs for R/Δ and τ with multiple-outputs (Radojčić et al. 2017b, and in addition for $S/\nabla^{2/3}$ and L_{WL}/L_P in Radojčić and Kalajdžić 2018) as opposed to those with a single-output, is a novel application of the ANN technique. Here too, the established reasoning that R/Δ and τ are only loosely related may be questioned, given that more than 92% of equation terms are common for the two different quantities. Note, however, that even a *loose connection* (in physical terms) between R/Δ and τ is often ignored, which is obviously wrong.

New ANN based models, whether multiple- or single-output, clearly exhibit the double hump for dynamic trim. This was not possible with regression-based models as these were stiffer. Double hump in dynamic trim curve is important and may indicate dynamic instability.

When comparing MMs for resistance prediction with those for a propeller's hydrodynamic characteristics, note that the dependent and independent variables for the resistance prediction vary from model to model, depending on speed range and hull type. This is not the case with the propeller's variables which are essentially predetermined. For both the extraction tool used was regression and ANN.

MMs are routinely used to represent a propeller's hydrodynamic characteristics (K_T and K_Q) and the original experimental results (the dataset) are rarely given, nor are they needed. Knowing this, there is no reason why contemporary MMs for resistance and trim prediction (those that are based on a single systematic series) should not also replace the original databases. It appears that the MMs for resistance prediction are not yet fully accepted as authentic representatives of the datasets upon which they are based.

Given the experience with resistance modeling, it may be anticipated that ANN with multiple outputs may be applied for modeling propellers' hydrodynamic characteristics K_T and K_Q (as K_T and K_Q are linked through the open water efficiency coefficient—$\eta_o = (K_T/K_Q) \cdot (J/2\pi)$).

In summary, reliable MMs for prediction of resistance and propeller efficiency, although important, are not sufficient for reliable HSC power modeling. The interaction of hull, propulsor, and engine is the principal issue for appropriate calm water power predictions. Although there is room for improvements, the authors believes that the integrated approach presented in Radojčić (1991) is fundamentally correct, and hence the subsequent MMs developed by them were actually routines which can be incorporated in this or similar comprehensive power prediction routines.

Last but not least, the side-effect benefit of modeling is data-smoothing over continuous multidimensional surface, naturally when overfitting is avoided. From that perspective, MMs that are based on the systematic series may produce better results than those that stem directly from the original database. In some cases MMs were even able to identify erroneous datasets, and irregularities with the model experiments conducted years before these MMs were developed. This is contrary to the conventional wisdom that a MM, in general, cannot be better than the data upon which it is based.

The book is arranged in two parts. The first part is presented as a text, while the second part, titled "Resistance and Propulsor Design Data with Examples", provides tabulated data for assessment of mathematical models of resistance and propeller efficiency as outlined in the first part. Worked examples are given in the second part. The relevant Excel files can be provided to the reader as Electronic Supplementary Material (ESM). The reader can access ESM via the Springer sites which are given in Chaps. 8, 9 and 10.

References

Bertram V, Mesbahi E. (2004) Estimating resistance and power of fast monohulls employing artificial neural nets. In: International conference on High Performance Marine Vehicles (HIPER), Rome

Blount DL (2014) Performance by design. ISBN 0-978-9890837-1-3

Blount DL, Fox DL (1976) Small craft power prediction. Mar Technol 13(1)

Blount DL, Fox DL (1978) Design considerations for propellers in cavitating environment. Mar Technol 15(2)

Blount DL, Hubble EN (1981) Sizing Segmental section commercially available propellers for small craft. In: Propellers '81 symposium, SNAME, Virginia Beach

Couser P, Mason A, Mason G, Smith CR, von Konsky BR (2004) Artificial neural network for hull resistance prediction. In: 3rd international conference on Computer and IT Applications in the Maritime Industries (COMPIT '04), Siguenza

Doust DJ (1960) Statistical analysis of resistance data for trawlers. Fishing Boats of the World: 2 Fishing News (Books) Ltd., London

Farlie-Clarke AC (1975) Regression analysis of ship data. Int Shipbuilding Prog 22(251)

Fung SC (1991) Resistance and powering prediction for transom stern hull forms during early stage ship design. SNAME Trans 99

Hadler JB (1966) The prediction of power performance of planing craft. SNAME Trans 74

Holtrop J, Mennen GGJ (1982) An approximate power prediction method. Int Shipbuilding Prog 29(335)

Koushan K (2005) Mathematical expressions of thrust and torque of newton-rader propeller series for high speed crafts using artificial neural networks. In: Proceedings of 8th international conference on Fast Sea Transportation (FAST 2005), St. Petersburg

Koushan K (2007) Mathematical expressions of thrust and torque of Gawn-Burrill propeller series for high speed crafts using artificial neural networks. In: Proceedings of 9th international conference on Fast Sea Transportation (FAST 2007), Shanghai

Mercier JA, Savitsky D (1973) Resistance of transom-stern craft in the pre-planing regime. Davidson Laboratory Report 1667

Müller-Graf B, Radojčić D, Simić A (2003) Resistance and propulsion characteristics of the VWS hard chine catamaran hull Series '89. SNAME Trans 110

Oosterveld MWC, van Oossanen P (1975) Further computer-analyzed data of the Wageningen B-screw series. Int Shipbuilding Prog 22(251)

Radojčić D (1985) An approximate method for calculation of resistance and trim of the planing hulls. University of Southampton, Ship Science Report No. 23. Paper presented on SNAME Symposium on Powerboats, September 1985

Radojčić D (1988) Mathematical model of segmental section propeller series for open-water and cavitating conditions applicable in CAD. In: Propellers '88 symposium, SNAME, Virginia Beach

Radojčić D (1991) An engineering approach to predicting the hydrodynamic performance of planing craft using computer techniques. RINA Trans 133

Radojčić D, Kalajdžić M (2018) Resistance and trim modeling of naples hard chine systematic series. RINA Trans Int J Small Craft Technol. https://doi.org/10.3940/rina.ijsct.2018.b1.211

Radojčić D, Rodić T, Kostić N (1997) Resistance and trim predictions for the NPL high speed round bilge displacement hull series. In: RINA conference on Power, Performance and Operability of Small Craft, Southampton

Radojčić D, Zgradić A, Kalajdžić M, Simić A (2014a) Resistance prediction for hard chine hulls in the pre-planing regime. Pol Mar Res 21(2(82)) (Gdansk)

Radojčić D, Morabito M, Simić A, Zgradić A (2014b) Modeling with regression analysis and artificial neural networks the resistance and trim of Series 50 experiments with V-bottom motor boats. J Ship Prod Des 30(4)

Radojčić DV, Zgradić AB, Kalajdžić MD, Simić AP (2017a) Resistance and trim modeling of systematic planing hull Series 62 (with 12.5, 25 and 30 degrees deadrise angles) using artificial neural networks, Part 1: the database. J Ship Prod Des 33(3)

Radojčić DV, Kalajdžić MD, Zgradić AB, Simić AP (2017b) Resistance and trim modeling of systematic planing hull Series 62 (with 12.5, 25 and 30 degrees deadrise angles) using artificial neural networks, Part 2: mathematical models. J Ship Prod Des 33(4)

Savitsky D (1964) Hydrodynamic design of planing hulls. Mar Technol 1(1)

Van Oortmerssen G (1971) A power prediction method and its application to small ships. Int Shipbuilding Prog 18

Part II
Resistance and Propulsor Design Data with Examples

Chapter 8
Resistance and Dynamic Trim Predictions

8.1 Programs—VBA Codes for Microsoft Excel

Two examples that explain the application of typical MMs for resistance and dynamic trim predictions are given.

The first example, for the Series NPL (speed dependent MM) developed by regression analysis, is given in Sect. 8.1.1. All necessary data for the Series NPL are given in Sect. 8.2.3. VBA codes for other regression based methods for estimation of hydrodynamic characteristics treated here are similar.

The second example, for the ANN based MMs with multiple outputs (ANN structure 5-1-8-5-2) for the Series 62 is given in Sect. 8.1.2. All necessary data for the Series 62 are given in Sect. 8.3.4. VBA codes for other ANN based methods, either with single or multiple outputs, for estimation of hydrodynamic characteristics are similar.

Both examples begin with the flow charts (Figs. 8.1 and 8.5) which describe the structure of the VBA macro code. VBA code follows, and then the tables that contain the required input data, MM's coefficients, and output data, all corresponding to a specific method, are presented. Comments, that explain the significant steps in the code are included too; they are marked with an apostrophe " ' " at the beginning of each line. Note that since VBA code reads the data required for calculation from the specified cells, each table containing the input data, MM coefficients, and output data must be placed at the exact position in the Excel Sheet 1, as indicated in Figs. 8.2, 8.3, 8.4 and 8.6, 8.7, 8.8, for the first and second example, respectively.

Electronic supplementary material The online version of this chapter (https://doi.org/10.1007/978-3-030-30607-6_8) contains supplementary material, which is available to authorized users.

D. Radojčić et al., *Power Prediction Modeling of Conventional High-Speed Craft*,
https://doi.org/10.1007/978-3-030-30607-6_8

VBA code for Froude expansion (or Froude extrapolation) is given in Sect. 8.1.3. An example of Froude expansion is given for the Series 62 (Fig. 8.9) for evaluation of total resistance for any displacement other than the standard 100000 lb. Note that standard conditions in this case are valid for ATTC-1947, $C_A = 0$ and sea water at 15 °C, i.e. $\rho = 1026$ kg/m^3 and $\nu = 1.1907 \times 10^{-6}$ m^2/s.

8.1.1 An Example of Regression Based Mathematical Models

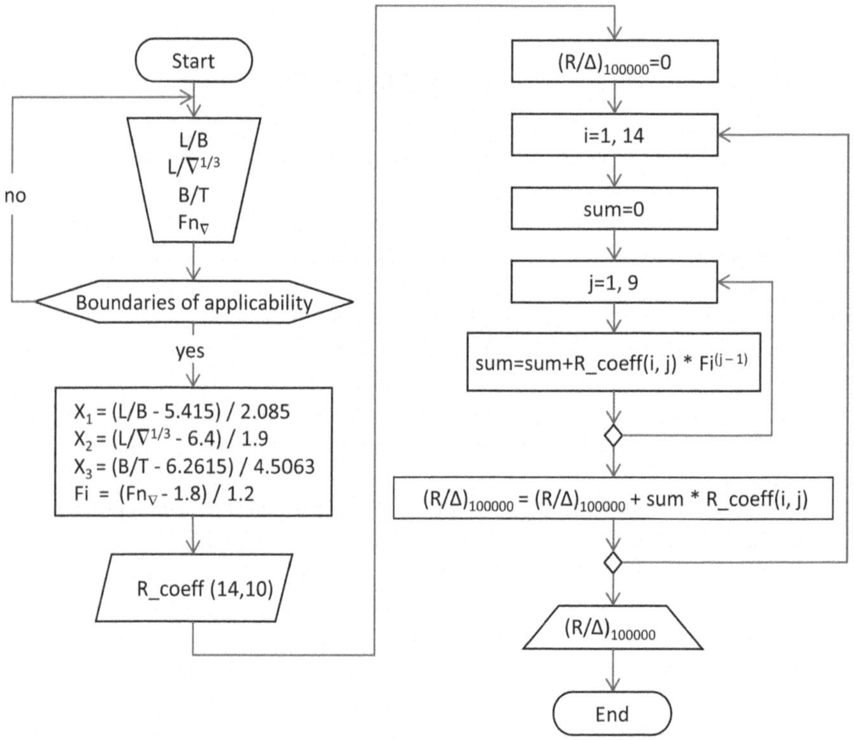

Fig. 8.1 Flow chart for regression based mathematical model (Example for resistance evaluation of the NPL series, speed dependent MM)

Example of VBA Macro Code for Regression Based Mathematical Model (NPL series, speed dependent MM)

```
Sub NPL_SD_Program()
Dim R_coeff(14, 10), T_coeff(15, 10), WS_coeff(13, 2)
'Clear previous results
Sheet1.Range("C63", "M66") = ""
'Normalization of Basic Independent Variables, as well as all
'other Independent Variables required for the calculation are
'prepared in the Excel worksheet table, and are given in the
'column "K". (see Fig. 8.3)
'Loading coefficient matrices for:
'(R/Δ)100000 , Dynamic Trim and Wetted Surface Coefficient
For i = 1 To 15
For j = 1 To 10
If i < 15 Then R_coeff(i, j) = Sheet1.Cells(i + 12, j + 1)
T_coeff(i, j) = Sheet1.Cells(i + 28, j + 1)
If i < 14 And j < 2 Then
   WS_coeff(i, 1) = Sheet1.Cells(i + 45, 2)
   WS_coeff(i, 2) = Sheet1.Cells(i + 45, 11)
   End If
Next j
Next i
'APPLICATION OF MATHEMATICAL MODELS
'Fn_vol is normalized according to the recommendation.
lok = 3
For Fn_vol = 1 To 3.01 Step 0.2
Fi = (Fn_vol – 1.8) / 1.2
'Calculation:  (R/Δ)100000
Resistance = 0
For i = 1 To 14
Sum = 0
For j = 1 To 9
Sum = Sum + R_coeff(i, j) * Fi ^ (j – 1)
Next j
Resistance = Resistance + Sum * R_coeff(i, j)
```

```
Next i
'Calculation:  Dynamic Trim
Dyn_Trim = 0
For i = 1 To 15
Sum = 0
For j = 1 To 9
Sum = Sum + T_coeff(i, j) * Fi ^ (j – 1)
Next j
Dyn_Trim = Dyn_Trim + Sum * T_coeff(i, j)
Next i
'Write results to Sheet1
Sheet1.Cells(63, lok) = Fn_vol
Sheet1.Cells(64, lok) = Resistance
Sheet1.Cells(65, lok) = Dyn_Trim
lok = lok + 1
Next Fn_vol
'Calculation:  Wetted Surface Coefficient
'WSC doesn't depend on a ship speed
WSC = 0
For i = 1 To 13
WSC = WSC + WS_coeff(i, 1) * WS_coeff(i, 2)
Next i
Sheet1.Cells(66, 3) = WSC
End Sub
```

Fig. 8.2 Sample of input data

	A	B	C	D
3				
4		**INPUT**		
5				
6		L/B	*4.17*	
7		$L/\nabla^{1/3}$	*6.402*	
8		B/T	*6*	
9				

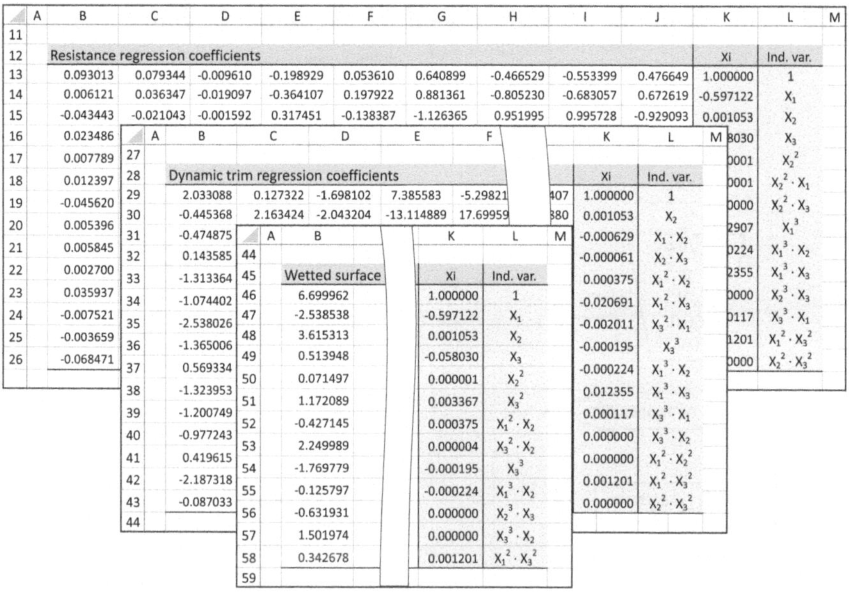

Fig. 8.3 Regression coefficients for resistance, dynamic trim and wetted surface (see Tables 8.6, 8.7 and 8.8; column "K" shows normalized independent variables)

OUTPUT												
Fn_∇	1.00	1.20	1.40	1.60	1.80	2.00	2.20	2.40	2.60	2.80	3.00
$(R/\Delta)_{100000}$	0.0307	0.0518	0.0672	0.0778	0.0868	0.0958	0.1054	0.1162	0.1290	0.1443	0.1618
τ [deg]	0.4191	0.9303	1.5612	1.9408	2.0408	2.0378	2.0510	2.0666	2.0909	2.3043	2.7197
WSC	8.1943										

Fig. 8.4 Sample of output data

8.1.2 An Example of ANN Based Mathematical Models

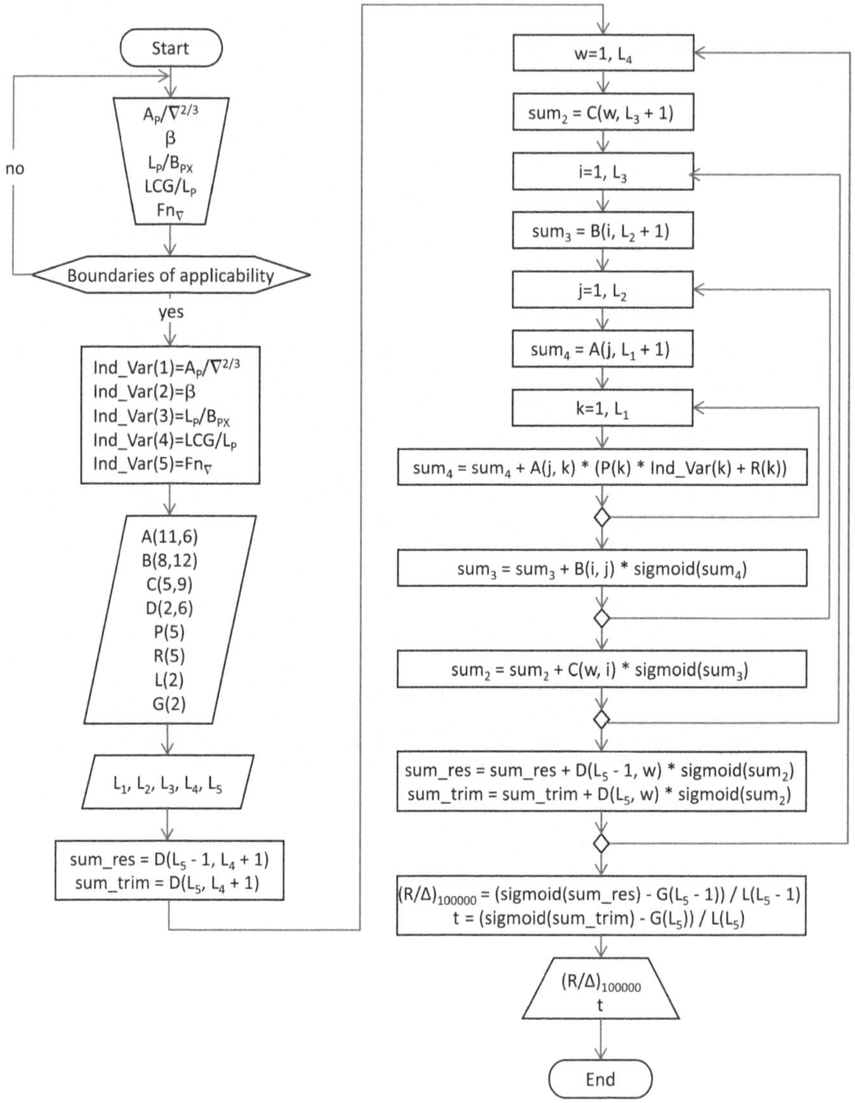

Fig. 8.5 Flow chart of ANN based mathematical models (example for the Series 62, multiple output MM)

Example of VBA Macro Code for ANN Based Mathematical Models (Series 62, multiple output MM)

```
Sub Program()
Dim A(11, 6), B(8, 12), C(5, 9), D(2, 6)
Dim P(5), R(5), L(2), G(2), Ind_Var(5)

'Clear all previous results
Sheet1.Cells.Range("C58", "M60") = ""
'Loading Independent Variables:
'Nomenclature:  Ind_Var(1)=Planing Area Coefficient
'               Ind_Var(2)=Deadrise angle
'               Ind_Var(3)=Length-Beam ratio
'               Ind_Var(4)=longitudinal center of gravity relative to transom
'               Ind_var(5)=Fn_vol
Ind_Var(1) = Sheet1.Cells(6, 3)
Ind_Var(2) = Sheet1.Cells(7, 3)
Ind_Var(3) = Sheet1.Cells(8, 3)
Ind_Var(4) = Sheet1.Cells(9, 3)

'Loading ANN configuration
L1 = Sheet1.Cells(7, 8): L2 = Sheet1.Cells(7, 9)
L3 = Sheet1.Cells(7, 10): L4 = Sheet1.Cells(7, 11)
L5 = Sheet1.Cells(7, 12)

'Loading ANN MM coefficients
For i = 1 To L2
  For j = 1 To L1 + 1
    A(i, j) = Sheet1.Cells(i + 13, j + 1)
  Next j
Next i
For i = 1 To L3
  For j = 1 To L2 + 1
    B(i, j) = Sheet1.Cells(i + 26, j + 1)
  Next j
Next i
For i = 1 To L4
  For j = 1 To L3 + 1
    C(i, j) = Sheet1.Cells(i + 36, j + 1)
  Next j
Next i
For i = 1 To L5
  For j = 1 To L4 + 1
    D(i, j) = Sheet1.Cells(i + 43, j + 1)
  Next j
Next i
```

```
For i = 1 To L1
   P(i) = Sheet1.Cells(i + 47, 2)
   R(i) = Sheet1.Cells(i + 47, 3)
Next i
For i = 1 To L5
   L(i) = Sheet1.Cells(i + 47, 4)
   G(i) = Sheet1.Cells(i + 47, 5)
Next i
'Application of ANN mathematical model
lok = 3
For Fn_vol = 1 To 4 Step 0.3
Ind_Var(5) = Fn_vol
sum_res = D(L5 – 1, L4 + 1)
sum_trim = D(L5, L4 + 1)
For w = 1 To L4
   sum2 = C(w, L3 + 1)
   For i = 1 To L3
      sum3 = B(i, L2 + 1)
      For j = 1 To L2
         sum4 = A(j, L1 + 1)
         For k = 1 To L1
         sum4 = sum4 + A(j, k) * (P(k) * Ind_Var(k) + R(k))
         Next k
      sum3 = sum3 + B(i, j) * sigmoid(sum4)
      Next j
   sum2 = sum2 + C(w, i) * sigmoid(sum3)
   Next i
sum_res = sum_res + D(L5 – 1, w) * sigmoid(sum2)
sum_trim = sum_trim + D(L5, w) * sigmoid(sum2)
Next w
Resistance = (sigmoid(sum_res) - G(L5 – 1)) / L(L5 – 1)
Dyn_Trim = (sigmoid(sum_trim) - G(L5)) / L(L5)

'Write results
Sheet1.Cells(58, lok) = Fn_vol
Sheet1.Cells(59, lok) = Resistance
Sheet1.Cells(60, lok) = Dyn_Trim
lok = lok + 1
Next Fn_vol

End Sub

'Calculation of the sigmoid function
Function sigmoid(x)
sigmoid = 1 / (1 + Exp(–x))
End Function
```

	A	B	C	D
3				
4		**INPUT**		
5				
6		$A_P/\nabla^{2/3}$	*7.024*	
7		β	*15*	
8		L_P/B_{PX}	*4.46*	
9		LCG/L_P	*0.4*	
10		Fn_∇	-	
11				

Fig. 8.6 Sample of input data

	A	B	C	D	E	F	G	H
12		MM Coefficients						
13		A_{1j}	A_{2j}	A_{3j}	A_{4j}	A_{5j}	a_j	
14		-0.1348189	-0.0076723	-0.2775522	3.9860210	-51.102770	5.0188620	
15		-0.4625220	-2.6474660	-4.3783800	-2.2981970	2.3711880	3.7986360	
16		-4.1258340	0.9317533	0.5740002	0.3302810	1.9288290	-0.8772002	
17		0.4276277						
18		0.3993703						
19		-2.7739500						
20		-0.3610690						
21		0.4993203						
22		-0.6632705						
23		0.6739060						
24		-1.9323510						
25								

	A	B	C	D	E	F	G	H	I	J	K	L	M	N
26		B_{1i}	B_{2i}	B_{3i}	B_{4i}	B_{5i}	B_{6i}	B_{7i}	B_{8i}	B_{9i}	B_{10i}	B_{11i}	b_i	
27		-1.037326	2.3060430	0.7746050	-0.5941365	-1.2888720	-0.7995334	-1.8261880	-3.6102430	-2.755224	-2.0167620	-0.5915413	3.1382240	
28		-16.05933	1.4665730	2.1381920	-0.0350787	1.8385600	0.1185768	1.6615480	-5.3048430	11.87204	-0.8111245	-1.5748320	0.1467929	
29		-3.242985											-1.7504580	
30		-8.723722											-14.483440	
31		-4.744526											-5.1791170	
32		-3.605005											-3.7377810	
33		-6.696820											1.5385620	
34		-12.77730											-5.3102150	
35														

	A	B	C	D	E	F	G	H	I	J	K
36		C_{1w}	C_{2w}	C_{3w}	C_{4w}	C_{5w}	C_{6w}	C_{7w}	C_{8w}	c_w	
37		-10.481130	2.8282740	1.2338810	-4.3893220	-2.9808120	-0.4040653	1.0968770	2.8939640	-2.8279170	
38		-12.570220	-3.6624040	-0.5793387	-0.3504248	-3.8378950	-3.0231180	-1.7668800	-0.0913801	2.1768760	
39		-1.820002	-3.2548080	-1.0896540	1.0606170	-2.7420750	-2.6956180	-1.4124410	0.5530158	2.2103580	
40		-4.307130	-4.1856740	2.7310460	-2.8798950	-4.9322180	-5.3763360	1.6572750	5.6926620	8.6448720	
41		5.824820	2.9226870	2.8063150	-2.7293670	0.6810613	1.1308260	0.6901268	-4.3622750	-2.8384770	
42											
43		D_{1V}	D_{2V}	D_{3V}	D_{4V}	D_{5V}	d_V				
44		-2.7865520	-1.1651470	-0.0683942	-12.662410	3.6777960	11.2581200				
45		-1.6608050	6.5392250	-10.209520	-8.4640710	-1.9419160	10.3451400				
46											
47		Pk	Rk	Lv	Gv						
48		0.1058824	0.0500000	2.0706787	0.0500000						
49		0.0300000	0.0500000	0.0481541	0.1602729						
50		0.1285714	0.0500000								
51		1.8442623	0.0500000								
52		0.1487603	0.0500000								
53											

Fig. 8.7 ANN coefficient for resistance and dynamic trim (see Table 8.28)

	A	B	C	D	E	F	G	H	I	J	K	L	M	N
55														
56		**OUTPUT**												
57														
58		Fn_∇	1.00	1.30	1.60	1.90	2.20	2.50	2.80	3.10	3.40	3.70	4.00	
59		$(R/\Delta)_{100000}$	0.0494	0.0818	0.0945	0.1026	0.1097	0.1160	0.1210	0.1253	0.1305	0.1370	0.1449	
60		τ [deg]	0.5160	2.0111	2.3670	2.6126	2.9787	3.3124	3.4324	3.3341	3.1148	2.8488	2.5718	
61														

Fig. 8.8 Sample of output data

8.1.3 Computation of New Design's Total Resistance

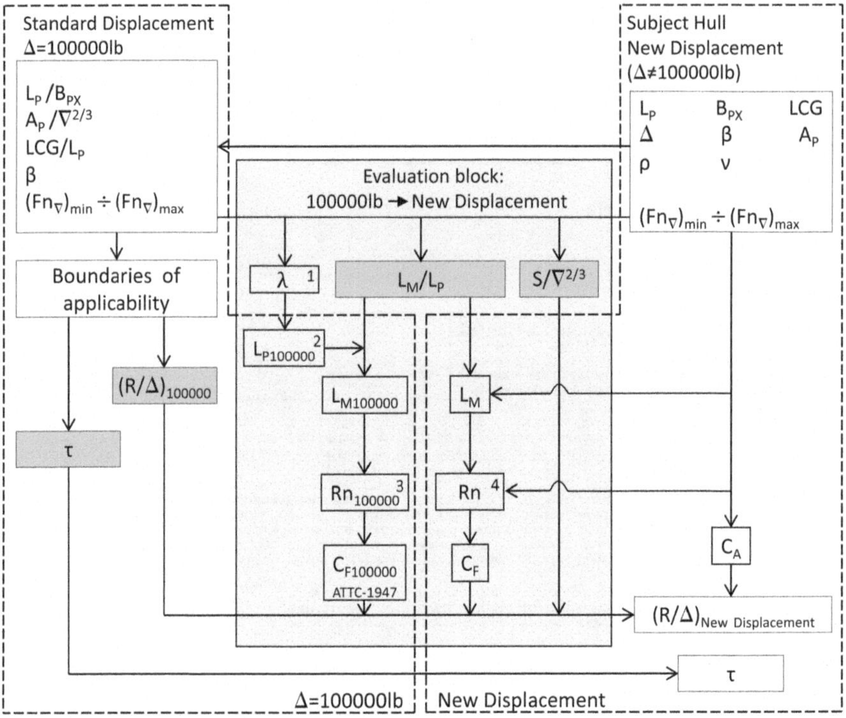

1) $$\lambda = \sqrt[3]{\frac{\Delta}{\Delta_{100000}} \cdot \frac{\rho_{100000}}{\rho}}$$

2) $$L_{P\,100000} = \frac{L_P}{\lambda}$$

4) $$Rn = \frac{Fn_{\nabla} \cdot \sqrt{g \cdot \nabla^{1/3}} \cdot L_M}{\nu}$$

3) $$Rn_{100000} = \frac{v \cdot L_M}{\nu} = \frac{Fn_{\nabla} \cdot \sqrt{g \cdot \nabla_{100000}^{1/3}} \cdot L_{M100000}}{\nu} = \frac{5.8887 \cdot Fn_{\nabla} \cdot L_M}{1.1907 \cdot 10^{-6}}$$

Fig. 8.9 Flow chart for computation of new design's total resistance (example for the Series 62)

When R/Δ is calculated for subject hull's displacement other than 100000 lb, C_A may be applied if desired. Namely:

$$(R/\Delta)_{New\,Displacement} = (R_T/\Delta)_{100000} + \left[(C_F + C_A)_{New\,Displacement} - C_{F100000}\right] \cdot 1/2 \cdot S/\nabla^{2/3} \cdot Fn_{\nabla}^2$$

Example of VBA Macro Code for Computation of New Design's Total Resistance for the Series 62

```
Sub Program()

'Clear all previous results
Sheet1.Cells.Range("D14", "D16") = ""
Sheet1.Cells.Range("H14", "H16") = ""
Sheet1.Cells.Range("L14", "L15") = ""

'Loading hull parameters
Ap_Vol = Sheet1.Cells(6, 3)
S_Vol = Sheet1.Cells(7, 3)
Lp_Bpx = Sheet1.Cells(8, 3)
Lm_Lp = Sheet1.Cells(6, 6)

'Loading Fn∇ and (R/Δ)100000
Fn_Vol = Sheet1.Cells(7, 6)
R_Disp_100000 = Sheet1.Cells(8, 6)

'Loading water characteristics and gravitational acceleration
rho = Sheet1.Cells(6, 9)
nu = Sheet1.Cells(7, 9)
ga = Sheet1.Cells(8, 9)

'Loading displacement other than 100000lb and correlation allowance coefficient
Disp = Sheet1.Cells(6, 12)
Ca = Sheet1.Cells(7, 12)

'Calculating L/∇ 1/3 according to the approximate method (Blount, 2014)
Lp_Vol = 1.1 * Sqr(Ap_Vol * Lp_Bpx)

'Calculating CF for Δ=100000 lb
Vol_100000 = 44.215816
Lp_100000 = Lp_Vol * Vol_100000 ^ (1 / 3)
Lm_100000 = Lm_Lp * Lp_100000
V_100000 = Fn_Vol * Sqr(ga * Vol_100000 ^ (1 / 3))
Rn_100000 = 0.5144 * V_100000 * Lm_100000 / 0.0000011907
Cf_100000 = 0.075 / (Log(Rn_100000) / Log(10) – 2) ^ 2
```

```
'Calculating CF for Δ≠100000 lb
Vol = Disp / rho
Lp = Lp_Vol * Vol ^ (1 / 3)
Lm = Lm_Lp * Lp
V = Fn_Vol * Sqr(ga * Vol ^ (1 / 3))
Rn = 0.5144 * V * Lm / nu
Cf = 0.075 / (Log(Rn) / Log(10) – 2) ^ 2
'Calculating R/Δ for Δ≠100000 lb
R_Disp = R_Disp_100000 + ((Cf + Ca) - Cf_100000) * 0.5 * S_Vol * Fn_Vol ^ 2

'Write results
Sheet1.Cells(14, 4) = Lp
Sheet1.Cells(15, 4) = Lm
Sheet1.Cells(16, 4) = Lp / Lp_Bpx
Sheet1.Cells(14, 8) = Ap_Vol * Vol ^ (2 / 3)
Sheet1.Cells(15, 8) = S_Vol * Vol ^ (2 / 3)
Sheet1.Cells(16, 8) = R_Disp
Sheet1.Cells(14, 12) = V
Sheet1.Cells(15, 12) = R_Disp * Disp * ga

End Sub
```

Note: All input and output data must be in the Sheet1 of Excel Workbook, and at the exact position as indicated in Figs. 8.10 and 8.11.

	A	B	C	D	E	F	G	H	I	J	K	L	M
1													
2		**Computation of Subject Hull Total Resistance**											
3													
4		**INPUT DATA**						**WATER**			**SUBJECT HULL**		
5													
6		$A_P/\nabla^{2/3}$	*5.9160*		L_M/L_P	*0.7500*		ρ [t/m^3]	*1.0259*		Δ [t]	*25*	
7		$S/\nabla^{2/3}$	6.6164		Fn_∇	*2.00*		ν [m^2/s]	*1.1907E-06*		C_A	*0.0002*	
8		L_P/B_{PX}	*4.0900*		$(R/\Delta)_{100000}$	*0.1080*		g [m/s^2]	*9.806*				
9													

Fig. 8.10 Sample of input data for the displacement of 100000 lb

	A	B	C	D	E	F	G	H	I	J	K	L	M
11													
12		**OUTPUT for Subject Hull**											
13													
14		L_P	[m]	15.687		A_P	[m^2]	49.7274		v	[m/s]	10.66	
15		L_M	[m]	11.766		S	[m^2]	55.6147		R_T	[kN]	27.45	
16		B_{PX}	[m]	3.836		R/Δ	[kN]	0.1120					
17													

Fig. 8.11 Sample of output data for subject hull

8.2 Recommended MMs for Semi-displacement Hull Forms

This sections contains tabulated data necessary for evaluation of resistance and dynamic trim (where available) of five selected MMs applicable for semi-displacement hull forms. Data for each series are presented as follows:

- **Name** of the Mathematical Model (MM)
- **Reference** (including cross-reference with other sections of this book)
- **Input** (independent variables, normalized in some cases) needed for resistance and dynamic trim predictions
- **Output** (dependent or target variables)
- **Body plan** of a series MM is based on
- **Boundaries of applicability** for a given MM presented in graphical format and/or as a set of equations.
- **Polynomial terms and regression coefficients** for resistance and dynamic trim MMs, as well as for wetted surface and other parameters needed for resistance evaluation.

Recommended MMs for semi-displacement hull forms are presented as follows:

- **Section** 8.2.1—**Mercier and Savitsky**. Random hull form MM for resistance prediction. Polynomial terms and regression coefficients for evaluation of $(R_T/\Delta)_{100,000}$ are given in Table 8.2.

- **Section** 8.2.2—**VTT**. Random hull form MMs for resistance prediction for *Round bilge* and *Hard chine* hulls. Polynomial terms and regression coefficients for evaluation of $(R_T/\Delta)_{100{,}000}$ are given in Table 8.3 (a) for round bilge, and (b) for hard chine hulls.
- **Section** 8.2.3—**NPL**. *Speed independent* and *Speed dependent* MMs for resistance and dynamic trim predictions based on the systematic NPL series. Their polynomial terms and regression coefficients are given in Tables 8.4, 8.5 and 8.6, 8.7, for $(R_T/\Delta)_{100{,}000}$ and τ, respectively. Note that the static waterline (Table 8.8) is used throughout.
- **Section** 8.2.4—**SKLAD**. MMs for resistance and dynamic trim predictions based on the systematic SKLAD series. Polynomial terms and regression coefficients for evaluation of C_R and τ are given in Tables 8.9 and 8.12 respectively. Note that running waterline, i.e. $(S) = f(v)$, (Table 8.10) is used throughout. Flowchart for evaluation of R_{TBH} from C_R, (S) and (M) is given in Fig. 8.20, while polynomial terms and regression coefficients for evaluation of (M) are given in Table 8.11.
- **Section** 8.2.5—**NTUA**. MMs for resistance and dynamic trim predictions based on the systematic double-chine NTUA series. Polynomial terms and regression coefficients for evaluation of C_R and τ are given in Table 8.13 (a) for C_R and (b) for τ. Note that static waterline (Table 8.13c) is used throughout.

Due to the differences amongst the MMs, stemming from the differences in the series and methods used for MM derivation, the MMs are not presented in an identical format (Table 8.14).

The main characteristics of the recommended MMs for semi-displacement hull forms are summarized and are given in the Table 8.1.

Table 8.1 Recommended MMs for evaluation of resistance and dynamic trim for semi-displacement hull forms

Name Reference	MMs main characteristics	Target value = f (input values)	Range of input parameters	Remarks
Mercier and Savitsky Mercier and Savitsky (1973) See 3.4.2 See 8.2.1	High-speed (Fn_∇=1–2), transom-stern, semi-displacement MM based on six round bilge series (NPL, Nordstrom, De Groot, SSPA, 64, 63, 62) and a single hard chine series (Series 62), in total 118 hull forms. Fig. 8.12	$(R_T/\Delta)_{100000}$ = f ($L_{WL}/\nabla^{1/3}$, ∇/B_X^3, i_e, A_T/A_X) for each Fn_∇ = 1.0, 1.1,...,2.0 $R_T/\Delta \neq$ f (LCG) LCG_{OPT} = (2–7)%L_{PP} aft of amidship $B_X=B_{WL}$	Constraints given in graphical format (scatter plots, see Fig. 8.13) Approximately: $L_{WL}/\nabla^{1/3}$=4–9 i_e=5–50 A_T/A_X=0–0.8 L_{WL}/B_X=2–7	PT: Regression analysis. I/F: 27/13 terms. SI MM. Corrections for $LCG \neq LCG_{OPT}$ and $R_T/\Delta \neq (R_T/\Delta)_{100000}$. Discrepancies between measured and calculated, less than 10% for 90% of cases. Schoenher's friction coef., C_A=0
VTT Lahtiharju et al. (1991) See 3.4.5 See 8.2.2	High-speed, transom-stern, round bilge semi-displacement (NPL, SSPA VTT). Fig. 8.14 (Fn_∇=1.8–3.2) Small draught, large C_B MM based on 65 hull forms ---------- High-speed, hard chine (Fn_∇=1.8–3.3) MM based on 13 hull forms.	$(R_T/\Delta)_{100000}$ = f ($\nabla^{1/3}$/T, L/T, B/L, A_T/A_X, B/T, C_X, L/ $\nabla^{1/3}$, B^3/∇, Fn_∇) ---------- $(R_T/\Delta)_{100000}$ = f (∇/T^3, $\nabla^{1/3}$/L, L/T, L/ $\nabla^{1/3}$, B/L, A_T/A_X, Fn_∇) $L=L_{WL}$, $B=B_{WL}$	L/ $\nabla^{1/3}$=4.47–8.3 B^3/∇=0.68–7.76 L/B=3.33–8.21 B/T=1.72–10.21 A_T/A_X=0.16–0.82 C_X =0.567–0.888 LCB=0.436L=const. ---------- L/ $\nabla^{1/3}$ =4.49–6.81 L/B=2.73–5.43 B/T=3.75–7.54 A_T/A_X=0.43–0.995	PT: Regression analysis. SI MM with 24 and 6 terms for round bilge and hard chine respectively Static wetted surface used for all speeds; ITTC-57 Given influence of variation of Δ, spray rails and wedges Predictions of seakeeping For Fn_∇=1.0–1.8 Mercier & Savitsky; Fn_∇>3.3 Savitsky Dynamic trim not given

(continued)

Table 8.1 (continued)

Name Reference	MMs main characteristics	Target value = f (input values)	Range of input parameters	Remarks
NPL Radojčić et al. (1997) See 3.4.7 See 8.2.3	High-speed, transom-stern, round bilge, semi-displacement. (Fn_∇=0.8–3.0). Fig. 8.16 SI and SD MMs, for Fn_∇=0.8–3.0 and Fn_∇=1.0–3.0, respectively MM based on solely on the NPL Series	$(R_T/\Delta)_{100000}$, τ, (S) = f (L/B, $L/\nabla^{1/3}$, B/T) and $(R_T/\Delta)_{100000}$, τ = f (L/B, $L/\nabla^{1/3}$, B/T, Fn_∇) $L=L_{WL}$, $B=B_{WL}$	Constraints given as 9 simple eq. of inequality type, see Fig. 8.17 Approximately: L/B=3.33–7.50 $L/\nabla^{1/3}$≈4.5–8.3 B/T=1.76–10.77	PT: Regression analysis. I/F: 27/14 for $(R/\Delta)_{100000}$, 27/15 for t and 27/13 for (S) S does not depend on speed. SD MMs have 14x9=126 and 15x9=135 terms for R/Δ and t respectively Propulsion coefficients in Bailey (1982)
SKLAD Radojčić et al. (1999) See 3.4.8 See 8.2.4	High-speed, transom-stern, round bilge, semi-displacement (Fn_∇=1.0–3.0). Fig. 8.18 MM based on solely on the SKLAD Series	C_R, $S/\nabla^{2/3}$, $L/\nabla^{1/3}$, τ = f (L/B, B/T, C_B) for each Fn_∇= 1.0, 1.25,..3.0 $L=L_{WL}$, $B=B_{WL}$	L/B=4–8 B/T=3–5 C_B=0.35–0.55 $L/\nabla^{1/3}$=4.5–9.7 Evaluated from $L/\nabla^{1/3} = (L/B)^{2/3}(B/T)^{1/3}(1/C_B)^{1/3}$	PT: Regression analysis. I/F: 100/22, 30, 21, and 19 terms for C_R, $S/\nabla^{2/3}$, $L/\nabla^{1/3}$, and τ respectively. All MMs are SI
NTUA Radojčić et al. (2001) See 3.4.11 See 8.2.5	High-speed, double-chine, wide-transom, pre-planing, semi-displacement. Fig. 8.21 (Fn_L = 0.5–0.9) MM based solely on the NTUA Series	C_R and τ = f(L/B, $L/\nabla^{1/3}$, B/T, Fn_L) (S) = f (B/T, $L/\nabla^{1/3}$) $L=L_{WL}$, $B=B_{WL}$	Odd-shaped 3-D space represented with 8 and 11 inequalities for C_R and τ respectively, see Fig 8.22. Approximately: L/B=4.3–7.5 $L/\nabla^{1/3}$=6.2–10.0 B/T=3.2–6.2	PT: Regression analysis. I/F: 100/31 and 100/28 for C_R and τ respectively SD MMs. For (S) - 6-term MM The independent variables and S are related to the static waterline

Legend: PT – Predictive Technique; I/F –Initial/Final MM; SI – Speed-Independent; SD – Speed-Dependent

8.2.1 Mercier and Savitsky

Reference Mercier and Savitsky (1973); see Sect. 3.4.2 (Fig. 8.12).

INPUT
$X_1 = (M)^{-1}$
$X_2 = \sqrt{2i_e}$
$X_3 = C_\Delta = \nabla/B_X^3$
$X_4 = A_T/A_X$

OUTPUT
$(R_T/\Delta)_{100000}$

INPUT
$X_1 = L_{WL}/\nabla^{1/3}$
$X_2 = B_X/T$

OUTPUT
(S)

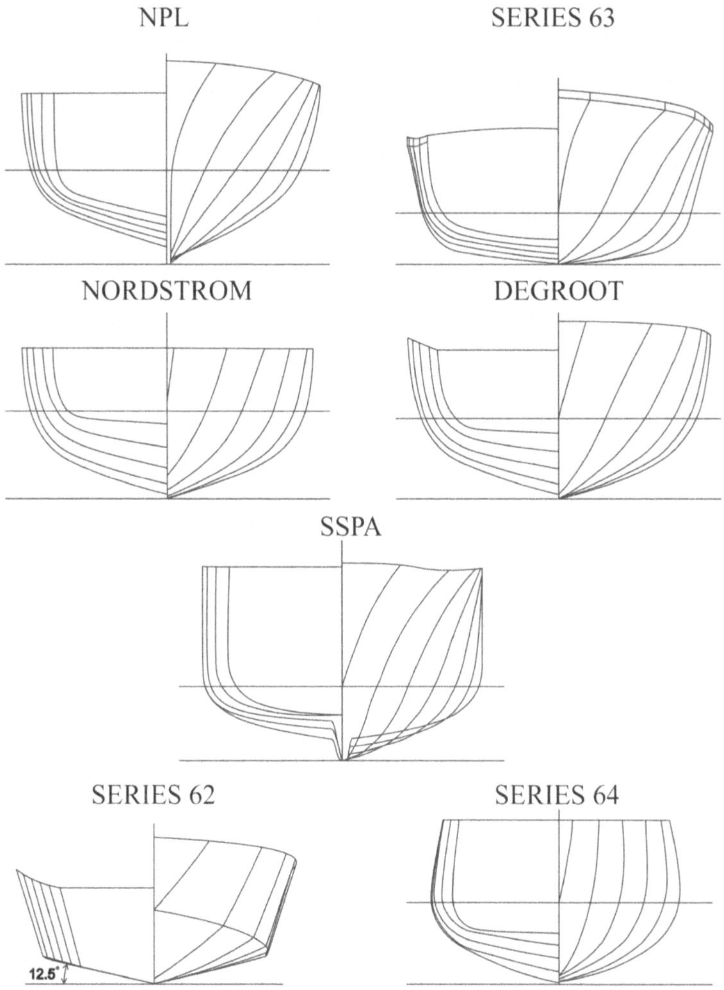

Fig. 8.12 Body plans of hull series used for the development of a MM. Note that the principal hull form parameters for the series 62 and 64 differ from the other series, see Fig. 8.13

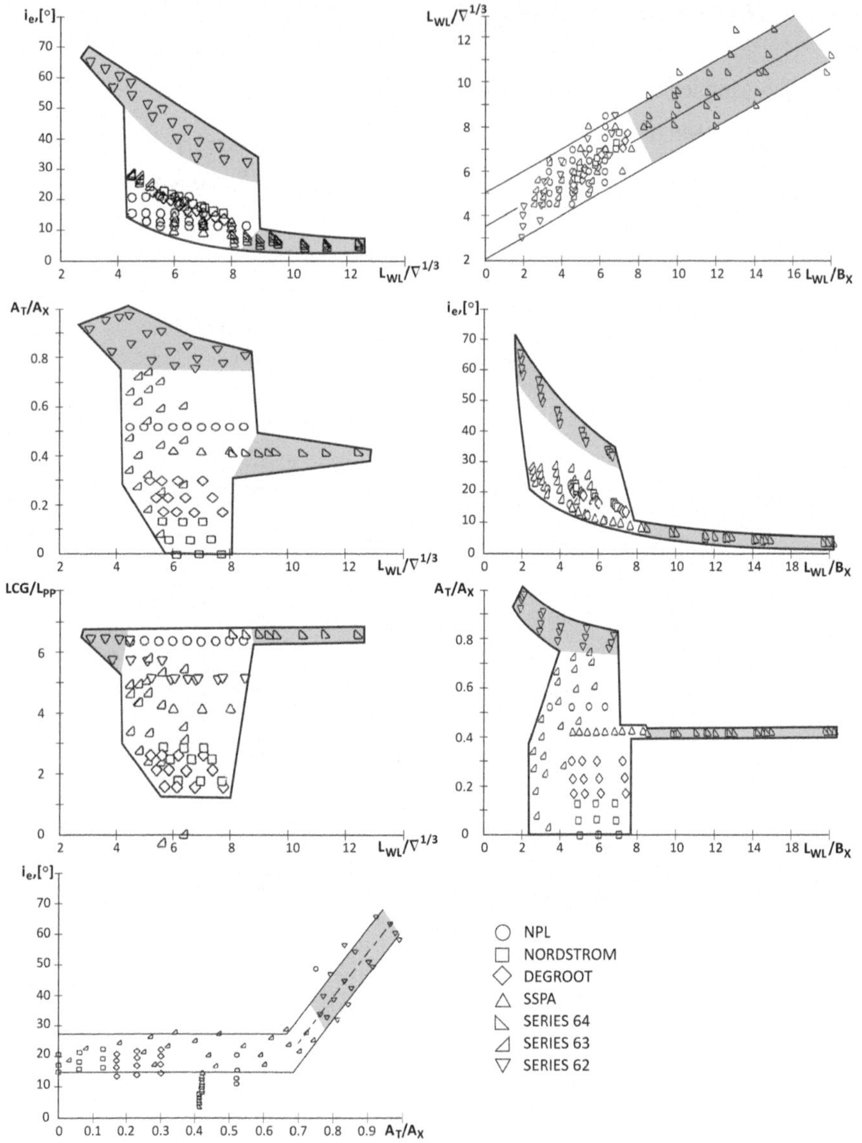

Fig. 8.13 Database distribution and original boundaries of applicability of a MM. Shaded areas represent data for series 62 and 64 which are outliers relative to the rest of the series but are included in the MM

Table 8.2 Polynomial terms and regression coefficients of the MM for resistance

$$(R_T/\Delta)_{100000} = a_0 + a_1X_1 + a_2X_2 + a_3X_4 + \cdots + a_{13}X_2^2X_4$$

		Fn_∇										
Coeff.	Multipl.	1	1.1	1.2	1.3	1.4	1.5	1.6	1.7	1.8	1.9	2
a_0	1	0.06473	0.10776	0.09483	0.03475	0.03013	0.03163	0.03194	0.04343	0.05036	0.05612	0.05967
a_1	X_1	−0.48680	−0.88787	−0.63720	0.00000	0.00000	0.00000	0.00000	0.00000	0.00000	0.00000	0.00000
a_2	X_2	−0.01030	−0.01634	−0.01540	−0.00978	−0.00664	0.00000	0.00000	0.00000	0.00000	0.00000	0.00000
a_3	X_4	−0.06490	−0.13444	−0.13580	−0.05097	−0.05540	−0.10543	−0.08599	−0.13289	−0.15597	−0.18661	−0.19758
a_4	X_1X_3	0.00000	0.00000	−0.16046	−0.21880	−0.19359	−0.20540	−0.19442	−0.18062	−0.17813	−0.18288	−0.20152
a_5	X_1X_2	0.10628	0.18186	0.16803	0.10434	0.09612	0.06007	0.06191	0.05487	0.05099	0.04744	0.04645
a_6	X_1X_4	0.97310	1.83080	1.55972	0.43510	0.51820	0.58230	0.52049	0.78195	0.92859	1.18569	1.30026
a_7	X_2X_3	−0.00272	−0.00389	−0.00309	−0.00198	−0.00215	−0.00372	−0.00360	−0.00332	−0.00308	−0.00244	−0.00212
a_8	X_3X_4	0.01089	0.01467	0.03481	0.04113	0.03901	0.04794	0.04436	0.04187	0.04111	0.04124	0.04343
a_9	X_4^2	0.00000	0.00000	0.00000	0.00000	0.00000	0.08317	0.07366	0.12147	0.14928	0.18090	0.19769
a_{10}	$X_1X_4^2$	−1.40962	−2.46696	−2.15556	−0.92663	−0.95276	−0.70895	−0.72057	−0.95929	−1.12178	−1.38644	−1.55127
a_{11}	$X_1^2X_3$	0.29136	0.47305	1.02992	1.06392	0.97757	1.19737	1.18119	1.01562	0.93144	0.78414	0.78282
a_{12}	$X_2X_4^2$	0.02971	0.05877	0.05198	0.02209	0.02413	0.00000	0.00000	0.00000	0.00000	0.00000	0.00000
a_{13}	$X_2^2X_4$	−0.00150	−0.00356	−0.00303	−0.00105	−0.00140	0.00000	0.00000	0.00000	0.00000	0.00000	0.00000

(S) can be estimated from the term at the foot of the table

$(S) = 2.262X_1(1 + 0.046X_2 + 0.00287X_2^2)$

8.2.2 VTT

Reference Lahtiharju et al. (1991); see Sect. 3.4.5 (Figs. 8.14 and 8.15).

INPUT	INPUT	OUTPUT
Round Bilge Vessels	**Hard Chine Craft**	
$X_1 = \nabla^{1/3}/T$	$X_1 = \nabla^{1/3}/T$	$(R_T/\Delta)_{100000}$
$X_2 = L/T$	$X_2 = L/T$	
$X_3 = B/L$	$X_3 = B/L$	
$X_4 = A_T/A_X$	$X_4 = A_T/A_X$	
$X_5 = B/T$	$X_7 = L/\nabla^{1/3}$	
$X_6 = C_X$	$X_9 = Fn_\nabla$	
$X_7 = L/\nabla^{1/3}$		
$X_8 = B^3/\nabla$		
$X_9 = Fn_\nabla$		

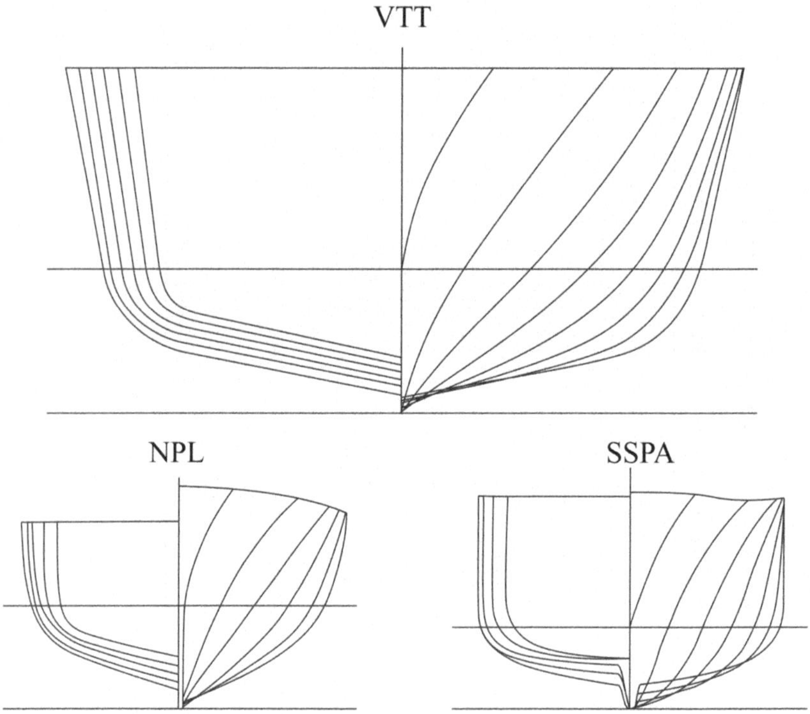

Fig. 8.14 Body plan of a VTT series together with two other series used for development of the MMs

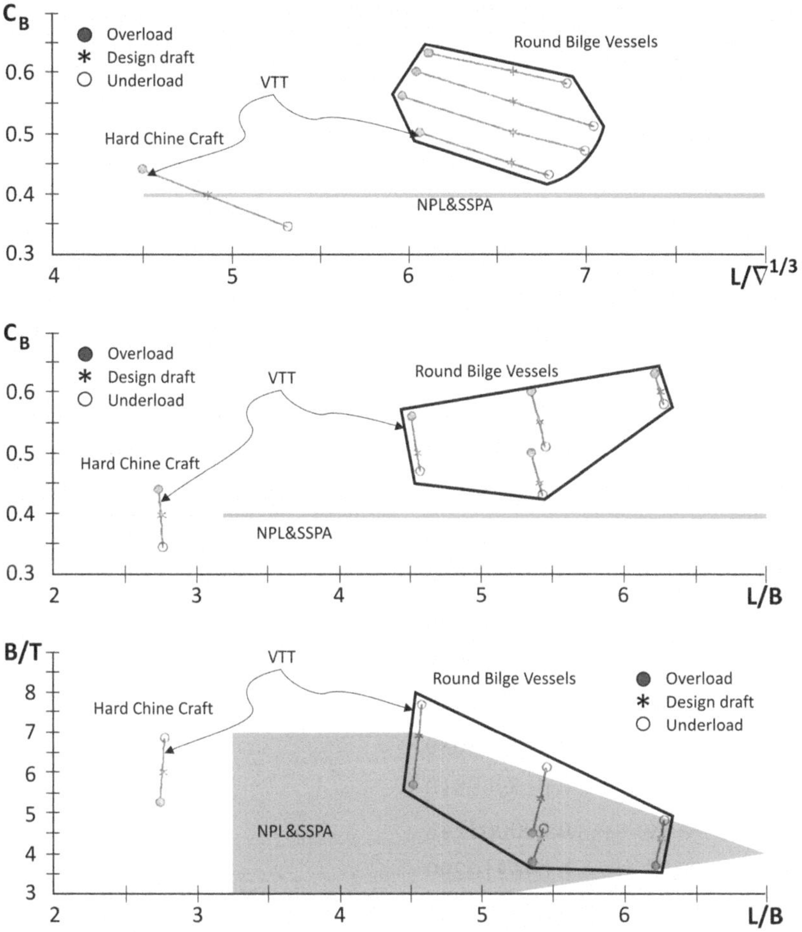

Boundaries of applicability of MMs given as a set of inequality equations.
Note: These equations do not exactly match the database distribution

Round Bilge	**Hard Chine**
$4.47 \le L\,/\,\nabla^{1/3} \le 8.3$	$4.494 \le L/\nabla^{1/3} \le 6.81$
$0.68 \le B^3/\nabla \le 7.76$	$2.73 \le L/B \le 5.43$
$3.33 \le L/B \le 8.21$	$3.75 \le B/T \le 7.54$
$1.72 \le B/T \le 10.21$	$0.43 \le A_T/A_X \le 0.995$
$0.16 \le A_T/A_X \le 0.82$	$1.8 \le Fn_\nabla \le 3.3$
$0.567 \le C_X \le 0.888$	
$1.8 \le Fn_\nabla \le 3.2$	
$LCG = 0.436L$	

Fig. 8.15 Database distribution and boundaries of applicability of MMs for the VTT series. Shaded areas represent data of NPL and SSPA series which are also used for the derivation of VTT series MMs

Table 8.3 Polynomial terms and regression coefficients of the MMs for resistance **a** round bilge, **b** hard chine

(a) Round Bilge

$$(R_T/\Delta)_{100000} = a_0 + a_1X_1^2 + a_2X_2^2 + \cdots + a_{24}X_4X_6^2X_9^4$$

Coeff.	Multipl.	a_i
a_0	1	0.085994790
a_1	X_1^2	−0.004033550
a_2	X_2^2	0.000050430
a_3	X_3X_4	0.503753960
a_4	$X_4X_5^2$	−0.004419540
a_5	$X_5^2X_6^2$	0.010066650
a_6	$X_2^2X_6^2$	−0.000229630
a_7	X_7X_9	−0.006485150
a_8	X_8X_9	0.017160910
a_9	$X_6^2X_8X_9$	−0.092915400
a_{10}	$X_2^2X_4X_9$	0.000070319
a_{11}	$X_4X_7X_9^2$	−0.010344410
a_{12}	$X_6^2X_7X_9^2$	0.023053080
a_{13}	$X_6^2X_8X_9^2$	0.015969810
a_{14}	$X_2^2X_9^3$	−0.000003248
a_{15}	$X_4X_9^3$	0.046510290
a_{16}	$X_6^2X_9^3$	−0.074689120
a_{17}	$X_8X_9^4$	0.001034050
a_{18}	$X_4X_8X_9^4$	−0.000694210
a_{19}	$X_6^2X_8X_9^4$	−0.003369480
a_{20}	$X_1X_6^2X_9^4$	−0.000124990
a_{21}	$X_3X_4X_9^4$	−0.053127110
a_{22}	$X_3X_6^2X_9^4$	0.117497870
a_{23}	$X_2^2X_4X_9^4$	0.000002201
a_{24}	$X_4X_6^2X_9^4$	−0.004705590

(b) Hard Chine

$$(R_T/\Delta)_{100000} = a_0 + a_1X_1^3 + \cdots + a_6X_3X_4X_9^2$$

Coeff.	Multipl.	a_i
a_0	1	−0.03546471
a_1	X_1^3	0.00129099
a_2	$X_7^{-1}X_9$	0.51603410
a_3	$X_2^2X_9$	−0.00010596
a_4	$X_7^2X_9^2$	−0.00090300
a_5	$X_7^3X_9^2$	0.00017501
a_6	$X_3X_4X_9^2$	−0.02784726

$(S) \approx 2.262X_7(1 + 0.046X_5 + 0.00287\,X_5^2)$ from Mercier and Savitsky

8.2.3 NPL

Reference Radojčić et al. (1997); see Sect. 3.4.7 (Figs. 8.16 and 8.17).

INPUT	OUTPUT
$X_1 = (L/B\text{-}5.415)/2.085$	$(R_T/\Delta)_{100000}$
$X_2 = ((M)\text{-}6.4)/1.9$	τ
$X_3 = (B/T\text{-}6.2615)/4.5063$	(S)
$X_4 = (Fn_\nabla\text{-}1.8)/1.2$	

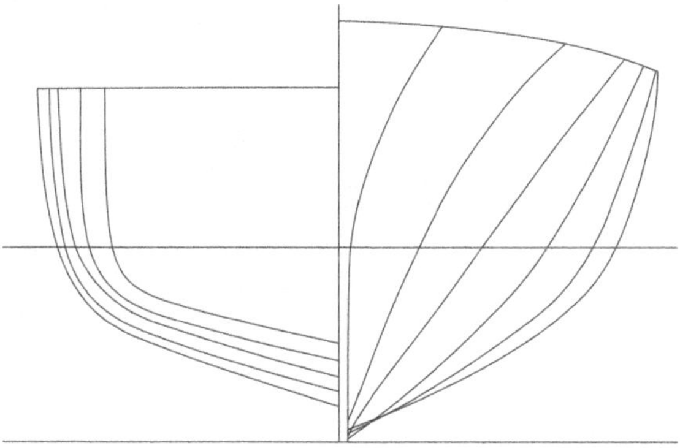

Fig. 8.16 Body plan of the parent hull of NPL series

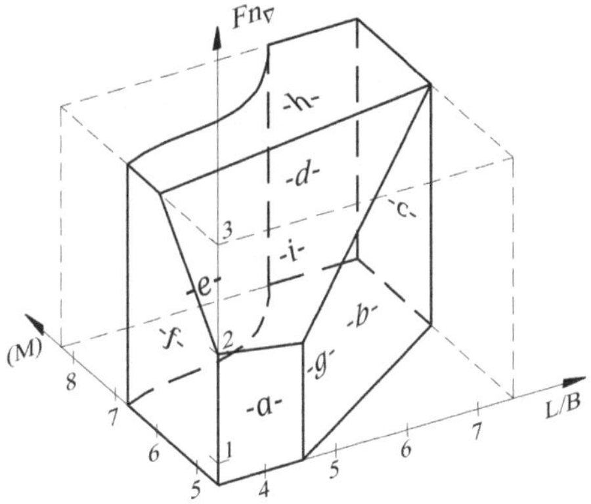

$-a-\ (M) \geq 4.5$

$-b-\ (M) \geq 0.6757 \cdot (L/B) + 1.4324$

$-c-\ L/B \leq 7.5$

$-d-\ (M) \leq 8.3$

$-e-\ (M) \leq 0.1534 \cdot (L/B)^3 - 1.9906 \cdot (L/B)^2 + 8.732 \cdot L/B - 5.9683$

$-f-\ L/B \geq 3.33$

$-g-\ Fn_\nabla \geq 0.8$

$-h-\ Fn_\nabla < 3.0$

$-i-\ Fn_\nabla \leq 0.714 \cdot (M) - 0.103 \cdot L/B - 0.872$

Fig. 8.17 Boundaries of applicability of NPL series MMs

Speed Independent Mathematical Model

Table 8.4 Polynomial terms and regression coefficients of the speed independent MM for resistance

$$(R_T/\Delta)_{100000} = a_0 + a_1X_1 + a_2X_2 + a_3X_3 + a_4X_2^2 + \cdots + a_{12}X_1^2X_3^2 + a_{13}X_2^2X_3^2$$

		Fn_∇											
Coeff.	Multipl.	0.8	1	1.2	1.4	1.6	1.8	2	2.2	2.4	2.6	2.8	3
a_0	1	0.012677	0.031092	0.057789	0.070863	0.080384	0.092592	0.105658	0.113350	0.118892	0.123105	0.120859	0.115058
a_1	X_1	−0.008102	0.000000	0.008445	0.004134	0.000962	0.005733	0.010769	0.007299	−0.001703	−0.014977	−0.041899	−0.077104
a_2	X_2	0.000061	−0.014184	−0.042841	−0.044967	−0.041259	−0.043200	−0.046379	−0.042913	−0.039950	−0.034003	−0.017179	0.005255
a_3	X_3	−0.002725	0.007983	0.015331	0.013614	0.014880	0.022656	0.031878	0.031266	0.023441	0.008990	−0.022888	−0.067378
a_4	X_2^2	−0.001185	0.002904	0.010337	0.013687	0.009803	0.008203	0.007646	0.013292	0.026694	0.043975	0.057718	0.071086
a_5	$X_1X_2^2$	0.004254	−0.003601	0.000000	0.005111	0.012468	0.012205	0.010466	0.000922	−0.018860	−0.043354	−0.046133	−0.048406
a_6	$X_2^2X_3$	−0.025702	0.000000	−0.000655	−0.013832	−0.038512	−0.043373	−0.050962	−0.050318	−0.036356	0.000000	0.051148	0.057769
a_7	X_1^3	0.001395	0.000062	0.000000	0.002578	0.004118	0.005143	0.006964	0.008068	0.009177	0.009824	0.009804	0.008532
a_8	$X_1^3X_2$	0.004913	0.005626	0.004361	0.004098	0.004899	0.004763	0.010023	0.012955	0.012276	0.006514	−0.008769	−0.034102
a_9	$X_1^3X_3$	−0.005901	0.002059	0.010917	0.001183	0.000667	0.002649	0.002493	−0.003659	−0.018867	−0.033644	−0.057179	−0.101116
a_{10}	$X_2^3X_3$	0.015476	0.000000	−0.003984	0.011616	0.030670	0.033934	0.044572	0.047574	0.025806	−0.015288	−0.047272	−0.062657
a_{11}	$X_1X_3^3$	−0.008887	0.002059	0.006095	0.001874	−0.004834	−0.006093	−0.009346	−0.010929	−0.015162	−0.020789	−0.019478	−0.051664
a_{12}	$X_1^2X_3^2$	−0.008427	0.005375	0.016608	0.006989	0.001094	−0.001804	−0.010127	−0.017531	−0.033868	−0.050023	−0.060285	−0.118198
a_{13}	$X_2^2X_3^2$	−0.033521	0.010811	0.008090	−0.023857	−0.057101	−0.064610	−0.081529	−0.080688	−0.063470	−0.014102	0.040157	0.073258

Table 8.5 Polynomial terms and regression coefficients of the speed independent MM for dynamic trim

$$\tau = b_0 + b_1X_2 + b_2X_1X_2 + b_3X_2X_3 + b_4X_1^2X_2 + b_5X_1^2X_3 + \cdots + b_{13}X_1^2X_3^2 + b_{14}X_2^2X_3^2$$

		Fn_∇											
Coeff.	Multipl.	**0.8**	**1**	**1.2**	**1.4**	**1.6**	**1.8**	**2**	**2.2**	**2.4**	**2.6**	**2.8**	**3**
b_0	1	0.175770	0.425175	0.883371	1.586455	1.899628	2.026007	2.058107	2.054187	2.069412	2.139490	2.357645	2.804328
b_1	X_2	−0.305797	−0.267412	−0.418518	−0.925061	−0.732979	−0.429611	−0.236105	−0.159138	−0.247707	−0.519239	−1.217875	−2.230391
b_2	X_1X_2	0.320664	0.041130	0.000000	0.042296	−0.272631	−0.472295	−0.567934	−0.598107	−0.404230	0.067575	1.151239	2.370417
b_3	X_2X_3	0.176390	0.241145	0.000000	−0.084784	−0.058048	0.161832	0.268277	0.368525	0.669915	1.201438	1.930230	2.947236
b_4	$X_1^2X_2$	−0.168558	−0.934109	−2.238618	−1.883572	−1.691190	−1.353073	−0.947598	−0.708352	−0.347415	0.178509	1.221190	2.742371
b_5	$X_1^2X_3$	−0.103548	−0.954868	−2.517245	−2.251547	−1.890948	−1.075607	−0.193149	0.622003	1.787335	3.090234	5.035156	6.999101
b_6	$X_1X_3^2$	0.349273	−0.937388	−4.491429	−4.304975	−3.724361	−2.552354	−1.283559	−0.064981	1.761040	3.810824	7.157361	11.191026
b_7	X_3^3	0.326051	−0.038224	−1.623344	−1.807736	−1.671105	−1.377836	−0.998759	−0.527308	0.283465	1.353872	3.053972	5.423357
b_8	$X_1^3X_2$	−1.040794	−0.806155	−0.703548	−1.350820	−0.537518	0.572009	1.628775	2.391011	3.030234	3.601990	3.710648	3.494198
b_9	$X_1^3X_3$	−0.383581	−1.348226	−2.781868	−2.624578	−2.221716	−1.349497	−0.344831	0.462487	1.541927	2.711225	4.260309	5.970641
b_{10}	$X_1X_3^3$	−0.078395	−0.557945	−2.635119	−2.260988	−1.924576	−1.229068	−0.401107	0.224152	1.045894	1.831750	3.596974	5.946926
b_{11}	$X_2X_3^3$	−0.358077	−0.434711	−1.523450	−2.023459	−1.530763	−1.018361	−0.408611	−0.118446	0.089195	0.194697	0.643298	1.373248
b_{12}	$X_1^2X_2^2$	0.794836	0.941649	1.417955	1.829837	1.260767	0.440716	−0.394656	−0.923245	−1.456637	−2.136660	−2.931875	−3.776398
b_{13}	$X_1^2X_3^2$	−0.342870	−1.516333	−4.571710	−4.392841	−3.747926	−2.234404	−0.422111	1.016714	2.912691	4.895521	8.005211	11.584463
b_{14}	$X_2^2X_3^2$	0.467308	0.865865	1.315406	1.042975	0.496254	−0.048087	−0.516796	−0.524839	−0.360649	−0.141175	−0.179117	−0.382556

Speed Dependent Mathematical Model

Table 8.6 Polynomial terms and regression coefficients of the speed dependent MM for resistance

$$(R_T/\Delta)_{100000} = \sum_{i=0}^{8} \alpha_i X_4^i + \left(\sum_{i=0}^{8} \alpha_{i+9} X_4^i\right) \cdot X_1 + \left(\sum_{i=0}^{8} \alpha_{i+18} X_4^i\right) \cdot X_2 + \cdots + \left(\sum_{i=0}^{8} \alpha_{i+108} X_4^i\right) \cdot X_1^2 X_3^2 + \left(\sum_{i=0}^{8} \alpha_{i+117} X_4^i\right) \cdot X_2^2 X_3^2$$

Multipl.		$X_4^0=1$		X_4^1		X_4^2		X_4^3		X_4^4		X_4^5		X_4^6		X_4^7		X_4^8
1	α_0	0.093013	α_1	0.079344	α_2	−0.009610	α_3	−0.198929	α_4	0.053610	α_5	0.640899	α_6	−0.466529	α_7	−0.553399	α_8	0.476649
X_1	α_9	0.006121	α_{10}	0.036347	α_{11}	−0.019097	α_{12}	−0.364107	α_{13}	0.197922	α_{14}	0.881361	α_{15}	−0.805230	α_{16}	−0.683057	α_{17}	0.672619
X_2	α_{18}	−0.043443	α_{19}	−0.021043	α_{20}	−0.001592	α_{21}	0.317451	α_{22}	−0.138387	α_{23}	−1.126365	α_{24}	0.951995	α_{25}	0.995728	α_{26}	−0.929093
X_3	α_{27}	0.023486	α_{28}	0.057679	α_{29}	−0.023726	α_{30}	−0.348106	α_{31}	0.179199	α_{32}	0.737798	α_{33}	−0.707577	α_{34}	−0.548689	α_{35}	0.562535
X_2^2	α_{36}	0.007789	α_{37}	−0.012854	α_{38}	0.048101	α_{39}	0.128945	α_{40}	0.017943	α_{41}	−0.038860	α_{42}	−0.182194	α_{43}	−0.067477	α_{44}	0.169689
$X_1X_2^2$	α_{45}	0.012397	α_{46}	0.000699	α_{47}	−0.041454	α_{48}	−0.078758	α_{49}	−0.411017	α_{50}	−0.020603	α_{51}	1.004690	α_{52}	0.100286	α_{53}	−0.614613
$X_2^2X_3$	α_{54}	−0.045620	α_{55}	−0.043439	α_{56}	0.100568	α_{57}	−0.136304	α_{58}	0.117069	α_{59}	0.787725	α_{60}	−0.349993	α_{61}	−0.584287	α_{62}	0.212039
X_1^3	α_{63}	0.005396	α_{64}	0.006646	α_{65}	0.001613	α_{66}	0.027995	α_{67}	−0.038633	α_{68}	−0.088732	α_{69}	0.106999	α_{70}	0.065021	α_{71}	−0.077782
$X_1^3X_2$	α_{72}	0.005845	α_{73}	0.015480	α_{74}	0.037527	α_{75}	−0.014682	α_{76}	−0.141940	α_{77}	−0.070899	α_{78}	0.174299	α_{79}	0.062046	α_{80}	−0.101799
$X_1^3X_3$	α_{81}	0.002700	α_{82}	0.023412	α_{83}	−0.052987	α_{84}	−0.415022	α_{85}	0.269027	α_{86}	1.022733	α_{87}	−0.841006	α_{88}	−0.773952	α_{89}	0.663990
$X_2^3X_3$	α_{90}	0.035937	α_{91}	0.044983	α_{92}	0.032679	α_{93}	0.147806	α_{94}	−0.993943	α_{95}	−1.042473	α_{96}	2.219790	α_{97}	0.948219	α_{98}	−1.455671
$X_1X_3^3$	α_{99}	−0.007521	α_{100}	−0.010052	α_{101}	0.052716	α_{102}	−0.110748	α_{103}	−0.257833	α_{104}	0.301729	α_{105}	0.394981	α_{106}	−0.194182	α_{107}	−0.220723
$X_1^2X_3^2$	α_{108}	−0.003659	α_{109}	−0.016928	α_{110}	0.006569	α_{111}	−0.297305	α_{112}	−0.143829	α_{113}	0.808866	α_{114}	0.059827	α_{115}	−0.591199	α_{116}	0.059515
$X_2^2X_3^2$	α_{117}	−0.068471	α_{118}	−0.066821	α_{119}	0.061252	α_{120}	−0.300738	α_{121}	0.829897	α_{122}	1.379352	α_{123}	−1.966275	α_{124}	−1.031222	α_{125}	1.236290

Table 8.7 Polynomial terms and regression coefficients of the speed dependent MM for dynamic trim

$$\tau = \sum_{i=0}^{8} \beta_i X_4^i + \left(\sum_{i=0}^{8} \beta_{i+9} X_4^i\right) \cdot X_2 + \left(\sum_{i=0}^{8} \beta_{i+18} X_4^i\right) \cdot X_1 X_2 + \cdots + \left(\sum_{i=0}^{8} \beta_{i+117} X_4^i\right) \cdot X_1^2 X_3^2 + \left(\sum_{i=0}^{8} \beta_{i+126} X_4^i\right) \cdot X_2^2 X_3^2$$

Multipl.		$X_4^0=1$		X_4^1		X_4^2		X_4^3		X_4^4		X_4^5		X_4^6		X_4^7		X_4^8
1	β_0	2.033088	β_1	0.127322	β_2	−1.698102	β_3	7.385583	β_4	−5.298216	β_5	−15.39244	β_6	17.085345	β_7	10.387316	β_8	−11.82641
X_2	β_9	−0.445368	β_{10}	2.163424	β_{11}	−2.043204	β_{12}	−13.114889	β_{13}	17.699596	β_{14}	25.43046	β_{15}	−40.013487	β_{16}	−17.534173	β_{17}	25.62888
X_1X_2	β_{18}	−0.474875	β_{19}	−1.209746	β_{20}	2.486439	β_{21}	3.532062	β_{22}	−9.678590	β_{23}	−1.98560	β_{24}	19.995371	β_{25}	0.638182	β_{26}	−10.93330
X_2X_3	β_{27}	0.143585	β_{28}	0.932906	β_{29}	−1.156282	β_{30}	−2.255990	β_{31}	11.399220	β_{32}	4.97365	β_{33}	−16.620768	β_{34}	−2.129468	β_{35}	7.66042
$X_1^2X_2$	β_{36}	−1.313364	β_{37}	1.397767	β_{38}	−0.172147	β_{39}	10.077473	β_{40}	−5.904343	β_{41}	−43.22277	β_{42}	42.104698	β_{43}	39.686211	β_{44}	−39.91394
$X_1^2X_3$	β_{45}	−1.074402	β_{46}	4.285762	β_{47}	1.742878	β_{48}	7.116343	β_{49}	−1.468058	β_{50}	−39.31659	β_{51}	35.581547	β_{52}	38.278200	β_{53}	−38.14810
$X_1X_3^2$	β_{54}	−2.538026	β_{55}	5.789688	β_{56}	1.867719	β_{57}	19.736459	β_{58}	−1.220753	β_{59}	−97.74601	β_{60}	76.166134	β_{61}	94.757031	β_{62}	−85.62398
X_3^3	β_{63}	−1.365006	β_{64}	1.504390	β_{65}	0.933899	β_{66}	9.069456	β_{67}	3.835651	β_{68}	−40.42210	β_{69}	23.516788	β_{70}	39.803665	β_{71}	−31.45430
$X_1^3X_2$	β_{72}	0.569334	β_{73}	7.844966	β_{74}	−2.930311	β_{75}	−28.830351	β_{76}	37.633001	β_{77}	61.19576	β_{78}	−84.867294	β_{79}	−43.279656	β_{80}	56.16172
$X_1^3X_3$	β_{81}	−1.323953	β_{82}	5.036429	β_{83}	1.245051	β_{84}	1.417878	β_{85}	4.007709	β_{86}	−24.91626	β_{87}	17.637162	β_{88}	26.233226	β_{89}	−23.36830
$X_1X_3^3$	β_{90}	−1.200749	β_{91}	3.403904	β_{92}	1.853146	β_{93}	13.498226	β_{94}	−13.727450	β_{95}	−68.02082	β_{96}	72.651723	β_{97}	66.400207	β_{98}	−68.91381
$X_2X_3^3$	β_{99}	−0.977243	β_{100}	3.779033	β_{101}	−1.710287	β_{102}	−8.874362	β_{103}	12.493018	β_{104}	−1.93510	β_{105}	−6.622121	β_{106}	11.688065	β_{107}	−6.46672
$X_1^2X_2^2$	β_{108}	0.419615	β_{109}	−5.617447	β_{110}	2.168412	β_{111}	17.290572	β_{112}	−24.321826	β_{113}	−29.82627	β_{114}	42.437218	β_{115}	17.202880	β_{116}	−23.53058
$X_1^2X_3^2$	β_{117}	−2.187318	β_{118}	8.724710	β_{119}	3.483912	β_{120}	5.944942	β_{121}	−1.689095	β_{122}	−63.17969	β_{123}	59.375699	β_{124}	67.577259	β_{125}	−66.46901
$X_2^2X_3^2$	β_{126}	−0.087033	β_{127}	−3.018830	β_{128}	3.801864	β_{129}	4.991406	β_{130}	−5.547206	β_{131}	3.75872	β_{132}	−5.683562	β_{133}	−7.913331	β_{134}	9.31614

Table 8.8 Polynomial terms and regression coefficients of the MM for Wetted surface

$$(S) = c_0 + c_1X_1 + c_2X_2 + c_3X_3 + c_5X_2^2 + \cdots + c_{26}X_1^2X_3^2$$

Coeff.	Multipl.	
c_0	1	6.699962
c_1	X_1	−2.538538
c_2	X_2	3.615313
c_3	X_3	0.513948
c_4	X_2^2	0.071497
c_5	X_3^2	1.172089
c_6	$X_1^2X_2$	−0.427145
c_7	$X_2X_3^2$	2.249989
c_8	X_3^3	−1.769779
c_9	$X_1^3X_2$	−0.125797
c_{10}	$X_2^3X_3$	−0.631931
c_{11}	$X_2X_3^3$	1.501974
c_{12}	$X_1^2X_3^2$	0.342678

Note Static waterline is used throughout

8.2.4 SKLAD

Reference Radojčić et al. (1999); see Sect. 3.4.8 (Figs. 8.18, 8.19 and 8.20).

INPUT	OUTPUT
$X_1 = L/B$	C_R
$X_2 = B/T$	(S)
$X_3 = C_B$	(M)
	τ

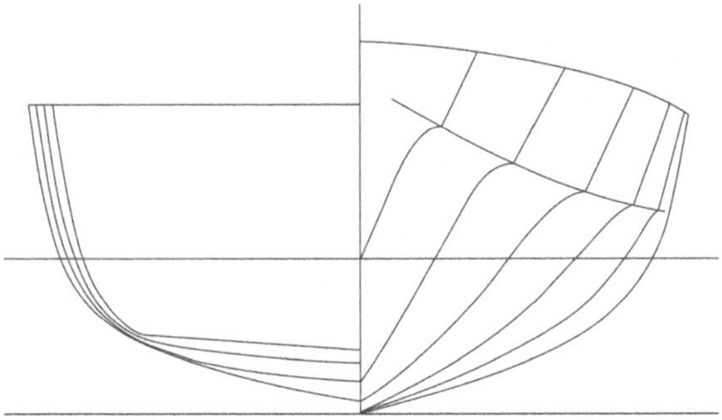

Fig. 8.18 Body plan of the parent hull of the SKLAD series

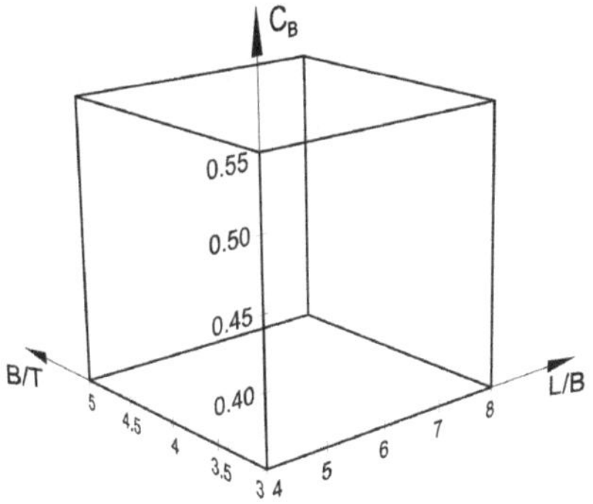

$4 \leq L/B \leq 8$
$3 \leq B/T \leq 5$
$0.35 \leq C_B \leq 0.55$

Fig. 8.19 Boundaries of applicability of SKLAD series MMs

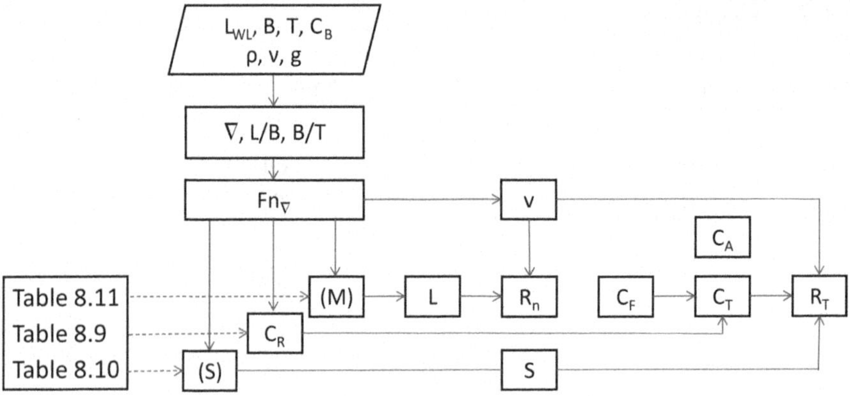

Fig. 8.20 Flow chart for evaluation of R_{TBH} from C_R, (S) and (M)

Table 8.9 Polynomial terms and regression coefficients of the MM for residuary resistance coefficient

$$10^3 \cdot C_R = a_0 + a_1 \cdot X_2^2 \cdot X_3 + a_2 \cdot X_3^3 + \cdots + a_{21} \cdot X_1^{1/3} \cdot X_2^{1/3} \cdot X_3^2$$

		Fn_∇							
Coeff.	Multipl.	1	1.25	1.5	1.75	2	2.25	2.5	3
a_0	1	96.8223198	103.9381866	110.5655308	71.3058851	93.7690256	115.4053829	61.2355975	76.1341171
a_1	$X_2^2X_3$	−2.0669480	−1.2278982	0.5872027	0.2057948	0.5984434	1.4291100	−1.0060833	−0.9078976
a_2	X_3^3	893.6445663	946.9998050	600.9090395	1259.3099391	506.0783293	283.8781596	1049.8083741	881.4630079
a_3	X_1^3	0.0109344	0.0174171	−0.0326842	0.0125131	−0.0121731	−0.0092164	0.0306355	0.0348260
a_4	$X_1^3X_2$	−0.0000495	−0.0023833	0.0192727	−0.0053168	0.0088797	0.0080694	−0.0133175	−0.0173138
a_5	$X_1^2X_2^2$	0.0137606	0.0210606	0.0004112	0.0462628	0.0156098	0.0101866	0.0497880	0.0519306
a_6	$X_1X_2X_3$	0.8715670	−4.1993409	−24.0776327	−15.3627479	−21.4294948	−17.0479806	−7.5885320	−3.4402352
a_7	$X_1X_2X_3^2$	9.8762251	26.5749514	31.4233606	33.7219409	19.2164161	5.5454734	20.1044928	13.0967680
a_8	$X_1X_2^3X_3$	0.0967935	0.1122517	0.1114783	0.0551373	0.0598996	0.0092344	0.0691665	0.0843002
a_9	$X_1^2X_2^3X_3$	−0.0163372	−0.0268178	−0.0094523	−0.0190360	0.0007145	0.0139293	−0.0210002	−0.0286189
a_{10}	$X_1^2X_3^3$	−1.6944782	−4.1287168	−3.3621228	−2.3523981	−1.6411920	−3.1483456	1.0054389	0.5876724
a_{11}	X_3^2	−339.8007885	−68.6127977	510.0083257	−539.1485747	213.2985704	210.1513827	−700.6978405	−654.2595284
a_{12}	$X_1^3X_2^2X_3^2$	−0.0191667	−0.0287483	−0.0304591	−0.0132764	−0.0053369	−0.0056433	0.0138415	0.0135857
a_{13}	$X_1^3X_2X_3^3$	0.1167621	0.1696262	0.1598483	0.0775111	0.0442892	0.0940339	−0.0911991	−0.0797247
a_{14}	$X_1^3X_2^3X_3$	0.0010030	0.0019994	0.0012553	0.0009020	−0.0004120	−0.0011889	−0.0001267	0.0004404
a_{15}	$X_1^{1/2}X_3$	−13.1481162	−51.6315189	−340.3783567	−59.1526677	−303.7267009	−308.1252917	−3.6166638	10.3084335
a_{16}	$X_1^{1/2}X_2^{1/2}X_3^3$	−121.9621802	−215.6609616	−178.1555441	−275.4644011	−136.9908259	−82.7606277	−223.7739956	−184.1230831
a_{17}	$X_1^{1/2}X_3^3$	−71.5325544	57.6006242	60.7219719	27.4766800	46.6008675	43.5704839	−1.7750139	−21.8789404
a_{18}	$X_1^{1/2}X_2^{1/3}$	−32.5608945	−29.2563972	−25.5054485	−21.6636306	−23.7092558	−29.6086847	−16.8034315	−17.9232221
a_{19}	$X_1^{1/2}X_2^{1/3}X_3$	54.5157933	142.5030600	819.0878327	196.1079105	711.2869865	768.5288342	0.0000000	0.0000000
a_{20}	$X_1^{1/3}X_2^{1/3}X_3$	−20.5288473	−111.2347584	−667.6456310	−102.0838597	−558.8027072	−665.3146674	44.6327609	−18.2683261
a_{21}	$X_1^{1/3}X_2^{1/3}X_3^2$	−16.2456722	−159.8058659	−331.8683190	−44.4746582	−179.2854206	−79.1653992	121.7294772	190.4342412

Table 8.10 Polynomial terms and regression coefficients of the MM for wetted surface

$$(S) = b_0 + b_1 \cdot X_1 \cdot X_2^2 + b_2 \cdot X_3^2 + \cdots + b_{29} \cdot X_1^{1/3} \cdot X_2^{2/3} \cdot X_3^2$$

		Fn_∇							
Coeff.	Multipl.	1	1.25	1.5	1.75	2	2.25	2.5	3
b_0	1	−20.3222043	−10.8220833	7.5356975	−30.0725537	17.0800003	6.8031675	8.6190277	17.2755056
b_1	$X_1X_2^2$	−0.4608138	−0.0671700	−0.1468507	0.4431587	0.3700828	0.2128709	0.1627703	0.1281514
b_2	X_3^3	33.0507625	−107.8139367	−160.1463058	0.0000000	−152.2769159	142.5114200	−111.8581759	0.0000000
b_3	$X_1X_2^3$	0.0320620	0.0017104	0.0091183	−0.0482034	−0.0706669	−0.0582786	−0.0323312	−0.0310084
b_4	$X_1^2X_2^2$	−0.0008098	−0.0226892	−0.0179048	0.0184048	−0.1271155	−0.0989131	−0.0896793	−0.1409993
b_5	$X_1X_2X_3$	5.0396726	2.8268146	−0.6209939	11.1077793	4.3840168	4.7719895	3.9933256	−9.4634858
b_6	$X_1^2X_2X_3$	−0.0999790	−0.1417253	−0.1960001	−0.1903931	−0.1120346	−0.0671769	−0.2872821	0.3100160
b_7	$X_1^2X_2^2X_3^2$	−0.0831839	−0.0129240	0.0451786	−0.1787343	0.0332116	−0.0422180	0.0730497	0.1530833
b_8	$X_2^3X_3^3$	−0.4857296	−0.4120887	−0.1460055	−0.6170326	−1.0706604	−1.0064388	0.0397205	0.3256037
b_9	$X_1^3X_3^3$	0.0473792	0.0533240	0.0233053	0.0223678	0.0044233	−0.0046020	−0.0015361	−0.0122720
b_{10}	$X_1^2X_2X_3^3$	0.4027168	0.2129878	0.4851503	0.4590570	0.5854488	0.8466962	0.5080877	−0.1999270
b_{11}	$X_1^3X_2^3X_3^3$	−0.0008906	0.0000033	−0.0006364	−0.0023135	−0.0027931	−0.0047260	−0.0017029	0.0008171
b_{12}	$X_2^3X_3^2$	0.2562829	0.2227769	−0.0126895	0.4699667	0.6867290	0.6098118	0.0289845	−0.0748263

(continued)

Table 8.10 (continued)

Coeff.	Multipl.	Fn_∇							
		1	**1.25**	**1.5**	**1.75**	**2**	**2.25**	**2.5**	**3**
b_{13}	$X_1^2X_2^3$	−0.0004158	−0.0005915	−0.0011976	0.0004667	0.0125057	0.0120290	0.0066721	0.0102015
b_{14}	$X_1^3X_2^2$	−0.0011251	−0.0009013	−0.0011138	−0.0010753	0.0029293	0.0022280	0.0029636	0.0025519
b_{15}	$X_1^3X_2^3X_3$	0.0008483	0.0017398	0.0018022	−0.0002252	−0.0008405	−0.0015728	−0.0000386	0.0008940
b_{16}	$X_1^3X_2^3X_3^2$	0.0001184	−0.0019315	−0.0020809	0.0030226	0.0019788	0.0045540	−0.0000724	−0.0033798
b_{17}	X_1^2	0.0450610	0.1014411	0.1055678	0.3199576	0.0074481	−0.0602615	0.0096783	−0.2380416
b_{18}	$X_1^{1/2}X_2^{1/2}X_3^3$	−143.7596456	−68.4260636	−102.4060911	−42.0191165	−137.9304649	−252.3175354	−27.2345608	13.6535455
b_{19}	$X_2^{1/2}X_3^2$	−24.7762923	−21.9403151	−136.4208675	142.6943104	−85.0785744	11.0914168	0.0000000	0.0000000
b_{20}	$X_2^{1/2}X_3^3$	−86.1780779	46.2360318	87.9955058	−103.6567150	−72.8348628	−444.1258458	0.0000000	−63.3710289
b_{21}	$X_2^{2/3}$	19.3064791	15.9269265	12.6744903	9.0139888	14.5171554	13.7079258	10.2622381	6.0938089
b_{22}	$X_1X_2^{1/3}X_3$	−12.2928131	−10.0878659	−3.8659302	−15.5731774	−0.4992810	3.6001809	1.1004028	1.6341905
b_{23}	$X_1^{1/3}X_2^{1/3}X_3$	110.7461046	106.4628058	0.0000000	189.3748495	0.0000000	0.0000000	0.0000000	0.0000000
b_{24}	X_1X_2	3.0389569	3.1733183	4.1945286	−3.8300759	5.1032541	3.8353508	3.6693245	8.2358601
b_{25}	$X_1^{1/3}X_2^{1/3}X_3^2$	0.0000000	0.0000000	166.8040085	−151.1244272	235.5192653	164.0899774	115.1075795	0.0000000
b_{26}	$X_1^{1/2}X_2^{2/3}$	−8.5009958	−11.4839564	−16.6736048	11.3123646	−22.5723538	−16.1302618	−14.4994276	−26.1421583
b_{27}	$X_1^{1/3}X_2^{2/3}X_3$	−74.1365525	−67.8966339	6.9338472	−145.0151116	−18.7380201	−24.1739323	−16.8751391	29.5524611
b_{28}	$X_1^{2/3}X_2^{2/3}X_3^2$	11.2908511	11.6971530	−20.6285934	28.7316397	−68.7401718	−70.4287216	−40.2954655	1.9302664
b_{29}	$X_1^{1/3}X_2^{2/3}X_3^3$	178.4800811	70.0208814	89.1311236	80.4165798	210.9135955	434.1094742	31.6515353	0.0000000

Table 8.11 Polynomial terms and regression coefficients of the MM for slenderness ratio

$$(M) = c_0 + c_1 \cdot X_2^2 \cdot X_3 + c_2 \cdot X_1^2 \cdot X_2 + \cdots + c_{20} \cdot X_1^3 \cdot X_2^3 \cdot X_3$$

		Fn_∇							
Coeff.	Multipl.	1	1.25	1.5	1.75	2	2.25	2.5	3
c_0	1	2.7147842	2.4131723	2.4919821	3.1094517	3.7168076	4.2478864	4.0315248	4.1595186
c_1	$X_2^2X_3$	0.1084159	0.1400754	0.1368889	−0.274898	−0.538303	−0.957328	−0.869128	−0.891656
c_2	$X_1^2X_2$	0.1174752	0.1669515	0.1496329	0.129631	0.0835509	0.0559054	0.0536775	0.024606
c_3	X_2^3	−0.020815	−0.039255	−0.008823	0.0012214	0.0209557	0.0341857	0.0561201	0.1350514
c_4	$X_1X_2^3$	0.0171647	0.0258062	0.0073781	−0.001093	−0.012472	−0.018085	−0.018059	−0.09263
c_5	$X_1^3X_2$	−0.008587	−0.012803	−0.010778	−0.011704	−0.008853	−0.006374	−0.003439	0.0010939
c_6	$X_1^2X_2^2$	−0.004503	−0.010477	−0.010363	0.0103247	0.0274227	0.0442724	0.0321514	0.0453312
c_7	$X_1^2X_2X_3$	−0.134682	−0.227868	−0.170697	−0.221691	−0.193745	−0.203335	−0.140373	−0.110202
c_8	$X_1X_2^3X_3$	−0.024084	−0.048342	−0.013126	0.0204465	0.0673502	0.0645402	0.0329156	0.2574829
c_9	$X_1X_2^3X_3^2$	−0.004478	0.0042342	0.0094231	0.0001916	−0.065449	−0.067576	0.0296763	−0.031173
c_{10}	$X_1^2X_2^3X_3$	0.0072843	0.010788	0.0057502	0.0022956	0.0014227	−0.001362	−0.009239	−0.03495
c_{11}	$X_2^3X_3^3$	−0.159663	−0.617235	−0.234906	−0.520283	−0.150194	−0.512061	−0.537317	0.3607485
c_{12}	$X_1^3X_3^3$	−0.111833	−0.15463	−0.087415	−0.203711	−0.216198	−0.230657	−0.157864	−0.065312
c_{13}	$X_1X_2^2X_3^3$	0.081302	0.2839137	0.0582071	0.2363007	0.2360265	0.4170054	0.3464805	0.176052
c_{14}	$X_2^3X_3^2$	0.1034529	0.3738507	0.0944764	0.2578312	0.1162081	0.4196101	0.2643682	−0.657283
c_{15}	$X_1^2X_2^3$	−0.004556	−0.00587	−0.002906	−0.005222	−0.006258	−0.006463	−0.002412	0.0056297
c_{16}	$X_1^3X_2^2$	0.0011616	0.0020341	0.0015099	0.000908	−0.000199	−0.001661	−0.002186	−0.004455
c_{17}	$X_1^3X_2X_3^2$	0.0739978	0.1073112	0.0659507	0.135823	0.1417216	0.1510289	0.0985996	0.0469201
c_{18}	$X_1^3X_2^2X_3^2$	−0.02424	−0.031309	−0.017845	−0.043142	−0.048802	−0.052582	−0.031638	−0.012929
c_{19}	$X_1^3X_2^2X_3^3$	0.0111925	0.0118577	0.0069871	0.0192738	0.0237294	0.0243721	0.0116343	0.0048947
c_{20}	$X_1^3X_2^3X_3$	0.0007953	0.0009471	0.0004821	0.0017321	0.0021298	0.0026561	0.0022107	0.0023326

Table 8.12 Polynomial terms and regression coefficients of the MM for dynamic trim

$$\tau = d_0 + d_1 \cdot X_2^2 \cdot X_3 + d_2 \cdot X_1^2 \cdot X_3 + \cdots + d_{18} \cdot X_1 \cdot X_2^{2/3} \cdot X_3^3$$

		Fn_∇							
Coeff.	Multipl.	**1**	**1.25**	**1.5**	**1.75**	**2**	**2.25**	**2.5**	**3**
d_0	1	11.4130900	−3.2607751	10.2096602	13.3085745	17.2397239	4.5440615	−14.3870919	−41.8552998
d_1	$X_2^2X_3$	−0.7951858	−1.8699231	−0.7138344	1.2178722	2.6250859	2.2015054	2.3088817	0.0000000
d_2	$X_1^2X_3$	3.4859281	6.9398210	2.1775328	1.2100264	3.6300596	4.4085896	6.0457764	17.0386412
d_3	X_3^3	301.7920806	60.6349588	111.6338575	76.5030320	213.5933559	221.4756056	291.4594590	113.6468208
d_4	X_3	−69.9613889	−196.9642890	−82.3664191	−82.0406376	−162.0247004	−152.8202149	−119.9843135	−386.8294246
d_5	$X_1^3X_3$	−0.0835811	−0.2429372	−0.0785838	−0.0614645	−0.1378837	−0.1581966	−0.2159598	−0.5661394
d_6	$X_1^2X_2X_3$	−0.3660518	0.0186235	0.0480117	0.1964759	0.0904302	0.0000000	−0.0823268	−0.1796438
d_7	$X_1X_2^3X_3^2$	−0.1860508	−0.0282358	−0.0206966	0.0274077	−0.0379986	−0.0361133	−0.0452606	0.0000000
d_8	$X_1^2X_2^3X_3$	0.0064777	−0.0003832	−0.0008200	−0.0024354	0.0000000	0.0000000	0.0000000	0.0000000
d_9	$X_1X_2X_3^3$	24.1360698	0.1194167	7.2268915	5.4658375	14.8710427	14.2496062	16.3188590	0.0000000
d_{10}	X_2^2	−0.2282975	0.5948589	0.3568662	0.0266102	−0.3345620	0.0000000	0.0000000	0.3634844
d_{11}	$X_2^3X_3^2$	0.7219893	0.3258176	0.1378759	−0.2915030	−0.2922701	−0.3935329	−0.3969477	−0.2146990
d_{12}	$X_1^{1/2}X_3$	106.3801708	271.0795078	90.9380974	75.8565448	163.5619967	156.3497628	154.8474777	600.7255381
d_{13}	$X_2^{1/2}X_3^3$	−180.4217905	−7.6347651	−34.3976055	−14.5885192	−86.5070522	−73.8337367	−115.6254817	−38.5903641
d_{14}	$X_1^{1/3}$	−9.6069864	30.5282765	9.7640007	−1.1860278	0.0000000	29.2150369	48.4046548	67.7812588
d_{15}	$X_1^{1/2}X_2^{1/3}$	1.6682977	−15.2035156	−7.5508487	−2.1958230	−3.0276817	−15.5961292	−19.8110262	−22.8860458
d_{16}	$X_1X_2^{2/3}X_3$	6.4985075	4.9446654	1.1003013	−4.0415709	−2.6805301	2.4634970	5.4061888	9.4720949
d_{17}	X_1X_3	−54.7083762	−120.0261438	−37.0074413	−19.6803795	−57.0320633	−67.5315593	−85.5267981	−280.2133516
d_{18}	$X_1X_2^{2/3}X_3^3$	−33.7594383	−2.9090450	−13.1893585	−10.5775990	−24.1131939	−25.6614819	−27.9377282	−1.9059874

8.2.5 NTUA

Reference Radojčić et al. (2001); see Sect. 3.4.11 (Figs. 8.21 and 8.22).

INPUT	OUTPUT
$X_1 = L/B$	C_R
$X_2 = (M)$	τ
$X_3 = B/T$	(S)
$X_4 = Fn$	

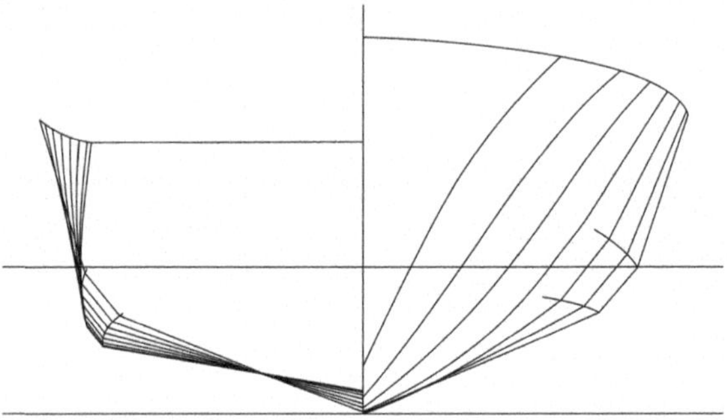

Fig. 8.21 Body plan of the parent hull of the NTUA systematic series

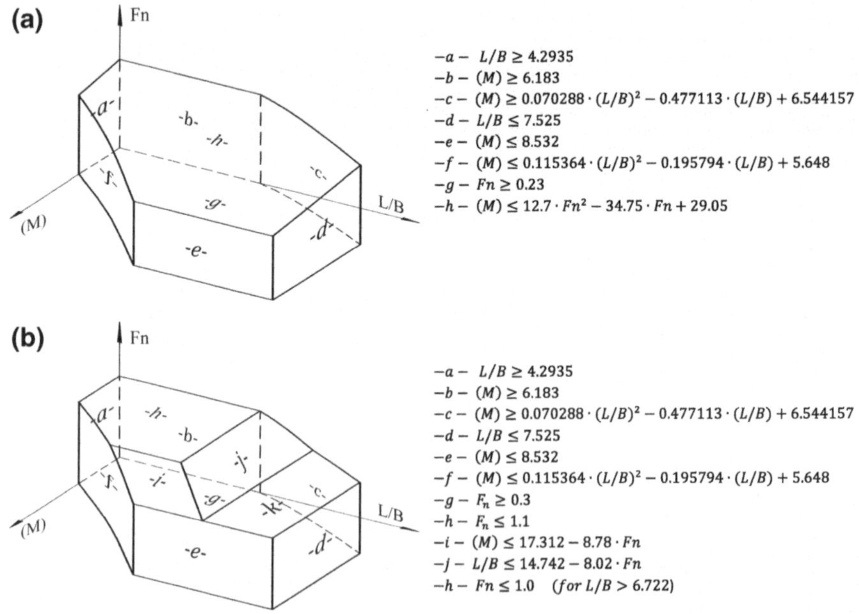

Fig. 8.22 Boundaries of applicability of NTUA series MMs for **a** C_R and **b** τ

Table 8.13 Polynomial terms and regression coefficients of the MMs for (a) C_R, (b) τ and (c) (S)

(a)

$$C_R = a_0 + \cdots + a_{30} \cdot X_2^2 \cdot X_4^3$$

Coeff.	Multipl.	a_i
a_0	1	81.947561
a_1	$X_2^2X_4$	−18.238522
a_2	X_2	−44.283380
a_3	X_3	−7.775629
a_4	X_1X_3	1.731934
a_5	$X_2^2X_3^2$	17.075124
a_6	$X_3^2X_4^3$	12.079902
a_7	$X_3X_4^5$	273.294648
a_8	$X_3^2X_4^5$	−16.121701
a_9	X_3^2	1.294730
a_{10}	$X_3^3X_4$	−0.187700
a_{11}	$X_3^3X_4^7$	−1.459234
a_{12}	$X_3^3X_4^6$	3.399356
a_{13}	$X_1X_3^2$	−0.235111
a_{14}	X_2^2	5.323100
a_{15}	$X_1^2X_2$	−0.021188
a_{16}	X_2X_4	108.448244
a_{17}	$X_2X_4^2$	−92.667206
a_{18}	$X_2^2X_3X_3^7$	0.176635
a_{19}	$X_3X_4^6$	−216.312999
a_{20}	$X_2X_3X_4^6$	−3.354160
a_{21}	$X_2^4X_4$	0.070018
a_{22}	$X_2X_3X_4$	1.145943
a_{23}	X_2^4	−0.017590
a_{24}	$X_2^4X_4^2$	−0.062264
a_{25}	$X_1X_3X_4^7$	−0.264296
a_{26}	$X_3X_4^3$	−105.059107
a_{27}	$X_3X_4^7$	55.703462
a_{28}	$X_3^3X_4^5$	−1.81086
a_{29}	$X_3^2X_4^7$	4.310164
a_{30}	$X_2^2X_4^3$	−1.240887

(b)

$$\tau = b_0 + \cdots + b_{27} \cdot X_2^2 \cdot X_4^5$$

Coeff.	Multipl.	b_i
b_0	1	1.444311
b_1	X_1X_4	−33.570197
b_2	$X_1X_2X_4$	0.246174
b_3	$X_1X_4^3$	−158.702695
b_4	$X_2^2X_4^4$	−4.977716
b_5	$X_2^2X_4^3$	7.433941
b_6	$X_1^2X_4^3$	4.833377
b_7	$X_2X_4^4$	8.684395
b_8	$X_1^2X_2X_4^2$	0.095234
b_9	$X_1^2X_4$	1.661479
b_{10}	$X_1X_4^5$	−18.319249
b_{11}	$X_2X_4^2$	−9.214203
b_{12}	$X_1X_2^2$	−0.006817
b_{13}	X_1^2	−0.123263
b_{14}	$X_2^2X_3X_4^5$	0.038370
b_{15}	$X_1X_4^4$	82.474887
b_{16}	$X_1^3X_4^5$	−0.066572
b_{17}	$X_1X_2X_4^2$	−1.861194
b_{18}	X_3X_4	1.562730
b_{19}	$X_2^2X_4^2$	−3.780954
b_{20}	X_2X_3	−0.091927
b_{21}	$X_2X_4^5$	−2.133727
b_{22}	X_1	2.371311
b_{23}	$X_1X_4^2$	128.709296
b_{24}	$X_1^2X_4^2$	−5.739856
b_{25}	$X_2^2X_3X_4^4$	−0.067986
b_{26}	$X_2^2X_4$	1.120683
b_{27}	$X_2^2X_4^5$	1.040239

(c)

$$(S) = c_0 + \cdots + c_5 \cdot X_2^2$$

Coeff.	Multipl.	c_i
c_0	1	2.400678
c_1	$X_2^2X_3$	0.002326
c_2	X_3^2	0.012349
c_3	X_2	0.689826
c_4	X_3^5	−0.000120
c_5	X_2^2	−0.018380

8.3 Recommended MMs for Semi-planing and Planing Hull Forms

This sections contains tabulated data necessary for evaluation of resistance and dynamic trim (where available) of five selected MMs applicable for semi-planing and planing hull forms. Data for each MM is presented as in Sect. 8.2, i.e. similar format is used. However, certain MMs are presented differently because ANN, rather than regression analysis, was used in derivation of some of them. So, ANN coefficients, as opposed to the polynomial terms and regression coefficients, are given. Note that no specific knowledge of ANN methodology is required – ANN coefficients are simply terms of the equations which are used as MMs for prediction of resistance and/or dynamic trim.

Recommended MMs for semi-planing and planing hull forms are presented as follows:

- **Section** 8.3.1—**62 & 65**. MMs for resistance and dynamic trim predictions. Polynomial terms and regression coefficients for evaluation of $(R_T/\Delta)_{100,000}$ and τ are given in Tables 8.15 and 8.16, respectively. Polynomial terms and regression coefficients for (*S*) and L/L_P are given in Tables 8.17 and 8.18, respectively.
- **Section** 8.3.2—**USCG & TUNS**. *Simple* and *Complex* MMs for resistance prediction based on two similar planing hull series. Polynomial terms and regression coefficients of a Simple MM for evaluation of $(R_T/\Delta)_{100,000}$, (*S*) and L_K/L are given in Table 8.19a–c respectively. ANN coefficients of a Complex MM for evaluation of $(R_T/\Delta)_{100,000}$ and (*S*) are given in Tables 8.20 and 8.21 respectively. Polynomial terms and regression coefficients of a Complex MM for evaluation of L_K/L are given in Table 8.22.
- **Section** 8.3.3—**Series 50**. MMs for resistance and baseline dynamic trim predictions based on the systematic Series 50. Polynomial terms and regression coefficients for evaluation of (R_R/Δ) and L_M/L_P are given in Tables 8.23 and 8.25 respectively. ANN coefficients for evaluation of τ_{BL} are given in Table 8.24. Note that wetted surface S is determined from a simple equation driven by the series' geometry.
- **Section** 8.3.4—**Series 62**. *Single* and *Multiple* output MMs for resistance and dynamic trim predictions derived by the ANN method, based on the systematic Series 62 with deadrise angles from 12.5° to 30°. Single output ANN coefficients for evaluation of $(R_T/\Delta)_{100,000}$ and τ are given in Tables 8.26 and 8.27 respectively. Multiple output ANN coefficients for evaluation of $(R_T/\Delta)_{100,000}$ and τ are given in Table 8.28. Polynomial terms and regression coefficients for evaluation of $A_P/\nabla^{2/3}$, L_M/L_P and (S) are given in Tables 8.29, 8.30 and 8.31 respectively.
- **Section** 8.3.5—**NSS series**. MMs with *Multiple* output only for resistance and dynamic trim predictions derived by the ANN method, based on the systematic NSS series. Multiple output ANN coefficients for evaluation of $(R_T/\Delta)_{100000}$ and τ, and $S/\nabla^{2/3}$ and L_{WL}/L_P are given in Tables 8.32 and 8.33 respectively.

Table 8.14 Recommended MMs for evaluation of resistance and dynamic trim for semi-planing and planing hull forms

Reference (series)	MMs main characteristics	Target value = f (input values)	Range of input parameters	Remarks
62 & 65 Radojčić (1985) See 3.4.4 See 8.3.1	High-speed, hard chine, planing hull forms. Fig. 8.23 SI and SD MMs for Fn_{∇}=1.0–4.0 MM based on the Series 62 (β=12.5 and 25 deg.), Series 65-B and one model DL-62-A Total 15 hull forms and 98 loadings	$(R_T/\Delta)_{100000}$ and τ – f ($A_P/\nabla^{2/3}$, LCG/L_P, L_P/B_{PA}, β) $S/\nabla^{2/3}$ and L/L_P = f($A_P/\nabla^{2/3}$, LCG/L_P, L_P/B_{PA}, β)	Constraints given in graphical format (scatter plots) or as 4-D ellipsoid (non-linear eq. replacing 60 bounds, see Fig. 8.24) Approximately: $A_P/\nabla^{2/3}$ – 4.25–9.5 LCG = (30–44.8)%L_P L_P/B_{PA}=2.36 6.73 β=13–37.4	PT: Regression analysis I/F: 27/18 terms for both $(R_T/\Delta)_{100000}$ and τ I/F: 27/8 terms for both $S/\nabla^{2/3}$ and L/L_P. SI and SD MMs (may be used independently). This MM replaces Radojčić (1984)
USCG & TUNS Radojčić et al. (2014a) See 3.4.13 See 8.3.2	Planing hull forms with wide transom (match the contemporary hulls). Fig. 8.25 Fn_{∇}=0.6–3.5 The applicability of the USCG Series (consisting of 4 models of 135–220 kg) is extended with TUNS Series (consisting of 9 models of 1 3.5 kg)	$(R_T/\Delta)_{100000}$ = f ($L/\nabla^{1/3}$, Fn_{∇}, L/B, LCG/L, β) $S/\nabla^{2/3}$ = f ($L/\nabla^{1/3}$, Fn_{∇}, L/B, LCG/L) L_K/L = f (Fn_{∇}, LCG/L) L = L_P, B = B_{PX}, β = β_{Bpx} $(R_T/\Delta)_{100000}$ = R/Δ	17 inequalities, see Fig. 8.26 Approximately: $L/\nabla^{1/3}$=3.9–6.9 L/B=2.5–4.7 LCG/L=0.27–0.41 β=12–24	PT: ANN (Regression for L_K/L) Final MM for R/Δ, $S/\nabla^{2/3}$ and L_K/L have 116, 23 and 16 terms respectively SD MMs L_K is representative length of wetted surface. Dynamic trim not modeled Simple MM f(Fn_{∇}, $L/\nabla^{1/3}$) only

(continued)

Table 8.14 (continued)

Reference (series)	MMs main characteristics	Target value = f (input values)	Range of input parameters	Remarks
Series 50 Radojčić et al. (2014b) See 3.4.14 See 8.3.3	Hulls with a warped bottom, negative keel angle and deep forefoot. Nowadays applicable for the semi-displacement regimes. Fig. 8.27 MM valid for $Fn_\nabla = 1.0–5.0$. MM based solely on the Series 50	R_R/Δ, τ_{BL}, L_M/L_P – f $(L_P/\nabla^{1/3}$, L_P/B_{PX}, LCG/L_P, $Fn_\nabla)$ $S=0.962(L_M/L_P)\cdot L_P\cdot B_{PX}$	34 inequalities, see Fig. 8.28 Approximately: $L_P/\nabla^{1/3} = 5.3–8.9$ $L_P/B_{PX} = 2.2–8.5$ $LCG/L_P = 0.15–0.43$	PT: ANN and Regression for τ_{BL}, and R_R/Δ and L_M/L_P respectively. SD MMs Final MMs for R_R/Δ, τ_{BL}, and L_M/L_P have 65, 91 and 16 terms respectively The "hydrodynamic trim" τ is approximately 1° less than the base line trim τ_{BL}
Series 62 β=12.5–30° Radojčić et al. (2017a, b) See 3.4.15 See 8.3.4	High-speed, hard chine, narrow-stern, planing hulls. Fig. 8.29 $Fn_\nabla = 1.0–4.0$ (to 6 for β=12.5) Total 14 hull forms and 280 loading cases with 2880 points out of which 10% are virtual MM based solely on the Series 62 with β=12.5, 25, and 30 deg.	$(R_T/\Delta)_{100000}$ and τ = f$(A_P/\nabla^{2/3}$, L_P/B_{PX}, LCG/L_P, β, $Fn_\nabla)$ $S/\nabla^{2/3}$ and L_M/L_P = f$(L_P/B_{PX}$, $L_P/\nabla^{1/3}$, LCG/L_P, $Fn_\nabla)$ $A_P/\nabla^{2/3}$ = f $(L_P/\nabla^{1/3}$, $L_P/B_{PX})$	Multidimensional cuboid, see Fig. 8.30. Approximately: β=12°–30° L_P/B_{PX}=2–7 $A_P/\nabla^{2/3}$=4.0–8.5 LCG/L_P=36.8–48.8	PT: ANN with multiple and single output. SD MMs The multiple output MMs for $(R_T/\Delta)_{100000}$ and τ have 225 terms each, out of these 217 terms are common. Single output MMs for R/Δ and τ are with 190 and 160 terms respectively For Froude expansion, separate MMs for evaluation of $A_P/\nabla^{2/3}$, L_M/L_P and $S/\nabla^{2/3}$ with 7, 11, and 25 terms respectively
NSS Radojčić and Kalajdžić (2018) See 3.4.18 See 8.3.5	Contemporary planing hull form with wide transom and slight bottom warp. Fig. 8.31 Fn_L=0.5–1.6 Fn_∇=1.1–4.2 MM is based solely on the NSS	$(R_T/\Delta)_{100000}$, τ, $S/\nabla^{2/3}$, L_{WL}/L_P = f $(L_P/\nabla^{1/3}$, L_P/B_{PX}, LCG/L_P, $Fn_\nabla)$ $L_{WL}=L_K$	Constraints given In graphical format (scatter plots), see Fig. 8.32 Approximately: $L_P/\nabla^{1/3}$ =5.0–7.8 L_P/B_{PX}=3.45–6.25 LCG/L_P=0.33–0.38	PT: ANN, multiple output. Final MMs for R/Δ, t, $S/\nabla^{2/3}$ and L_{WL}/L_P have 122 terms each, out of these 113 are common for R/Δ & t and $S/\nabla^{2/3}$ & L_{WL}/L_P. SD MMs Representative length of wetted surface is L_{KEEL}. Influence of interceptors not modeled

Legend: PT – Predictive Technique; I/F –Initial/Final MM; SI – Speed-Independent; SD – Speed-Dependent

8.3.1 62 & 65

Reference Radojčić (1985); see Sect. 3.4.4 (Figs. 8.23 and 8.24).

INPUT

$X_1 = (A_P/\nabla^{2/3} - 6.875)/2.625$

$X_2 = (100(LCG/L_P) - 37.4)/7.4$

$X_3 = (L_P/B_{PA} - 4.545)/2.185$

$X_4 = (\beta - 25.2)/12.2$

OUTPUT

$(R_T/\Delta)_{100000}$

τ

$S/\nabla^{2/3}$

L/L_P

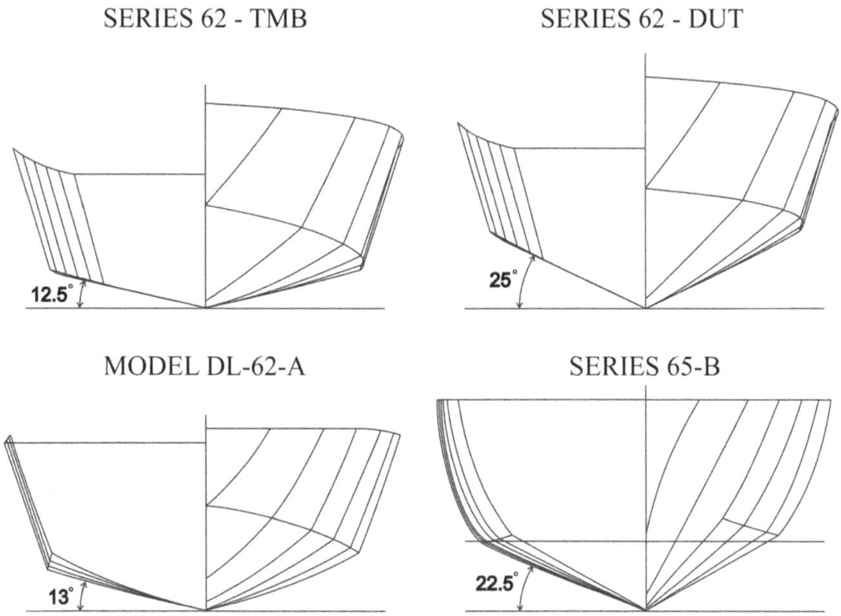

Fig. 8.23 Body plans of hull series used for the development of a MM

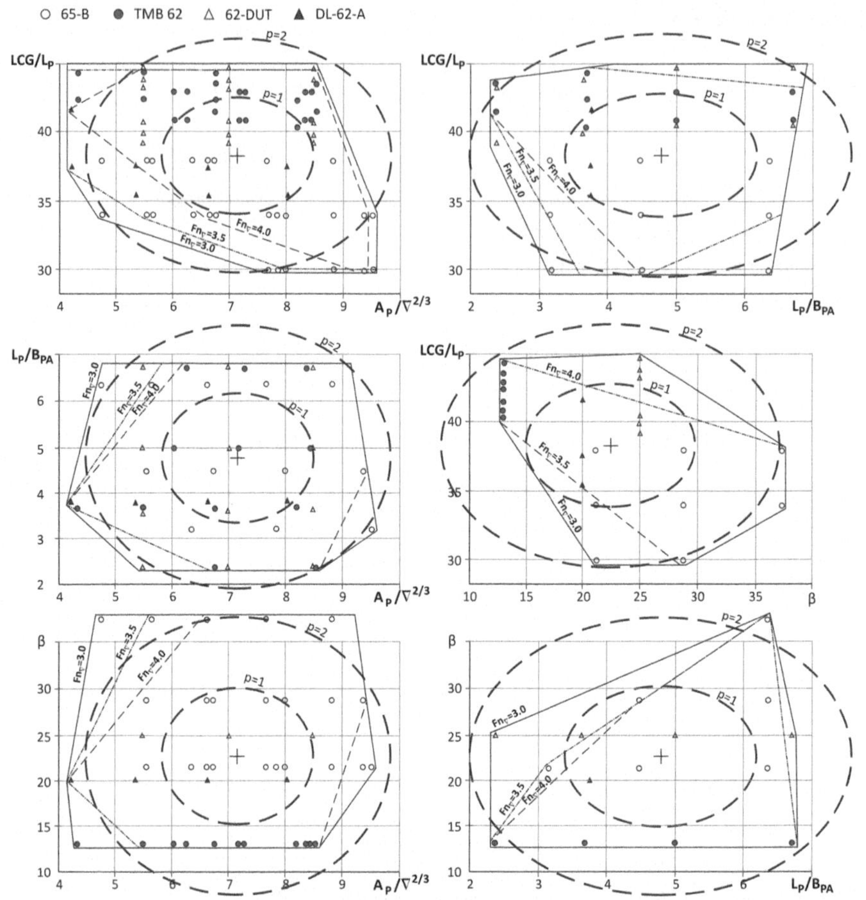

$$1 - p^{-2} \cdot (-3.6982X_1^2 + 0.7766X_1 - 2.8727X_2^2 + 0.7182X_2 - 2.3669X_3^2 + 0.5207X_3 - 2.6015X_4^2 - 1.1967X_4 - 0.2519) \geq 0$$

Fig. 8.24 Database distribution and boundaries of applicability of MMs. Boundaries are given in a graphical format as a set of linear bounds, and as a single non-linear equation, i.e. 4D ellipsoid obtained with the coefficient p = 1 and p = 2

Table 8.15 Polynomial terms and regression coefficients of the MM for resistance

$$(R_T/\Delta)_{100000} = a_0 + a_1 \cdot X_1 + a_2 \cdot X_2 + a_3 \cdot X_3 + a_4 \cdot X_4 \ldots + a_{16} \cdot X_3^2 \cdot X_4 + a_z \cdot z$$

		Fn_∇							
Coeff.	Multipl.	1.00	1.25	1.50	1.75	2.00	2.50	3.00	3.50
a_0	1	0.06166659	0.09218460	0.10261366	0.10997288	0.12342421	0.13344753	0.13738033	0.12710047
a_1	X_1	−0.02068559	−0.04135239	−0.03581788	−0.03368683	−0.03039523	−0.02033408	−0.00486718	–
a_2	X_2	−0.01159428	−0.01133362	−0.00927631	−0.00721882	−0.00996893	–	0.01517530	0.01344670
a_3	X_3	−0.03102352	−0.05630887	−0.04981934	−0.04533287	−0.04212487	−0.02674611	−0.01721230	0.02592634
a_4	X_4	–	–	–	–	–	–	0.01429118	0.02900173
a_5	$X_1 X_2$	–	–	–	–	–	0.01028233	0.01811335	0.04055339
a_6	$X_1 X_3$	0.00286041	0.01238261	0.00914594	0.00966439	0.01108910	–	–	–
a_7	$X_3 X_4$	–	–	–	–	–	−0.00838570	−0.01831473	−0.01584742
a_8	X_1^2	0.00905674	0.01448677	0.01237267	0.01881432	0.02114185	0.01695077	0.01440470	0.03768682
a_9	X_2^2	0.00630047	0.01698212	0.01670406	0.01666895	0.01498626	0.01453727	0.00948421	–
a_{10}	X_3^2	0.02153703	0.03570836	0.02497512	0.01610927	0.00450583	−0.00453941	–	0.01635087
a_{11}	X_4^2	–	–	–	–	–	–	0.00765337	0.02470056
a_{12}	$X_1 X_4^2$	–	–	–	−0.00584924	−0.01213373	−0.01539055	−0.02448441	−0.03109249
a_{13}	$X_2 X_1^2$	–	–	–	–	0.01025070	0.01179381	–	−0.03995866
a_{14}	$X_2 X_3^2$	−0.01066115	−0.01426742	−0.00970227	−0.00776786	–	–	−0.00778079	−0.02575325
a_{15}	$X_2 X_4^2$	–	–	–	–	−0.00927555	−0.02265445	−0.02389701	–
a_{16}	$X_3 X_4^2$	–	0.00417371	0.00535018	0.00798569	0.01115392	0.01116712	–	−0.04082650
z	0 or 1	−0.00990906	−0.00499868	–	–	–	0.01462743	0.02486347	0.00541199

Note z = 1 for Series 65-B; z = 0 for Series DMB 62, Series 62-DUT and Model 62-A

Table 8.16 Polynomial terms and regression coefficients of the MM for dynamic trim

$$\tau = b_0 + b_1 \cdot X_1 + b_2 \cdot X_2 + b_3 \cdot X_3 + b_4 \cdot X_4 + \cdots + b_{16} \cdot X_2^2 \cdot X_4 + b_z \cdot z$$

Coeff.	Multipl.	Fn_∇ 1.00	1.25	1.50	1.75	2.00	2.50	3.00	3.50	4.00
b_0	1	1.33336293	2.75105083	3.20884172	3.46368920	3.92282114	4.33912398	3.84085312	4.02473013	4.24066972
b_1	X_1	−1.03702812	−1.86276316	−1.79648922	−2.36844448	−2.75880036	−2.48144444	−1.97280091	−1.38826006	−1.28313404
b_2	X_2	−1.83432596	−1.53407422	−1.55056152	−1.72094149	−1.84237867	−1.53976972	−0.90049877	−0.83941923	−0.44692903
b_3	X_3	−1.15780847	−2.57565899	−2.73103753	−2.87897170	−2.67156966	−1.63892185	−0.75972903	1.02117918	1.33437093
b_4	X_4	−0.44247566	−0.36603100	−0.38527222	−0.48195065	−0.42806926	−0.73197975	−0.74778298	–	0.83158531
b_5	$X_1\ X_3$	0.48222430	1.14962838	1.32316858	1.30002617	1.13565883	0.30211957	−0.45415533	−0.41309091	–
b_6	$X_2\ X_3$	1.56022120	0.78308941	0.77159688	0.74495377	0.60675291	–	−0.48898871	–	–
b_7	$X_2\ X_4$	–	–	–	–	–	−0.53871035	−0.65202352	−1.02696360	−0.98125790
b_8	X_1^2	0.23129493	0.72788950	0.78113230	1.08787427	1.48532225	0.83576886	0.81571059	0.57099631	−0.13408536
b_9	X_2^2	0.81705244	0.99026200	0.93398783	0.89195140	0.82089541	0.67384787	0.50817590	−0.53188085	−0.79474929
b_{10}	X_3^2	0.77721697	1.63732229	1.66100134	1.45160997	0.61935881	−0.73224333	−0.63049648	−0.24371396	–
b_{11}	X_4^2	0.42095809	0.48076851	0.61161560	0.72207309	0.77346695	0.32902463	–	–	–
b_{12}	$X_1\ X_3^2$	–	−0.76619380	−1.02118319	−0.45775147	–	0.46066770	–	–	–
b_{13}	$X_2\ X_1^2$	–	–	–	–	–	–	–	−1.01829318	–
b_{14}	$X_2\ X_3^2$	−1.31998028	−1.06647051	−0.75930774	−0.46488590	–	0.74883439	0.68629671	–	–
b_{15}	$X_3\ X_4^2$	–	–	–	–	–	0.24236948	0.77829394	−1.01220609	−1.31954182
b_{16}	$X_4\ X_2^2$	1.77654067	0.98510573	0.82727403	0.70090751	0.31589615	1.05978194	0.89846908	–	–
z	0 or 1	–	–	–	–	−0.64480930	1.13525183	–		–

Note z = 1 for Series 65-B; z = 0 for Series DMB 62, Series 62-DUT and Model 62-A

Table 8.17 Polynomial terms and regression coefficients of the MM for wetted surface

$$(S) = c_0 + c_1 \cdot X_1 + c_2 \cdot X_2 + c_3 \cdot X_3 + c_4 \cdot X_4 + \cdots + c_7 \cdot X_2 \cdot X_3^2$$

		Fn_∇								
Coeff.	Multipl.	1.00	1.25	1.50	1.75	2.00	2.50	3.00	3.50	4.00
c_0	1	7.35972955	7.28287556	7.20829266	7.15098528	7.01829675	5.95292073	5.43200915	5.36244782	5.02788170
c_1	X_1	1.67293911	1.90314232	1.88377274	1.97128608	2.03566157	2.11515615	2.00972331	2.02278043	2.01557671
c_2	X_2	−0.07193080	−0.17774343	−0.17756813	0.01462322	0.16029332	0.62862102	0.83801974	1.34709626	1.70218866
c_3	X_3	0.42800156	0.72564445	0.99430103	1.29217163	1.47999022	1.47732914	1.26920476	0.72279653	0.55645074
c_4	X_4	−0.06055862	−0.11250616	−0.02734152	0.13028786	0.29805380	0.72808374	0.85896318	1.61946479	1.58873964
c_5	X_2^2	0.27473585	−0.03968004	−0.22836400	−0.58283712	−0.81367931	−0.63393784	−0.53871323	−0.16624795	−0.37730699
c_6	X_3^2	−0.38915391	−0.49279052	−0.60402762	−0.63793414	−0.71261690	0.09803725	−0.00352813	−0.46934336	−0.74794949
c_7	$X_2 X_3^2$	0.44976935	0.58469101	0.54501203	0.28694207	0.24704447	−0.26483977	−0.38300992	0.29587827	0.47600574

Table 8.18 Polynomial terms and regression coefficients of the MM for wetted length

$$L/L_P = d_0 + d_1 \cdot X_1 + d_2 \cdot X_2 + d_3 \cdot X_3 + d_4 \cdot X_4 + \cdots + d_7 \cdot X_2 \cdot X_4^2$$

		Fn_∇								
Coeff.	Multipl.	1.00	1.25	1.50	1.75	2.00	2.50	3.00	3.50	4.00
d_0	1	0.77418144	0.76501529	0.75219908	0.73615807	0.70128520	0.61213431	0.56207094	0.53520799	0.50483383
d_1	X_1	–0.12184031	–0.10417762	–0.09481551	–0.07336621	–0.05133490	–0.01425025	–	–	–
d_2	X_2	0.18434727	0.13168583	0.10375325	0.10912236	0.11752245	0.14618187	0.15420571	0.15817110	0.12210944
d_3	X_3	0.02339267	0.04775750	0.07262345	0.09796173	0.11686546	0.11621771	0.09202312	0.05459982	0.04477227
d_4	X_4	–0.25569737	–0.19439791	–0.14539163	–0.10503104	–0.05033986	–	0.02384494	0.03461336	0.02768969
d_5	$X_1\ X_4$	–0.01290730	–0.09672286	–0.11005719	–0.11534376	–0.11283571	–0.07581007	–0.03928807	–0.03873253	–0.31366527
d_6	X_3^2	–	–0.02082503	–0.04265106	–0.06489656	–0.05404190	–	–	–	–
d_7	$X_2\ X_4^2$	–0.36472950	–0.23899868	–0.15158548	–0.12144172	–0.06682152	–0.04426719	–0.01844410	–0.02929236	0.01197709

8.3.2 USCG & TUNS

Reference Radojčić et al. (2014a); see Sect. 3.4.13 (Figs. 8.25 and 8.26).

INPUT	OUTPUT
$X_1 = L/B$	$(R_T/\Delta)_{100000}$
$X_2 = L/\nabla^{1/3}$	$S/\nabla^{2/3}$
$X_3 = Fn_\nabla$	L_K/L
$X_4 = LCG/L$	
$X_5 = \beta$	

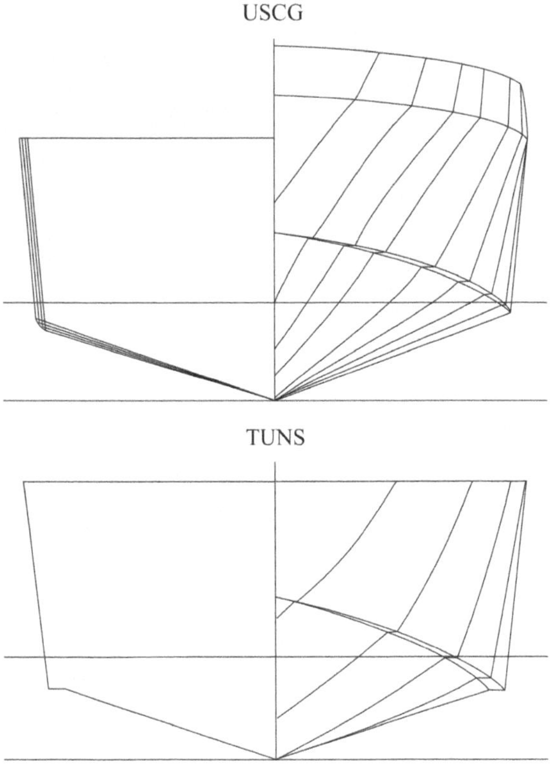

Fig. 8.25 Body plans of hull series used for the development of a MM

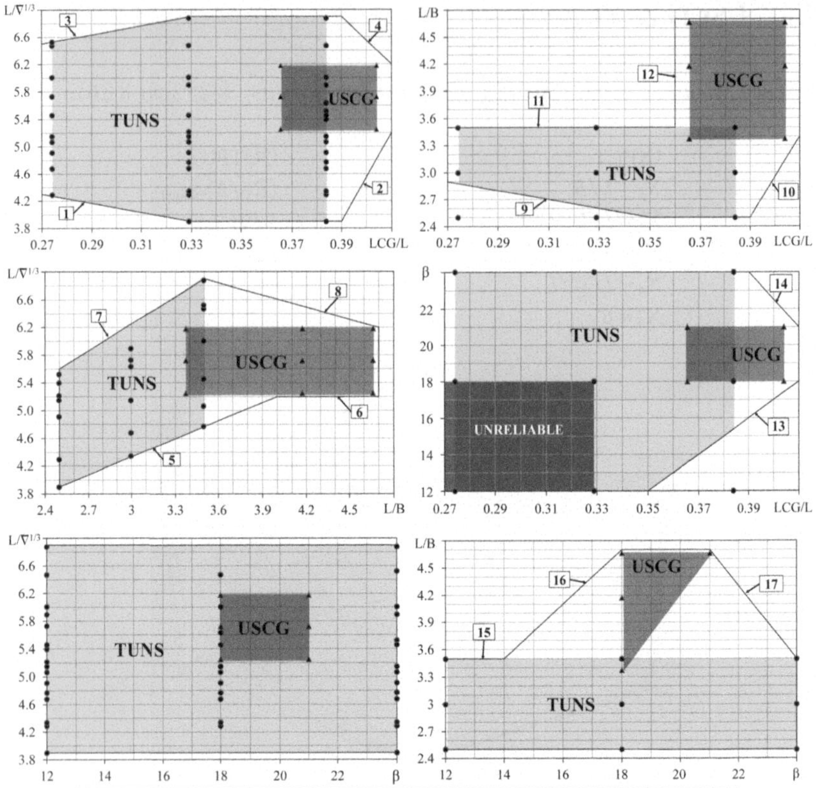

No.		Boundaries
1	$0.27 \le LCG/L < 0.33$	$L/\nabla^{1/3} \ge 6.1 - 6.66667 \cdot LCG/L$
2	$0.39 \le LCG/L \le 0.41$	$L/\nabla^{1/3} \ge 65 \cdot LCG/L - 21.45$
3	$0.27 \le LCG/L < 0.33$	$L/\nabla^{1/3} \le 6.66667 \cdot LCG/L + 4.7$
4	$0.39 \le LCG/L \le 0.41$	$L/\nabla^{1/3} \le 20.55 - 35 \cdot LCG/L$
5	$2.5 \le L/B < 4.0$	$L/\nabla^{1/3} \ge 0.866667 \cdot L/B + 1.73333$
6	$4.0 \le L/B \le 4.7$	$L/\nabla^{1/3} \ge 5.2$
7	$2.5 \le L/B \le 3.5$	$L/\nabla^{1/3} \le 1.3 \cdot L/B + 2.35$
8	$3.5 < L/B \le 4.7$	$L/\nabla^{1/3} \le 8.94167 - 0.583333 \cdot L/B$
9	$0.27 \le LCG/L \le 0.35$	$L/B \ge 4.25 - 5 \cdot LCG/L$
10	$0.39 \le LCG/L \le 0.41$	$L/B \ge 45 \cdot \mathrm{LCG/L} - 15.05$
11	$0.27 \le LCG/L < 0.36$	$L/B \le 3.5$
12	$3.5 < L/B \le 4.7$	$LCG/L \ge 0.36$
13	$0.35 \le LCG/L \le 0.41$	$\beta \ge 100 \cdot LCG/L - 23$
14	$0.39 \le LCG/L \le 0.41$	$\beta \le 82.5 - 150 \cdot LCG/L$
15	$12° \le \beta \le 14°$	$L/B \le 3.5$
16	$14° < \beta \le 18°$	$L/B \le 0.3 \cdot \beta - 0.7$
17	$21° \le \beta \le 24°$	$L/B \le 13.1 - 0.4 \cdot \beta$

Fig. 8.26 Database distribution and boundaries of applicability of MMs. Boundaries are given as a set of 17 inequality equations

Simple Mathematical Model

Table 8.19 Polynomial terms and regression coefficients of the MMs for (a) Resistance, (b) wetted surface and (c) wetted length

(a)

$$(R_T/\Delta)_{100000} = a_0 + \cdots + a_{19} \cdot X_2^4 \cdot X_3^3$$

Coeff.	Multipl.	a_i
a_0	1	15.1435300
a_1	X_2	−12.5431467
a_2	X_2^2	3.6580101
a_3	X_2^3	−0.4573261
a_4	X_2^4	0.0209140
a_5	X_3	−37.3113557
a_6	$X_2 X_3$	30.6066779
a_7	$X_2^2 X_3$	−8.9006694
a_8	$X_2^3 X_3$	1.1119496
a_9	$X_2^4 X_3$	−0.0508810
a_{10}	X_3^2	22.1656778
a_{11}	$X_2 X_3^2$	−18.0560600
a_{12}	$X_2^2 X_3^2$	5.2443776
a_{13}	$X_2^3 X_3^2$	−0.6562651
a_{14}	$X_2^4 X_3^2$	0.0301221
a_{15}	X_3^3	−3.5683179
a_{16}	$X_2 X_3^3$	2.8994541
a_{17}	$X_2^2 X_3^3$	−0.8432151
a_{18}	$X_2^3 X_3^3$	0.1057838
a_{19}	$X_2^4 X_3^3$	−0.0048694

(b)

$$(S) = b_0 + \cdots + b_{15} \cdot X_2^3 \cdot X_3^3$$

Coeff.	Multipl.	b_i
b_0	1	58.8177152
b_1	X_2	−33.1423777
b_2	X_2^2	6.7794188
b_3	X_2^3	−0.4432887
b_4	X_3	−49.7163001
b_5	$X_2 X_3$	29.8350864
b_6	$X_2^2 X_3$	−6.1030996
b_7	$X_2^3 X_3$	0.4226973
b_8	X_3^2	18.0264798
b_9	$X_2 X_3^2$	−11.3511782
b_{10}	$X_2^2 X_3^2$	2.3295698
b_{11}	$X_2^3 X_3^2$	−0.1612880
b_{12}	X_3^3	−2.1175915
b_{13}	$X_2 X_3^3$	1.3944044
b_{14}	$X_2^2 X_3^3$	−0.2876721
b_{15}	$X_2^3 X_3^3$	0.0197989

(c)

$$L_K/L = c_0 + \cdots + c_{15} \cdot X_2^3 \cdot X_3^3$$

Coeff.	Multipl.	c_i
c_0	1	3.2719934
c_1	X_2	−0.9891214
c_2	X_2^2	0.1543888
c_3	X_2^3	−0.0086702
c_4	X_3	−1.1483516
c_5	$X_2 X_3$	0.0626019
c_6	$X_2^2 X_3$	0.0482171
c_7	$X_2^3 X_3$	−0.0049667
c_8	X_3^2	0.7839794
c_9	$X_2 X_3^2$	−0.3171667
c_{10}	$X_2^2 X_3^2$	0.0437025
c_{11}	$X_2^3 X_3^2$	−0.0021325
c_{12}	X_3^3	−0.1630761
c_{13}	$X_2 X_3^3$	0.0910387
c_{14}	$X_2^2 X_3^3$	−0.0165489
c_{15}	$X_2^3 X_3^3$	0.0010024

Complex Mathematical Model

Table 8.20 ANN coefficients of the MM for resistance

$$(R_T/\Delta)_{100000} = \frac{sig\left(d_v+\sum_{w=1}^{3}\left(D_{wv}\cdot sig\left(c_w+\sum_{i=1}^{5}\left(C_{iw}\cdot sig\left(b_i+\sum_{j=1}^{7}\left(B_{ji}\cdot sig\left(a_j+\sum_{k=1}^{5}\left(A_{kj}\cdot(P_k\cdot X_k+R_K)\right)\right)\right)\right)\right)\right)\right)\right)-G}{H}$$

j	A_{1j}	A_{2j}	A_{3j}	A_{4j}	A_{5j}	a_j
1	1.7456920	2.7293300	8.8455260	5.4907050	−9.3073300	−9.857926
2	0.2898082	−3.1160330	20.4518000	−0.8953395	0.0587176	−2.315033
3	−0.7191356	−3.2455800	2.2043970	1.0204330	1.1992110	−2.723165
4	0.5621026	2.9209610	8.5967350	1.6920190	0.3731644	−3.539546
5	−7.5173060	−1.3333580	−0.0657259	−1.0204240	−0.3872136	2.180475
6	−6.2340650	4.1786120	8.8794550	2.3562540	2.3852490	−7.769617
7	0.3437763	0.9243057	23.8714800	0.3581851	0.1575263	−2.576811

i	B_{1i}	B_{2i}	B_{3i}	B_{4i}	B_{5i}	B_{6i}	B_{7i}	b_i
1	−2.4406510	2.1080950	7.6929170	−6.196037	−5.0275680	3.0244980	10.194260	−7.880680
2	−3.0829330	1.8681110	3.5004800	−12.575160	−1.8650630	3.8124340	13.984750	−5.292542
3	0.0122627	−4.3228210	0.4912690	−15.314440	0.6553449	13.0049100	10.002760	−4.638958
4	1.3330090	9.3863860	−2.6382690	−3.542744	−24.9670200	−0.3168464	1.446800	−7.829987
5	1.0186670	−0.2119331	0.1647189	−4.278055	0.3412170	0.4457378	7.245709	−4.301029

w	C_{1w}	C_{2w}	C_{3w}	C_{4w}	C_{5w}	c_w
1	−7.0169640	1.0933530	−5.2393960	0.4377735	−5.5052590	−0.3100626
2	−3.2096000	−3.6988810	−11.5074100	−6.9135450	−6.0141480	11.2548400
3	2.9190180	−9.8329010	0.7894529	−17.4901300	6.3501410	5.8545430

v	D_{1v}	D_{2v}	D_{3v}	d_v
1	−6.9155950	−9.0677310	−0.9125847	10.0196700

k	P_k	R_k
L/B	0.4334425	−1.0323493
$L/\nabla^{1/3}$	0.3021148	−1.1270393
Fn_∇	0.1665957	−0.0434102
LCG/L	6.5312046	−1.7402032
β	0.0750000	−0.8500000

v	H	G
$(R_T/\Delta)_{100000}$	3.1461161	0.0266244

Table 8.21 ANN coefficients of the MM for wetted surface

$$(S) = \frac{\mathrm{sig}\left(b_i + \sum_{j=1}^{2}\left(B_{ji} \cdot \mathrm{sig}\left(a_j + \sum_{k=1}^{4}\left(A_{kj} \cdot (P_k \cdot X_k + R_k)\right)\right)\right)\right) - G}{H}$$

j	A_{1j}	A_{2j}	A_{3j}	A_{4j}	a_j
1	−1.0502660	4.8476660	−1.5388710	−1.5804480	−0.4047123
2	2.0970270	−1.0619980	−3.1135780	−0.3070406	1.6046860

i	B_{1i}	B_{2i}	b_i
1	−2.7913260	−3.2475410	2.4859060

k	P_k	R_k
L/B	0.4157620	−0.9881993
Fn_∇	0.2589555	−0.0951964
$L/\nabla^{1/3}$	0.3021148	−1.1270393
LCG/L	6.9230769	−1.8476154

v	H	G
(S)	0.1007964	−0.2070596

Table 8.22 Polynomial terms and regression coefficients of the MM for wetted length

$$L_K/L = a_0 + a_1 \cdot X_4 + a_2 \cdot X_4^2 + a_3 \cdot X_4^3 + a_4 \cdot X_3 + \cdots + a_{15} \cdot X_3^3 \cdot X_4^3$$

Coeff.	Multipl.	a_i
a_0	1	−12.0970
a_1	X_4	120.7900
a_2	X_4^2	−359.6600
a_3	X_4^3	347.6000
a_4	X_3	16.9360
a_5	$X_3 X_4$	−171.0200
a_6	$X_3 X_4^2$	541.8500
a_7	$X_3 X_4^3$	−548.6600
a_8	X_3^2	−5.4136
a_9	$X_3^2 X_4$	57.0400
a_{10}	$X_3^2 X_4^2$	−188.9300
a_{11}	$X_3^2 X_4^3$	198.9800
a_{12}	X_3^3	0.4443
a_{13}	$X_3^3 X_4$	−5.0552
a_{14}	$X_3^3 X_4^2$	17.8950
a_{15}	$X_3^3 X_4^3$	−19.8610

8.3.3 Series 50

Reference Radojčić et al. (2014b); see Sect. 3.4.14 (Figs. 8.27 and 8.28).

INPUT	OUTPUT
$X_1 = 0.9/3.638261 \cdot (L_P/\nabla^{1/3} - 5.291901) + 0.05$	R_R/Δ
$X_2 = 0.9/6.316902 \cdot (L_P/B_{PX} - 2.198484) + 0.05$	L_M/L_P
$X_3 = 0.9/0.289 \cdot (LCG/L_P - 0.146) + 0.05$	
$X_4 = 0.9/5.05671 \cdot (Fn_\nabla - 1.055012) + 0.05$	

INPUT	OUTPUT
$X_1 = L_P/\nabla^{1/3}$	τ_{BL}
$X_2 = L_P/B_{PX}$	
$X_3 = LCG/L_P$	
$X_4 = Fn_\nabla$	

NOTE: Surface area $S = 0.962 \left(\frac{L_M}{L_P}\right) L_P B_{PX}$ [m²]
(determined by geometry of the hulls)

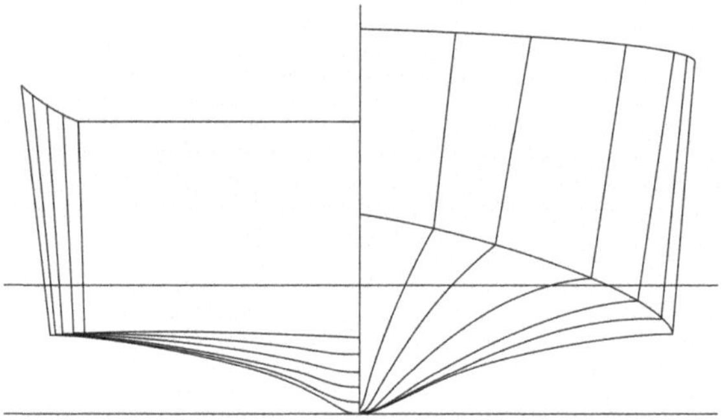

Fig. 8.27 Body plan of the parent hull of the Series 50

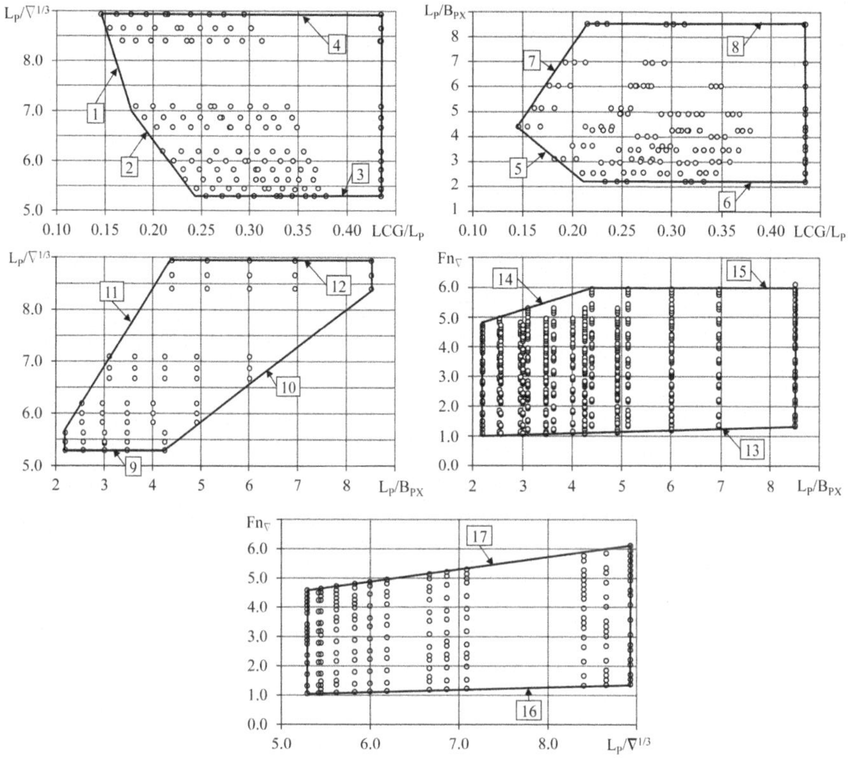

No.		Boundaries
1	$0.145 \le LCG/L_P < 0.175$	$L_P/\nabla^{1/3} \ge 17.8952 - 62.2581 \cdot LCG/L_P$
2	$0.175 \le LCG/L_P \le 0.240$	$L_P/\nabla^{1/3} \ge 11.5341 - 25.909 \cdot LCG/L_P$
3	$0.240 \le LCG/L_P < 0.435$	$L_P/\nabla^{1/3} \ge 5.290$
4	$0.145 \le LCG/L_P \le 0.435$	$L_P/\nabla^{1/3} \le 8.930$
5	$0.145 \le LCG/L_P \le 0.210$	$L_P/B_{PX} \ge 9.0762 - 32.0290 \cdot LCG/L_P$
6	$0.210 \le LCG/L_P \le 0.435$	$L_P/B_{PX} \ge 2.190$
7	$0.145 \le LCG/L_P \le 0.215$	$L_P/B_{PX} \le 59.5652 \cdot LCG/L_P - 4.2965$
8	$0.215 \le LCG/L_P \le 0.435$	$L_P/B_{PX} \le 8.510$
9	$2.190 \le L_P/B_{PX} \le 4.250$	$L_P/\nabla^{1/3} \ge 5.290$
10	$4.250 \le L_P/B_{PX} \le 8.510$	$L_P/\nabla^{1/3} \ge 0.7268 \cdot L_P/B_{PX} + 2.1646$
11	$2.190 \le L_P/B_{PX} \le 4.400$	$L_P/\nabla^{1/3} \le 1.4615 \cdot L_P/B_{PX} + 2.4992$
12	$4.400 \le L_P/B_{PX} \le 8.510$	$L_P/\nabla^{1/3} \le 8.930$
13	$2.190 \le L_P/B_{PX} \le 8.510$	$Fn_\nabla \ge 0.0555 \cdot L_P/B_{PX} + 0.8779$
14	$2.190 \le L_P/B_{PX} \le 4.400$	$Fn_\nabla \le 0.5045 \cdot L_P/B_{PX} + 3.7250$
15	$4.400 \le L_P/B_{PX} \le 8.510$	$Fn_\nabla \le 6.0$
16	$5.290 < L_P/\nabla^{1/3} \le 8.930$	$Fn_\nabla \ge 0.0879 \cdot L_P/\nabla^{1/3} + 0.5849$
17	$5.290 \le L_P/\nabla^{1/3} \le 8.930$	$Fn_\nabla \le 0.4234 \cdot L_P/\nabla^{1/3} + 2.3305$

Fig. 8.28 Database distribution and boundaries of applicability of MMs. Boundaries are given as a set of 17 inequality equations

Table 8.23 Polynomial terms and regression coefficients of the MM for residuary resistance

$$R_R/\Delta = a_0 + a_1 \cdot X_4^{1/4} + \cdots + a_{39} \cdot X_1^{1/4} \cdot X_2^{1/2} \cdot X_3 + \cdots + a_{65} \cdot X_1^2 \cdot X_3^2 \cdot X_4^{1/2}$$

Coeff.	Multipl.	a_i	Coeff.	Multipl.	a_i
a_0	1	–6.916756	a_{33}	$X_1^{1/4} X_3^2$	–3.174796
a_1	$X_4^{1/4}$	37.952889	a_{34}	$X_1^{1/4} X_3^2 X_4^{1/2}$	13.360202
a_2	$X_4^{1/2}$	–51.703283	a_{35}	$X_1^{1/4} X_3^2 X_4$	–17.140057
a_3	X_4	25.190336	a_{36}	$X_1^{1/4} X_3^2 X_4^2$	5.387301
a_4	X_4^2	–3.558576	a_{37}	$X_1^{1/4} X_2^{1/2} X_4$	–3.224661
a_5	X_3	6.711970	a_{38}	$X_1^{1/4} X_2^{1/2} X_4^2$	4.720016
a_6	$X_3 X_4^{1/4}$	–45.369194	a_{39}	$X_1^{1/4} X_2^{1/2} X_3$	10.942077
a_7	$X_3 X_4^{1/2}$	67.507315	a_{40}	$X_1^{1/4} X_2^{1/2} X_3 X_4^{1/4}$	–24.076455
a_8	$X_3 X_4$	–35.787648	a_{41}	$X_1^{1/4} X_2^{1/2} X_3 X_4$	33.013553
a_9	$X_3 X_4^2$	4.366035	a_{42}	$X_1^{1/4} X_2^{1/2} X_3 X_4^2$	–26.532165
a_{10}	X_3^2	2.231629	a_{43}	$X_1^{1/4} X_2^{1/2} X_3^2$	–3.872807
a_{11}	$X_3^2 X_4^{1/2}$	–7.641255	a_{44}	$X_1^{1/4} X_2^{1/2} X_3^2 X_4^{1/2}$	21.486989
a_{12}	$X_3^2 X_4$	7.135076	a_{45}	$X_1^{1/4} X_2^{1/2} X_3^2 X_4$	–33.261115
a_{13}	$X_2^{1/2} X_4$	4.155461	a_{46}	$X_1^{1/4} X_2^{1/2} X_3^2 X_4^2$	20.428120
a_{14}	$X_2^{1/2} X_4^2$	–5.724032	a_{47}	$X_1^{1/4} X_2 X_3^2 X_4^{1/4}$	–0.775084
a_{15}	$X_2^{1/2} X_3$	18.303913	a_{48}	$X_1^{1/4} X_2 X_3^2 X_4^{1/2}$	1.347047
a_{16}	$X_2^{1/2} X_3 X_4^{1/4}$	–103.117971	a_{49}	$X_1 X_4^{1/4}$	–0.926208
a_{17}	$X_2^{1/2} X_3 X_4^{1/2}$	156.441838	a_{50}	$X_1 X_4^{1/2}$	1.416676
a_{18}	$X_2^{1/2} X_3 X_4$	–108.176883	a_{51}	$X_1 X_3 X_4^{1/4}$	–1.221706
a_{19}	$X_2^{1/2} X_3 X_4^2$	43.832920	a_{52}	$X_1 X_3^2$	0.823968
a_{20}	$X_2^{1/2} X_3^2$	–21.460088	a_{53}	$X_1 X_2^{1/2} X_3^2$	–4.584800
a_{21}	$X_2^{1/2} X_3^2 X_4^{1/4}$	113.420767	a_{54}	$X_1 X_2^{1/2} X_3^2 X_4^{1/4}$	18.916114
a_{22}	$X_2^{1/2} X_3^2 X_4^{1/2}$	–164.838625	a_{55}	$X_1 X_2^{1/2} X_3^2 X_4^{1/2}$	–21.826423
a_{23}	$X_2^{1/2} X_3^2 X_4$	105.054460	a_{56}	$X_1 X_2^{1/2} X_3^2 X_4$	6.949923
a_{24}	$X_2^{1/2} X_3^2 X_4^2$	–37.931912	a_{57}	$X_1 X_2 X_3^2$	0.099882
a_{25}	$X_1^{1/4} X_4^{1/4}$	–9.131407	a_{58}	$X_1^2 X_4$	–0.151425
a_{26}	$X_1^{1/4} X_4^{1/2}$	19.633711	a_{59}	$X_1^2 X_3$	4.941884
a_{27}	$X_1^{1/4} X_4$	–14.537889	a_{60}	$X_1^2 X_3 X_4^{1/4}$	–13.632337
a_{28}	$X_1^{1/4} X_4^2$	2.850167	a_{61}	$X_1^2 X_3 X_4^{1/2}$	11.119057
a_{29}	$X_1^{1/4} X_3 X_4^{1/4}$	19.994687	a_{62}	$X_1^2 X_3 X_4$	–1.847766
a_{30}	$X_1^{1/4} X_3 X_4^{1/2}$	–43.672852	a_{63}	$X_1^2 X_3^2$	–4.624755
a_{31}	$X_1^{1/4} X_3 X_4$	34.991251	a_{64}	$X_1^2 X_3^2 X_4^{1/4}$	10.767824
a_{32}	$X_1^{1/4} X_3 X_4^2$	–8.566126	a_{65}	$X_1^2 X_3^2 X_4^{1/2}$	–6.413523

Table 8.24 ANN coefficients of the MM for dynamic trim

$$\tau_{BL} = \frac{\mathrm{sig}\left(c_W + \sum_{i=1}^{4}\left(C_{iw}\cdot \mathrm{sig}\left(b_i + \sum_{j=1}^{8}\left(B_{ji}\cdot \mathrm{sig}\left(a_j + \sum_{k=1}^{4}\left(A_{kj}\cdot (P_k\cdot X_k + R_k)\right)\right)\right)\right)\right)\right) - G}{H}$$

j	A_{1j}	A_{2j}	A_{3j}	A_{4j}	a_j
1	−1.6191150	−2.32547900	−2.96805700	12.5204500	1.4575500
2	1.1651560	−4.26778300	0.87332730	1.6141680	−2.1720760
3	−0.3954377	2.22601100	−1.88776900	−4.1061690	0.3364133
4	−3.5358610	0.06969679	−0.04187283	−0.8056500	−0.5813576
5	3.5991000	−1.97274500	−3.24267300	−7.8127540	−2.1192560
6	−12.9202300	0.53439580	−10.26596000	−0.4580198	5.4269850
7	6.5522910	5.88211200	0.80276450	1.3732150	−2.0152160
8	−0.9626058	−2.64446900	−2.25809100	15.4689900	−1.9805980

i	B_{1i}	B_{2i}	B_{3i}	B_{4i}	B_{5i}	B_{6i}	B_{7i}	B_{8i}	b_i
1	−5.0961540	−3.2049040	−5.7697620	−5.2860470	−2.8687000	−0.9393527	−1.1365520	7.8473890	6.7543800
2	−17.6980200	7.0609170	−0.3278437	5.1920420	−1.4596790	8.3570150	−1.9821370	6.8562210	0.4168208
3	−10.6568300	3.7257140	−0.9556682	−3.5100510	−0.3494906	−2.9589420	−1.2496820	1.7426410	10.5056300
4	−1.3054560	−1.5834450	−3.3944200	−6.8907960	0.3421457	4.1135250	−1.8178670	1.2753590	2.5923960

w	C_{1w}	C_{2w}	C_{3w}	C_{4w}	c_w
1	−3.1824090	−8.0000370	−4.0781820	−2.6389610	6.2952590

k	P_k	R_k
$L_P/\nabla^{1/3}$	0.2473709	−1.2590625
L_P/B_{PX}	0.1424905	−0.2632632
LCG/L_P	3.1141869	−0.4046713
Fn_∇	0.1779813	−0.1377724

v	H	G
τ_{BL}	0.0703125	0.0078125

Table 8.25 Polynomial terms and regression coefficient of the MM for wetted length

$$L_M/L_P = b_0 + b_1 \cdot X_4^{1/4} + \cdots + b_{10} \cdot X_2^2 \cdot X_3^2 \cdot X_4^{1/4} + \cdots + b_{16} \cdot X_1^2 \cdot X_4$$

Coeff.	Multipl.	b_i	Coeff.	Multipl.	b_i
b_0	1	0.584272	b_9	$X_2\,X_3\,X_4^{1/2}$	−1.036234
b_1	$X_4^{1/4}$	−0.620672	b_{10}	$X_2^2\,X_3^2\,X_4^{1/4}$	0.953004
b_2	X_4^2	0.848228	b_{11}	$X_1^{1/4}\,X_3\,X_4^{1/4}$	0.382653
b_3	X_3	0.439793	b_{12}	$X_1^{1/4}\,X_3^2$	−0.088835
b_4	$X_3\,X_4^{1/4}$	1.279120	b_{13}	$X_1^{1/4}\,X_2^{1/2}\,X_4^2$	0.795342
b_5	$X_3\,X_4^{1/2}$	−2.125064	b_{14}	$X_1^{1/4}\,X_2^2\,X_3^2$	−0.896709
b_6	$X_3^2\,X_4^{1/2}$	0.188852	b_{15}	$X_1\,X4^2$	−0.739886
b_7	$X_2^{1/2}\,X_4^2$	−0.897188	b_{16}	$X_1^2\,X_4$	0.169817
b_8	$X_2\,X_3\,X_4^{1/4}$	1.320159			

8.3.4 Series 62

Reference Radojčić et al. (2017); see Sect. 3.4.15 (Figs. 8.29 and 8.30).

INPUT	OUTPUT
$X_1 = A_P/\nabla^{2/3}$	$(R_T/\Delta)_{100000}$
$X_2 = \beta$	τ
$X_3 = L_P/B_{PX}$	
$X_4 = LCG/L_P$	
$X_5 = Fn_\nabla$	

INPUT	OUTPUT
$X_1 = 0.9(L_P/\nabla^{1/3} - 3.05)/5.48 + 0.05$	$S/\nabla^{2/3}$
$X_2 = 0.9(L_P/B_{PX} - 2.0)/5.01 + 0.05$	L_M/L_P
$X_3 = 0.9(LCG/L_P - 0.3511)/0.1369 + 0.05$	$A_P/\nabla^{2/3}$
$X_4 = 0.9(Fn_\nabla - 0.88)/5.23 + 0.05$	

NOTE: Parameter $L_P/\nabla^{1/3}$ may be used instead of $A_P/\nabla^{2/3}$ and vice versa, as they are interrelated. Approximately $L_P/\nabla^{1/3} = 1.1 \cdot \sqrt{(A_P/\nabla^{2/3} \cdot L_P/\nabla^{1/3})}$ as per Blount (2014). For precise relationship see Table 8.29.

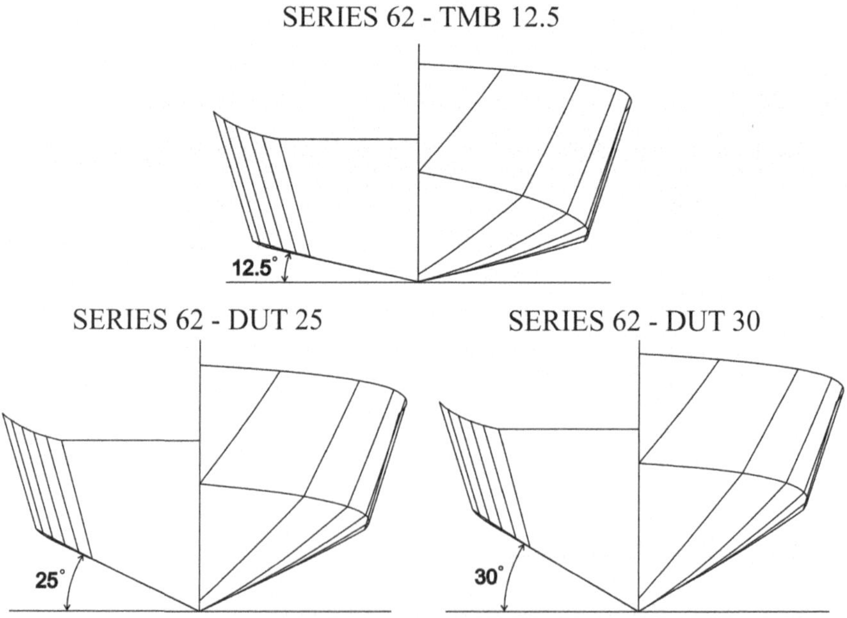

Fig. 8.29 Body plans of the parent hulls of subseries 62 with deadrise angles 12.5°, 25° and 30°

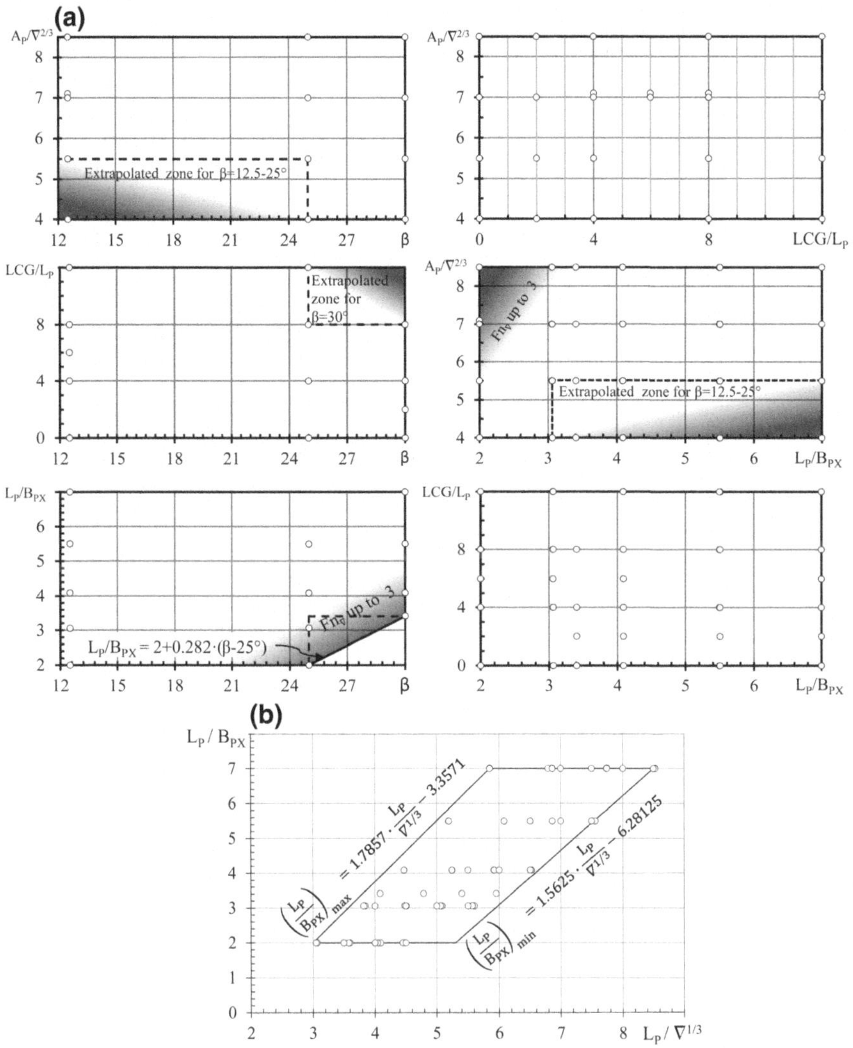

Fig. 8.30 Database distribution and boundaries of applicability of MMs for **a** $(R_T/\Delta)_{100,000}$ and τ, and **b** $A_P/\nabla^{2/3}$, L_M/L_P and $S/\nabla^{2/3}$

Single Output

Table 8.26 ANN coefficients of the MM for resistance

$$(R_T/\Delta)_{100000} = \frac{\mathrm{sig}\left(d_v + \sum_{w=1}^{3}\left(D_{wv}\cdot \mathrm{sig}\left(c_w + \sum_{i=1}^{7}\left(C_{iw}\cdot \mathrm{sig}\left(b_i + \sum_{j=1}^{11}\left(B_{ji}\cdot \mathrm{sig}\left(a_j + \sum_{k=1}^{5}\left(A_{kj}\cdot (P_k\cdot X_k + R_k)\right)\right)\right)\right)\right)\right)\right)\right) - G}{H}$$

j	A_{1j}	A_{2j}	A_{3j}	A_{4j}	A_{5j}	a_j
1	−0.8953071	0.8692768	−4.7686240	−1.1079190	9.2061240	−2.1124090
2	−1.5405570	0.3500689	−4.2759300	−2.3067310	−1.0030320	2.4277080
3	0.3453476	−0.0049233	0.6422353	−2.1320090	6.1046770	−1.7066510
4	1.7825570	1.3610870	−0.8890811	0.6342062	4.7003190	−5.1683690
5	−1.2542430	−2.2489860	−2.2436170	−0.3778002	11.9060800	0.8990684
6	−3.3724320	4.3842810	0.1963734	−3.3333470	−2.0846420	−1.2670800
7	−1.8988970	−2.0034850	0.1667324	1.6199720	0.0064824	1.5954960
8	0.9573705	1.3396530	1.3731320	2.3323730	−2.8576140	−1.3796650
9	1.3187180	2.7784720	−7.8972430	0.7002557	0.7720946	−0.7854592
10	1.1887340	−2.4052050	1.2648970	0.7803378	−5.8422660	0.1389654
11	−0.1580780	0.3736366	0.1123325	−0.2578748	−49.2733600	2.8607880

i	B_{1i}	B_{2i}	B_{3i}	B_{4i}	B_{5i}	B_{6i}	B_{7i}	B_{8i}	B_{9i}	B_{10i}	B_{11i}	b_i
1	−4.6437600	4.1446310	3.4275960	−0.6230085	−0.7022514	0.3030230	1.5060330	0.8447835	2.1267640	0.8477411	−6.3509450	−4.9093410
2	−5.1399190	−8.1142070	−4.7752640	−8.0000640	4.4795430	8.0141620	−4.1592040	−0.7853532	11.7619400	−3.3671720	−6.3914490	−6.1837660
3	−0.2668565	−0.8623143	−3.5619770	5.6276760	1.9989060	−0.7182558	−0.2423009	−3.6738990	0.5044518	−6.6138120	1.8863820	1.7233500
4	−0.6802325	0.5959017	1.9988190	−4.1171680	2.5590320	−0.0593718	−0.1853874	3.9956520	−2.3192120	−0.5018532	−4.7479050	−3.0574160
5	−7.4609430	−1.2534840	−3.2511810	−3.1709260	3.0848880	0.6152146	−0.9464355	3.9562660	0.2374572	−0.2996799	−5.1985010	−3.0923370
6	−0.6157955	−0.0848251	−3.6492890	4.1921580	8.8131410	3.8060420	1.1645430	−0.7344435	−0.2260222	−0.2168286	−2.2656090	−8.6381780
7	−2.5060510	3.3721840	1.6767890	6.4910810	−1.7542240	0.1150844	1.6736070	−2.9776620	1.1050280	−0.4623301	0.1154653	2.6849580

(continued)

Table 8.26 (continued)

w	C_{1w}	C_{2w}	C_{3w}	C_{4w}	C_{5w}	C_{6w}	C_{7w}	c_w
1	−0.3752839	−9.2892360	0.5501294	−1.2299960	2.7512290	−4.4879880	−0.8110411	2.6199090
2	−10.1823900	−1.5401370	−6.7694470	−0.6879896	−8.2779170	−3.9376570	−0.0731905	10.8929500
3	−2.0982320	−3.4298510	−2.7347090	−1.3847540	−3.8894430	2.0274030	−6.3145600	7.8656930

v	D_{1v}	D_{2v}	D_{3v}	d_v
1	−2.6426030	−7.5461400	−3.0005930	9.9164240

k	P_k	R_k
$A_P/\nabla^{2/3}$	0.2000000	−0.7500000
β	0.0514286	−0.5928571
L_P/B_{PX}	0.1800000	−0.3100000
LCG/L	6.5693431	−2.2558394
Fn_∇	0.1679104	−0.0658582

v	L_v	G_v
$(R_T/\Delta)_{100000}$	2.1325847	0.0230932

Table 8.27 ANN coefficients of the MM for dynamic trim

$$\tau = \frac{\mathrm{sig}\left(d_v + \sum_{w=1}^{3}\left(D_{wv}\cdot \mathrm{sig}\left(c_w + \sum_{i=1}^{5}\left(C_{iw}\cdot \mathrm{sig}\left(b_i + \sum_{j=1}^{11}\left(B_{ji}\cdot \mathrm{sig}\left(a_j + \sum_{k=1}^{5}\left(A_{kj}\cdot (P_k\cdot X_k + R_k)\right)\right)\right)\right)\right)\right)\right)\right) - G}{H}$$

j	A_{1j}	A_{2j}	A_{3j}	A_{4j}	A_{5j}	a_j
1	−3.4492220	0.6310675	1.1905780	−1.3575430	−4.1485210	0.8722453
2	−0.8492672	0.0661794	−3.4550510	0.2499540	−0.5284106	0.4412649
3	0.4586774	−0.4420829	0.9486749	0.5034463	−38.8782000	2.8810010
4	−0.0470687	−0.4667896	0.5517367	−2.1696190	−8.2071860	0.6869534
5	0.9114847	1.0125350	1.9974450	0.7818773	−6.8750270	0.7216546
6	−0.1913945	0.6899573	4.3440070	1.5357960	−19.2195500	0.5476219
7	0.5051449	−0.2937532	0.0833985	−3.2630980	1.3037720	0.0877794
8	−2.3953230	−1.3161350	−2.2536810	−2.0134050	−2.1285080	1.8148740
9	0.0149306	−0.6897116	−0.6168764	1.2792000	−34.6604900	4.1287740
10	0.8241501	−21.4137800	11.8292200	0.5723563	0.1774283	16.8573700
11	0.9066686	0.4675484	−2.1144510	1.0989180	4.0938930	−1.9485680

i	B_{1i}	B_{2i}	B_{3i}	B_{4i}	B_{5i}	B_{6i}	B_{7i}	B_{8i}	B_{9i}	B_{10i}	B_{11i}	b_i
1	−0.0239567	3.0932320	−9.0294150	11.2363000	0.8773211	−2.3694060	−3.4714430	−5.4702330	4.9118150	−3.0352150	−1.8290510	1.8177600
2	2.1742880	6.2769100	−13.4009000	−1.1237480	1.6166490	−1.0753720	0.6387935	−5.9491080	2.3083680	−4.9426420	0.6164492	2.6787350
3	4.0613520	3.5635640	−7.3242030	−1.7176130	−7.4597110	−7.4434450	1.0348500	−1.9399860	−5.0507800	−3.7413790	0.3508019	8.0311540
4	−0.8599226	−1.2800770	1.6735850	11.6958500	−0.5515437	3.0834550	−1.1009040	−3.4225850	−4.6215220	0.2014546	−1.3737640	3.6216240
5	1.8460310	3.2741470	−9.0812330	1.9338690	−2.7308470	−0.7307843	−2.2696510	2.7261490	2.0294720	3.0257060	−3.7928420	−2.1601720

(continued)

Table 8.27 (continued)

w	C_{1w}	C_{2w}	C_{3w}	C_{4w}	C_{5w}	c_w
1	−2.9055990	−1.5828060	−2.0270280	−7.4149200	−3.0381230	10.5337300
2	−10.4940000	−1.5774530	−4.2350510	−0.1853783	−5.8421700	12.9098300
3	−0.9955937	−1.3356660	−0.9657380	−7.6183680	−7.7233020	5.6019720

v	D_{1v}	D_{2v}	D_{3v}	d_v
1	−2.0900740	−7.5094650	−5.1224900	8.4527260

k	P_k	R_k
$A_P/\nabla^{2/3}$	0.2000000	−0.7500000
β	0.0514286	−0.5928571
L_P/B_{PX}	0.1800000	−0.3100000
LCG/L	6.5693431	−2.2558394
Fn_∇	0.1679104	−0.0658582

v	H	G
1	0.0481541	0.1602729

Multiple Output

Table 8.28 ANN coefficients of the MMs for resistance and dynamic trim

$$(R_T/\Delta)_{100000}, \tau = \frac{\mathrm{sig}\left(d_v + \sum_{w=1}^{5}\left(D_{wv}\cdot \mathrm{sig}\left(c_w + \sum_{i=1}^{8}\left(C_{iw}\cdot \mathrm{sig}\left(b_i + \sum_{j=1}^{11}\left(B_{ji}\cdot \mathrm{sig}\left(a_j + \sum_{k=1}^{5}\left(A_{kj}\cdot (P_k\cdot X_k + R_k)\right)\right)\right)\right)\right)\right)\right)\right) - G_{1,2}}{H_{1,2}}$$

j	A_{1j}	A_{2j}	A_{3j}	A_{4j}	A_{5j}	a_j
1	−0.1348189	−0.0076723	−0.2775522	3.9860210	−51.1027700	5.0188620
2	−0.4625220	−2.6474660	−4.3783800	−2.2981970	2.3711880	3.7986360
3	−4.1258340	0.9317533	0.5740002	0.3302810	1.9288290	−0.8772002
4	0.4276277	24.9427300	−15.9999900	−1.1010020	5.7635070	−17.2446300
5	0.3993703	2.5567490	4.1169740	−3.9867760	36.0846400	−8.1654170
6	−2.7739500	−0.1677839	−6.8352360	−0.0920398	−4.7806060	8.0008130
7	−0.3610690	−2.6026650	1.2987690	−2.3065480	−4.5788530	6.6255860
8	0.4993203	0.4236119	1.8421170	−3.2782560	−8.3202350	3.9722700
9	−0.6632705	0.1599472	−1.0277510	−6.4186220	2.3015750	3.7007440
10	0.6739060	−1.9705250	7.8950650	−4.8384850	1.9820560	1.6988550
11	−1.9323510	−1.4435490	−5.7124600	−9.2345910	16.4392800	8.8538460

i	B_{1i}	B_{2i}	B_{3i}	B_{4i}	B_{5i}	B_{6i}	B_{7i}	B_{8i}	B_{9i}	B_{10i}	B_{11i}	b_i
1	−1.0373260	2.3060430	0.7746050	−0.5941365	−1.2888720	−0.7995334	−1.8261880	−3.6102430	−2.7552240	−2.0167620	−0.5915413	3.1382240
2	−16.0593300	1.4665730	2.1381920	−0.0350787	1.8385600	0.1185768	1.6615480	−5.3048430	11.8720400	−0.8111245	−1.5748320	0.1467929
3	−3.2429850	−7.3641040	−6.2394800	−2.2290960	8.9277110	−3.9129150	−2.5996480	0.5317162	−2.7476220	−8.7133150	4.0550770	−1.7504580
4	−8.7237220	−0.9508612	3.6166210	1.4410360	3.7705990	1.8351820	3.3766650	−1.5381720	−1.1865670	−0.7550833	8.3972550	−14.4834400
5	−4.7445260	−9.1662120	0.3310782	3.7112300	−3.1729500	4.2862620	−0.9947226	4.3514310	−10.3105000	1.2045130	2.8764090	−5.1791170
6	−3.6050050	1.4773420	3.6873220	−0.5288379	−1.7603580	−0.3749855	2.5213510	3.1052740	−3.7475370	−1.7986580	2.4222280	−3.7377810
7	−6.6968200	−2.5532780	−0.8749955	0.9327890	1.3859110	2.6257200	−0.4362571	−3.0893150	−4.1500940	2.2058950	−4.0140100	1.5385620
8	−12.7773000	4.2003320	2.3674180	−0.5768666	−0.1976727	−1.3412960	−0.1038523	0.4837250	−4.0861120	−1.9748190	5.5389870	−5.3102150

(continued)

Table 8.28 (continued)

w	C_{1w}	C_{2w}	C_{3w}	C_{4w}	C_{5w}	C_{6w}	C_{7w}	C_{8w}	c_w
1	−10.4811300	2.8282740	1.2338810	−4.3893220	−2.9808120	−0.4040653	1.0968770	2.8939640	−2.8279170
2	−12.5702200	−3.6624040	−0.5793387	−0.3504248	−3.8378950	−3.0231180	−1.7668800	−0.0913801	2.1768760
3	−1.8200020	−3.2548080	−1.0896540	1.0606170	−2.7420750	−2.6956180	−1.4124410	0.5530158	2.2103580
4	−4.3071300	−4.1856740	2.7310460	−2.8798950	−4.9322180	−5.3763360	1.6572750	5.6926620	8.6448720
5	5.8248200	2.9226870	2.8063150	−2.7293670	0.6810613	1.1308260	0.6901268	−4.3622750	−2.8384770

v	D_{1v}	D_{2v}	D_{3v}	D_{4v}	D_{5v}	d_v	
1	−2.7865520	−1.1651470	−0.0683942	−12.6624100	3.6777960	11.2581200	← $(R_T/\Delta)_{100000}$
2	−1.6608050	6.5392250	−10.2095200	−8.4640710	−1.9419160	10.3451400	← τ

k	P_k	R_k
$A_P/\nabla^{2/3}$	0.1058824	0.0500000
β	0.0300000	0.0500000
L_P/B_{PX}	0.1285714	0.0500000
LCG/L	1.8442623	0.0500000
Fn_∇	0.1487603	0.0500000

v	L_v	G_v
$(R_T/\Delta)_{100000}$	2.0706787	0.0500000
τ	0.0481541	0.1602729

MMs for $A_p/\nabla^{2/3}$, L_M/L_P and (S)

Table 8.29 Polynomial terms and regression coefficients of the MM for planing area coefficient

$$A_P/\nabla^{2/3} = 5.068 \cdot \left(c_0 + c_1 \cdot X_2^{1/2} + c_2 \cdot X_2^{3/2} + \cdots + c_6 \cdot X_1^{3/2} - 0.05\right) + 4$$

Coeff.	Multipl.	c_i
c_0	1	0.054055
c_1	$X_2^{1/2}$	0.408490
c_2	$X_2^{3/2}$	1.000686
c_3	$X_1^{1/2}$	−0.730380
c_4	$X_1^{1/2} X_2^{1/2}$	−5.826540
c_5	X_1	7.155628
c_6	$X_1^{3/2}$	−1.072560

Table 8.30 Polynomial terms and regression coefficients of the MM for wetted length

$$L_M/L_P = 0.8172 \cdot (a_0 + a_1 \cdot X_4^{1/4} + \cdots + a_{10} \cdot X_1^2 \cdot X_3^{1/2} \cdot X_4 - 0.05) + 0.3579$$

Coeff.	Multipl.	a_i
a_0	1	1.540554
a_1	$X_4^{1/4}$	−2.011497
a_2	X_4^2	0.236221
a_3	X_3	0.457643
a_4	$X_2 X_3^{1/2} X_4^{1/4}$	1.032858
a_5	$X_2 X_3^{1/2} X_4$	−1.146167
a_6	$X_1^{1/4} X_3^{1/4} X_4^{1/4}$	0.024093
a_7	$X_1 X_3^{1/4} X_4^{1/4}$	2.550234
a_8	$X_1 X_3^{1/2}$	−1.983606
a_9	$X_1^2 X_3^{1/4} X_4$	−1.671383
a_{10}	$X_1^2 X_3^{1/2} X_4$	1.392485

Table 8.31 Polynomial terms and regression coefficients of the MM for wetted surface

$$(S) = 9.419 \cdot (b_0 + b_1 \cdot X_4^{1/3} + \cdots + b_{24} \cdot X_1^3 \cdot X_2^2 \cdot X_4 - 0.05) + 1.859$$

Coeff.	Multipl.	b_i
b_0	1	0.336686
b_1	$X_4^{1/3}$	3.765763
b_2	$X_4^{1/2}$	−7.030424
b_3	X_4	4.321102
b_4	$X_3 X_4^{1/3}$	0.230037
b_5	$X_3 X_4$	−1.512141
b_6	X_2	−1.209557
b_7	$X_2 X_4^{1/2}$	−1.567582
b_8	$X_2 X_3^2$	−0.460442
b_9	X_2^2	0.439141
b_{10}	$X_2^2 X_4$	24.319819
b_{11}	$X_1^{1/3} X_4$	−2.311625
b_{12}	$X_1^{1/3} X_3 X_4^{1/2}$	3.433392
b_{13}	$X_1^{1/3} X_2 X_4^{1/2}$	2.898156
b_{14}	$X_1^{1/3} X_2 X_3^2 X_4^{1/3}$	0.390216
b_{15}	$X_1^{1/3} X_2^2 X_4$	−42.530085
b_{16}	$X_1^{1/2} X_3 X_4^{1/3}$	−1.681586
b_{17}	$X_1 X_4^{1/3}$	10.485111
b_{18}	$X_1 X_4^{1/2}$	−9.431033
b_{19}	$X_1 X_2^2 X_4$	18.317865
b_{20}	X_1^2	−0.823152
b_{21}	$X_1^2 X_2 X_4$	4.327095
b_{22}	$X_1^3 X_4$	5.416627
b_{23}	$X_1^3 X_2 X_4$	−15.668252
b_{24}	$X_1^3 X_2^2 X_4$	6.243324

8.3.5 NSS

Reference Radojčić and Kalajdžić (2018), see Sect. 3.4.18 (Figs. 8.31 and 8.32).

INPUT	OUTPUT
$X_1 = L_P/\nabla^{1/3}$	$(R_T/\Delta)_{100000}$
$X_2 = L_P/B_{PX}$	τ
$X_3 = LCG/L_P$	$S/\nabla^{2/3}$
$X_4 = Fn_\nabla$	L_{WL}/L_P

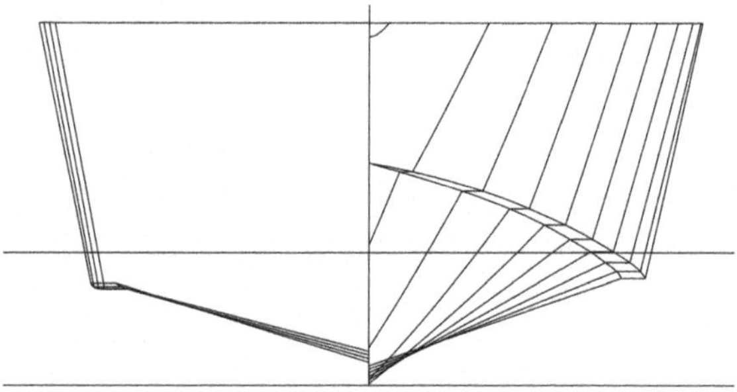

Fig. 8.31 Body plan of the parent hull of the NSS series

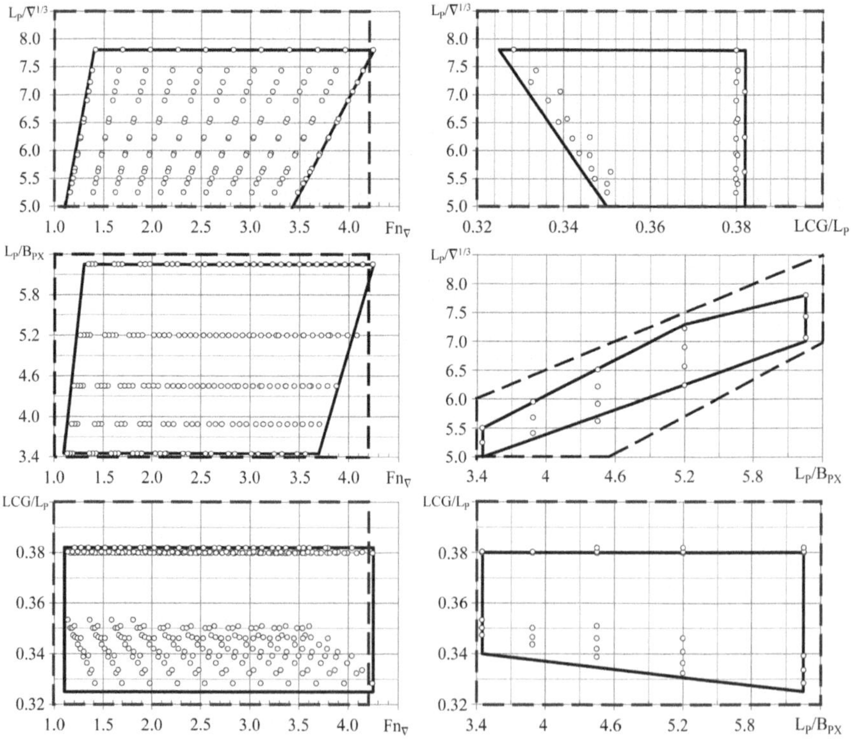

Fig. 8.32 Database distribution and boundaries of applicability of MMs (full lines). Dashed lines show extended boundaries which should be treated with caution

Table 8.32 ANN coefficients of the MMs for resistance and dynamic trim

$$(R_T/\Delta)_{100000}, \tau = \frac{\mathrm{sig}\left(c_w + \sum_{i=1}^{6}\left(C_{iw}\cdot \mathrm{sig}\left(b_i + \sum_{j=1}^{9}\left(B_{ji}\cdot \mathrm{sig}\left(a_j + \sum_{k=1}^{4}\left(A_{kj}\cdot (P_k\cdot X_k + R_k)\right)\right)\right)\right)\right)\right) - G_{1,2}}{H_{1,2}}$$

j	A_{1j}	A_{2j}	A_{3j}	A_{4j}	a_j
1	1.6841110	3.0981180	1.5282840	−16.4245400	3.1890700
2	2.4092060	0.3142931	0.3908146	0.2556609	−1.8597140
3	−1.9152880	−7.1345370	−3.6484370	−0.6078711	1.1340150
4	−1.4699330	−10.6498300	−0.0225162	−0.5634111	10.0365500
5	4.1680880	−4.5602380	0.6937186	2.7010170	−2.0246550
6	1.3626750	0.1168425	0.7784404	−10.0020100	−0.6886486
7	0.5479282	2.5084270	−1.1066880	−3.2642590	−0.3612035
8	−0.9420888	−0.2510580	0.0049670	−7.9977910	2.1865540
9	−3.1893870	1.2750320	−0.5223626	−3.9322540	4.7179180

i	B_{1i}	B_{2i}	B_{3i}	B_{4i}	B_{5i}	B_{6i}	B_{7i}	B_{8i}	B_{9i}	b_i
1	−0.4082681	0.0891881	1.4520660	0.0838715	1.1443510	−4.0656230	1.7600280	3.3407880	2.1149690	−2.0033580
2	0.6056641	1.8613140	0.3999820	0.1692302	0.0305501	2.5074690	−0.4054057	−0.8660036	1.2468480	−3.5325410
3	−0.1796350	−2.7916050	−2.9371210	0.3929720	−1.9068460	−6.4689190	−0.7239615	−2.9223100	−2.3012630	3.3583560
4	−5.3774030	−4.4911580	−2.2997850	−1.4914010	−3.7586970	−7.1752740	0.5064505	8.1555560	5.5388670	−4.1627820
5	−0.2961807	−4.7618480	−2.7215400	−0.5896583	2.0540800	0.1658212	−1.8644930	−1.6496020	−8.5582700	2.9999680
6	−6.2256920	2.9541990	−2.5990310	0.7695088	−3.3201360	−4.2225760	−1.7524090	7.9941520	−3.7219060	−0.1329643

(continued)

Table 8.32 (continued)

w	C_{1w}	C_{2w}	C_{3w}	C_{4w}	C_{5w}	C_{6w}	c_w	
1	−2.2633980	−7.7446910	0.2705126	4.3702590	7.6477400	0.0098969	3.2038660	← $(R_T/\Delta)_{100000}$
2	7.3747910	−1.3781800	4.4610630	8.6496370	−1.1401120	6.7151070	−6.6791630	← τ

k	P_k	R_k
$L_P/\nabla^{1/3}$	0.3214286	−1.5571429
L_P/B_{PX}	0.3215434	−1.0596463
LCG/L	16.6666698	−5.4166677
Fn_∇	0.2893891	−0.2770096

v	L_v	G_v
$(R_T/\Delta)_{100000}$	6.4766839	−0.2948834
τ	0.1573427	−0.1309441

Table 8.33 ANN coefficients of the MMs for wetted surface and wetted length

$$(S), L_{WL}/L_P = \frac{sig\left(c_w + \sum_{i=1}^{6}\left(C_{iw}\cdot sig\left(b_i + \sum_{j=1}^{9}\left(B_{ji}\cdot sig\left(a_j + \sum_{k=1}^{4}\left(A_{kj}\cdot(P_k\cdot X_k + R_k)\right)\right)\right)\right)\right)\right) - G_{1,2}}{H_{1,2}}$$

j	A_{1j}	A_{2j}	A_{3j}	A_{4j}	a_j
1	−1.0290940	−8.3437940	−0.7758193	0.3231824	1.9491680
2	4.2965780	−6.9091490	−1.0277140	8.8388290	−1.9010410
3	5.1826030	−1.5968630	5.0166010	1.0523230	−6.5190030
4	−15.3890100	9.5143390	0.6286043	4.9452410	−2.6309080
5	10.4899200	−4.8726330	−1.6352600	−0.1959135	−4.5200380
6	−0.9499520	−2.1286730	−0.9494023	8.1789890	−0.9933102
7	−26.5298400	0.8592815	−1.6622860	0.6121404	9.7718200
8	1.6415560	−8.1853450	−0.8103762	1.7085260	0.0527342
9	−12.0271000	−2.0909870	−0.9701849	0.9512810	1.7652420

i	B_{1i}	B_{2i}	B_{3i}	B_{4i}	B_{5i}	B_{6i}	B_{7i}	B_{8i}	B_{9i}	b_i
1	4.1591200	−0.9906697	13.9657000	−2.6740540	6.1056830	−4.2079160	−2.5134440	2.3902560	2.6879430	7.9379350
2	1.0908680	−1.9158720	0.3214091	2.3244010	−3.6995700	−1.0235730	−4.3936600	1.6758450	−6.1349780	4.2691960
3	5.9825890	−1.3113770	0.1117310	−0.0907498	−2.4306030	0.6541541	1.3742260	−3.7154990	−2.3982970	−0.7658195
4	−12.5662400	1.0044840	10.1118400	0.2574254	14.9457800	−0.5477254	7.7494610	−9.2629270	12.7814500	3.2341290
5	2.4613880	1.2783540	0.3146927	−3.3574090	1.8264710	−1.5419090	−10.0103900	−1.0326580	0.1740037	−2.0128590
6	−3.9455390	0.4092577	0.7670850	0.1192115	−0.9617954	−2.9109090	−0.9420338	4.2584630	0.7633737	2.8597680

(continued)

Table 8.33 (continued)

w	C_{1w}	C_{2w}	C_{3w}	C_{4w}	C_{5w}	C_{6w}	c_w	
1	5.2640740	−1.3256100	1.4416850	6.9203660	0.8226122	5.7311380	−15.1649800	← L_{WL}/L_P
2	3.5917060	2.1159010	4.8052780	−0.4571030	12.4226700	1.2936160	−8.0888610	← $S/V^{2/3}$

k	P_k	R_k
$L_P/\nabla^{1/3}$	0.3214286	−1.5571429
L_P/B_{PX}	0.3215434	−1.0596463
LCG/L	16.6666698	−5.4166677
Fn_∇	0.2893891	−0.2770096

v	H_v	G_v
L_{WL}/L_P	2.0529197	−1.0735219
$S/V^{2/3}$	0.1524332	−0.3873496

8.4 Some Typical Examples

Two worked examples are given in this section. The first one is for a typical semi-displacement mega yacht of 500 t, and the second one is for a hard chine yacht of 100000 lb (45.4 t), presented respectively, in Sect. 8.4.1 (Tables 8.34, 8.35, 8.36, 8.37 and 8.38 with final comparison in Fig. 8.33), and Sect. 8.4.2 (Tables 8.39, 8.40, 8.41 and 8.42 with final comparison in Fig. 8.34). The purpose of numerical examples is to compare the MMs for prediction of bare hull resistance and dynamic trim, presented in Sects. 8.2 and 8.3. A comprehensive procedure for evaluation of bare hull resistance for the semi-displacement NPL hull form is given in Sect. 10.3.1, hence step by step explanations for each MM is omitted here because the procedure described is same as that.

Note that some discrepancies among the comparable MMs (used in the examples) are expected, due to the fact that they represent different hull forms, i.e. their secondary hull form parameters are different, although their input variables are identical.

8.4.1 An Example of Semi-displacement Hull Forms

Resistance and dynamic trim predictions of five MMs applicable for semi-displacement hull forms (4 round bilge, 1 double-chine) are shown in Fig. 8.33. The hull dimensions used are typical for a 500t mega yacht (L_{WL} = 52 m, B_{WL} = 9.3 m, T = 2.55 m, LCG = 23.4 m), and hence the main non-dimensional parameters are $L_{WL}/\nabla^{1/3} = 6.6$ and $L_{WL}/B_{WL} = 5.6$ (as per Blount 2014). Other assumed variables used as input for some MMs, were: $A_T/A_X = 0.52$ and $C_B = 0.396$ according to the NPL series, $C_X = 0.573$ according to the VTT series, and $i_e = 15°$ an average value. R/Δ is valid for $C_A = 0$ and 500 t, hence here $R/\Delta = (R/\Delta)_{500t}$.

For each MM the density of water, kinetic viscosity and frictional resistance coefficient are taken as suggested in the original papers (see notes for Tables 8.34, 8.35 and 8.36), while for evaluation of the new design's total resistance (see Sect. 8.1.3) the following values were adopted: $\rho = 1.025$ t/m^3, $g = 9.81$ m/s^2, $\nu = 1.1907E{-}6$ m^2/s and C_F according to ITTC-1957.

Table 8.34 Results for the Mercier and Savitsky MM

	Fn_∇	[-]	1.0	1.1	1.2	1.3	1.4	1.5	1.6	1.7	1.8	1.9	2.0
Δ = 100000 lb (45.359t)	$(R_T/\Delta)_{100000}$	[-]	0.0269	0.0377	0.0473	0.0549	0.0613	0.0665	0.0706	0.0747	0.0784	0.0829	0.0868
	(S)	[-]	6.9935	6.9935	6.9935	6.9935	6.9935	6.9935	6.9935	6.9935	6.9935	6.9935	6.9935
	$Rn_{100000}\ 10^{-8}$	[-]	1.1560	1.2716	1.3872	1.5029	1.6185	1.7341	1.8497	1.9653	2.0809	2.1965	2.3121
	$C_{F100000}\ 10^{3}$	[-]	2.0181	1.9920	1.9687	1.9476	1.9284	1.9108	1.8945	1.8794	1.8653	1.8521	1.8398
Δ = 500t	Fn_L	[-]	0.389	0.428	0.467	0.506	0.545	0.584	0.623	0.662	0.701	0.740	0.778
	v	[m/s]	8.793	9.672	10.552	11.431	12.310	13.190	14.069	14.948	15.827	16.707	17.586
	V	[kn]	17.094	18.803	20.512	22.222	23.931	25.641	27.350	29.059	30.769	32.478	34.187
	$Rn\ 10^{-8}$	[-]	3.8401	4.2241	4.6081	4.9921	5.3761	5.7601	6.1441	6.5281	6.9121	7.2961	7.6801
	$C_F\ 10^{3}$	[-]	1.7300	1.7084	1.6891	1.6716	1.6557	1.6410	1.6275	1.6149	1.6032	1.5923	1.5820
	S	[m²]	433.37	433.37	433.37	433.37	433.37	433.37	433.37	433.37	433.37	433.37	433.37
	$(R_T/\Delta)_{500t}$	[-]	0.0259	0.0365	0.0459	0.0533	0.0594	0.0643	0.0682	0.072	0.0754	0.0796	0.0832
	R_T	[kN]	127.2	178.8	225.0	261.5	291.6	315.6	334.3	353.1	370.0	390.7	408.0
	P_E	[kW]	1118.8	1729.6	2374.3	2989.1	3589.5	4162.7	4703.6	5278.5	5856.3	6526.7	7175.2

Note for Δ = 100000 lb: $C_{F100000}$ according to Schoenherr friction line

$$Rn_{100{,}000} = \frac{Fn_\nabla (L/\nabla^{1/3})\sqrt{32.2 \times 100000/64}}{1.2817 \times 10^{-5}} \text{ (in Impeial system)}$$

Table 8.35 Results for the VTT MM

	Fn_∇	[-]	1.8	2	2.2	2.4	2.6	2.8	3	3.2
Δ = 100000 lb (45.359t)	$(R_T/\Delta)_{100000}$	[-]	0.0786	0.0855	0.0944	0.1055	0.119	0.1349	0.1535	0.1751
	(S)	[-]	6.9935	6.9935	6.9935	6.9935	6.9935	6.9935	6.9935	6.9935
	$Rn_{100000}\ 10^{-8}$	[-]	2.0720	2.3023	2.5325	2.7627	2.9929	3.2232	3.4534	3.6836
	$C_{F100000}\ 10^{3}$	[-]	1.8798	1.8529	1.829	1.8076	1.7883	1.7706	1.7544	1.7395
Δ = 500t	Fn_L	[-]	0.7003	0.7782	0.856	0.9338	1.0116	1.0894	1.1672	1.2451
	v	[m/s]	15.82	17.58	19.33	21.09	22.85	24.61	26.36	28.12
	V	[kn]	30.75	34.17	37.58	41.00	44.42	47.83	51.25	54.67
	$Rn\ 10^{-8}$	[-]	6.912	7.68	8.448	9.216	9.984	10.752	11.52	12.288
	$C_F\ 10^{3}$	[-]	1.6032	1.582	1.5632	1.5462	1.5309	1.5169	1.5041	1.4922
	S	[m²]	433.37	433.37	433.37	433.37	433.37	433.37	433.37	433.37
	$(R_T/\Delta)_{500t}$	[-]	0.0755	0.0817	0.0899	0.1003	0.1129	0.1279	0.1457	0.1662
	R_T	[kN]	370.4	400.8	441.1	491.8	553.7	627.6	714.4	815.3
	P_E	[kW]	5858.2	7044.9	8527.9	10372.5	12650.6	15442.0	18834.9	22926.9

Note for Δ = 100000 lb: $C_{F100000}$ according to ITTC-1957

$$Rn_{100000} = \frac{Fn_\nabla (L/\nabla^{1/3})\sqrt{32.2 \times 100000/64}}{1.2817 \times 10^{-5}} \text{ (in Impeial system)}$$

Table 8.36 Results for the NPL MM

	Fn_∇	[–]	0.8	1.0	1.2	1.4	1.6	1.8	2.0	2.2	2.4	2.6	2.8	3.0
Δ = 100000 lb (45.359t)	$(R_T/\Delta)_{100000}$	[–]	0.0137	0.025	0.0452	0.0588	0.0678	0.0756	0.0835	0.0917	0.1013	0.1138	0.1294	0.1495
	(S)	[–]	7.3535	7.3535	7.3535	7.3535	7.3535	7.3535	7.3535	7.3535	7.3535	7.3535	7.3535	7.3535
	τ	[deg]	0.093	0.380	1.123	1.782	2.124	2.206	2.174	2.105	1.986	1.810	1.660	1.545
	$Rn_{100000}\ 10^{-8}$	[–]	0.9232	1.154	1.3847	1.6155	1.8463	2.0771	2.3079	2.5387	2.7695	3.0003	3.2311	3.4619
	$C_{F100000}\ 10^3$	[–]	2.1077	2.0408	1.9885	1.9459	1.91	1.8792	1.8523	1.8284	1.807	1.7877	1.7701	1.7539
Δ = 500t	Fn_L	[–]	0.311	0.389	0.467	0.545	0.623	0.701	0.78	0.86	0.93	1.01	1.09	1.17
	v	[m/s]	7.034	8.793	10.552	12.310	14.069	15.827	17.586	19.345	21.103	22.862	24.620	26.379
	V	[kn]	13.675	17.094	20.512	23.931	27.350	30.769	34.187	37.61	41.02	44.44	47.86	51.28
	$Rn\ 10^{-8}$	[–]	3.0721	3.8401	4.6081	5.3761	6.1441	6.9121	7.6801	8.4481	9.2162	9.9842	10.752	11.52
	$C_F\ 10^3$	[–]	1.7820	1.7300	1.6891	1.6557	1.6275	1.6032	1.5820	1.5632	1.5462	1.5309	1.5169	1.5041
	S	[m²]	456.78	456.78	456.78	456.78	456.78	456.78	456.78	456.78	456.78	456.78	456.78	456.78
	$(R_T/\Delta)_{500t}$	[–]	0.0130	0.0240	0.0438	0.0569	0.0654	0.0727	0.0799	0.0874	0.0963	0.1079	0.1227	0.1418
	R_T	[kN]	63.87	117.75	215.01	279.34	320.88	356.40	391.74	428.64	472.20	529.19	601.87	695.76
	P_E	[kW]	449.3	1035.4	2268.7	3438.7	4514.4	5640.8	6889.1	8292.0	9964.9	12098.2	14818.2	18353.4
	τ	[deg]	0.093	0.380	1.123	1.782	2.124	2.206	2.174	2.105	1.986	1.810	1.660	1.545

Note for Δ = 100000 lb: $C_{F100000}$ according to ITTC-1957, see Table 10.4

Table 8.37 Results for the SKLAD MM

	Fn_∇	[–]	1.00	1.25	1.50	1.75	2.00	2.25	2.50	3.00
MM	$C_R\ 10^3$	[–]	6.5621	7.5062	6.4489	4.7492	3.8867	3.1703	2.4814	1.9042
	(S)	[–]	7.5369	7.6269	7.7477	7.8544	7.9847	8.126	8.1948	8.3602
	(M)	[–]	6.5958	6.5862	6.5485	6.5171	6.5249	6.5036	6.4444	6.4898
	τ	[deg]	0.525	1.274	1.926	2.067	1.890	1.544	1.361	1.485
Δ = 500t	Fn_L	[–]	0.389	0.487	0.584	0.681	0.779	0.876	0.973	1.168
	v	[m/s]	8.793	10.991	13.190	15.388	17.586	19.784	21.983	26.379
	V	[kn]	17.094	21.367	25.641	29.914	34.187	38.461	42.734	51.281
	$Rn\ 10^{-8}$	[–]	3.8401	4.8001	5.7601	6.7201	7.6801	8.6401	9.6002	11.5202
	$C_F\ 10^3$	[–]	1.7300	1.6801	1.6410	1.6090	1.5820	1.5588	1.5384	1.5041
	$C_T\ 10^3$	[–]	8.2920	9.1863	8.0899	6.3582	5.4687	4.7290	4.0198	3.4083
	S	[m²]	467.76	473.33	480.63	487.1	495.09	503.82	508.03	518.34
	$(R_T/\Delta)_{500t}$	[–]	0.0313	0.0549	0.0707	0.0766	0.0875	0.0974	0.1031	0.1284
	R_T	[kN]	153.7	269.2	346.7	375.8	429.1	478.0	505.8	630.0
	P_E	[kW]	1351.4	2958.9	4572.2	5783.2	7546.9	9455.9	11117.8	16619.5
	τ	[deg]	0.525	1.274	1.926	2.067	1.890	1.544	1.361	1.485

Table 8.38 Results for the NTUA MM

	Fn_∇	[–]	0.51	0.77	1.03	1.29	1.54	1.80	2.06	2.31	2.57	2.83
MM	$C_R\ 10^3$	[–]	0.0036	0.0073	0.0087	0.0081	0.0065	0.0047	0.0034	0.0028	0.0027	0.0031
	(S)	[–]	7.0447	7.0447	7.0447	7.0447	7.0447	7.0447	7.0447	7.0447	7.0447	7.0447
	τ	[deg]	0.310	0.271	0.883	1.540	1.962	2.110	2.090	2.061	2.143	2.325
Δ = 500t	Fn_L	[–]	0.20	0.30	0.40	0.50	0.60	0.70	0.80	0.90	1.00	1.10
	v	[m/s]	4.5	6.8	9.0	11.3	13.6	15.8	18.1	20.3	22.6	24.8
	V	[kn]	8.8	13.2	17.6	22.0	26.3	30.7	35.1	39.5	43.9	48.3
	$Rn\ 10^{-8}$	[–]	1.9906	2.9859	3.9812	4.9765	5.9718	6.9672	7.9625	8.9578	9.9531	10.9484
	$C_F\ 10^3$	[–]	1.8903	1.7888	1.7218	1.6723	1.6334	1.6016	1.5748	1.5517	1.5315	1.5135
	$C_T\ 10^3$	[–]	5.4568	9.1062	10.4050	9.7897	8.1089	6.2728	4.9354	4.3119	4.2386	4.5820
	S	[m²]	436.54	436.54	436.54	436.54	436.54	436.54	436.54	436.54	436.54	436.54
	$(R_T/\Delta)_{500t}$	[–]	0.0051	0.0191	0.0387	0.0569	0.0679	0.0715	0.0735	0.0813	0.0986	0.1290
	R_T	[kN]	24.9	93.5	190.0	279.3	333.1	350.8	360.5	398.6	483.7	632.7
	P_E	[kW]	112.5	633.7	1716.4	3154.1	4514.4	5545.5	6513.0	8101.8	10924.8	15718.8
	τ	[deg]	0.310	0.271	0.883	1.540	1.962	2.110	2.090	2.061	2.143	2.325

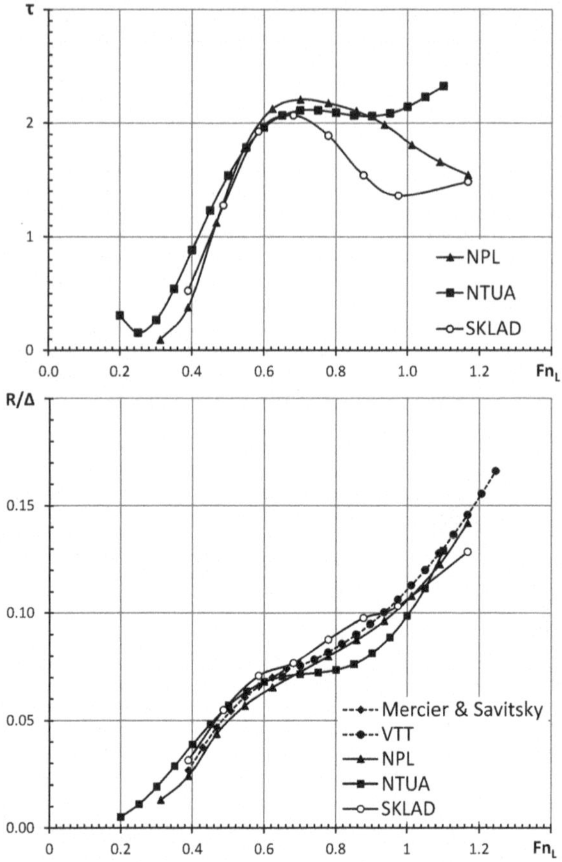

Fig. 8.33 The Predictions of five MMs (4 round bilge, 1 double-chine-NTUA) applicable for a typical 500 t mega yacht (L_{wl} = 52 m, B_{wl} = 9.3 m, T = 2.55 m, LCG = 23.4 m)

8.4.2 An Example of Semi-planing and Planing Hull Forms

Predictions of four MMs applicable for hard chine hull forms are shown in Fig. 8.34. The hull dimensions are typical for a 100000 lb (45.4 t) yacht (L_P = 23 m, B_{PX} = 4.1 m, LCG = 8.6 m), and hence the main non-dimensional parameters are $L_P/\nabla^{1/3} = 6.5$ and L_P/B_{PX} = 5.6 (as per Blount 2014). C_A = 0 for all cases. Note that these parameters are beyond the bounds of 62 & 65 MM, hence it is excluded in this example. ρ = 1.025 t/m^3, g = 9.81 m/s^2, ν = 1.1907E−6 m^2/s, C_F according to ITTC-1957 and C_A = 0 for all MMs throughout this example.

Additional elaboration is needed for the Series 62 MMs because the input parameters are not the same as for other MMs. That is, (a) the planing area coefficient $A_P/\nabla^{2/3}$ is used instead of the slenderness ratio $L_P/\nabla^{1/3}$ (relationship

$A_P/\nabla^{2/3} = f(L_P/\nabla^{1/3}, L_P/B_{PX})$ is given in Sect. 8.3.4), and (b) in order to cover the deadrise span of other series, two Series 62 curves are evaluated, for $\beta = 16$ and 22.5° (Tables 8.41 (a) and (b) respectively), see Fig. 8.34.

Note that the MM for Series 50 predicts the baseline trim (τ_{BL}). Hydrodynamic trim (shown in Fig. 8.34) is obtained as τ_{BL}-2° (for this particular example 2° static trim is assumed).

Table 8.39 Results for the USCG & TUNS MM

	Fn_∇	[–]	0.60	1.00	1.40	1.80	2.20	2.60	3.00	3.60
MM	$(R_T/\Delta)_{100000}$	[–]	0.0132	0.0451	0.0741	0.0861	0.097	0.1106	0.1288	0.1372
	L_K/L	[–]	0.9618	0.9384	0.9007	0.8555	0.8096	0.7698	0.7429	0.7416
	(S)	[–]	6.442	6.496	6.512	6.465	6.323	6.055	5.649	4.867
Δ = 100000 lb (45.359t)	v	[m/s]	3.53	5.89	8.25	10.60	12.96	15.31	17.67	21.20
	V	[kn]	6.87	11.45	16.03	20.61	25.19	29.77	34.35	41.22
	L_K	[m]	22.12	21.58	20.72	19.68	18.62	17.70	17.09	17.06
	S	[m^2]	80.60	81.27	81.47	80.88	79.10	75.75	70.67	60.89
	R_T	[kN]	5.9	20.1	33.0	38.3	43.2	49.2	57.3	61.1
	P_E	[kW]	20.7	118.3	272.1	406.3	559.6	754.1	1013.0	1294.9

Table 8.40 Results for the Series 50 MM

	Fn_∇	[–]	1.16	1.94	2.73	3.12	3.51	3.91	4.30	5.08
MM	R_R/Δ	[–]	0.0554	0.0942	0.1069	0.1032	0.0967	0.0894	0.0834	0.0803
	τ_{BL}	[deg]	3.284	4.740	6.664	6.792	6.366	5.933	5.562	4.9251
	L_M/L_P	[–]	3.4164	2.7824	2.3887	2.2436	2.1249	2.0301	1.9586	1.8885
Δ = 100000 lb (45.359t)	v	[m/s]	6.81	11.44	16.07	18.38	20.69	23.01	25.32	29.95
	V	[kn]	13.24	22.24	31.23	35.73	40.23	44.72	49.22	58.21
	L_M	[m]	19.92	16.22	13.93	13.08	12.39	11.84	11.42	11.01
	S	[m^2]	78.58	63.99	54.94	51.60	48.87	46.69	45.05	43.44
	Rn 10^{-8}	[–]	1.1398	1.5587	1.8794	2.0195	2.1533	2.2872	2.4286	2.7697
	C_F 10^3	[–]	2.0444	1.9557	1.9053	1.8865	1.8699	1.8546	1.8395	1.8070
	R_F	[kN]	3.82	8.39	13.85	16.85	20.05	23.48	27.22	36.07
	R_R	[kN]	24.68	41.92	47.57	45.95	43.03	39.81	37.10	35.73
	R_T	[kN]	28.50	50.31	61.42	62.80	63.08	63.29	64.32	71.80
	P_E	[kW]	194.2	575.5	986.6	1154.1	1305.2	1456.1	1628.5	2150.2
	$(R_T/\Delta)_{100000}$	[–]	0.0640	0.1130	0.1379	0.1410	0.1416	0.1421	0.1444	0.1612
	τ	[deg]	1.284	2.740	4.664	4.792	4.366	3.933	3.562	2.925

Table 8.41 Results for the Series 62 MM (multiple output) (a) β = 16° (b) β = 22.5°

(a)	Fn_∇	[–]	1.00	1.50	2.00	2.50	3.00	3.50	4.00
MM	$(R_T/\Delta)_{100000}$	[–]	0.0444	0.0903	0.109	0.1264	0.1355	0.1402	0.1472
	τ	[deg]	0.5502	2.2143	2.7265	3.5481	3.7373	3.2996	2.7822
	L_M/L_P	[–]	0.9107	0.8183	0.7495	0.6924	0.6433	0.6004	0.5628
	(S)	[–]	7.007	6.517	5.791	5.105	4.502	3.983	3.545
Δ = 100000 lb (45.359t)	v	[m/s]	5.89	8.83	11.78	14.72	17.67	20.61	23.56
	V	[kn]	11.45	17.17	22.90	28.62	34.35	40.07	45.80
	L_M	[m]	20.95	18.82	17.24	15.93	14.80	13.81	12.94
	S	[m²]	24.79	23.06	20.49	18.06	15.93	14.09	12.54
	R_T	[kN]	19.7	40.2	48.5	56.3	60.3	62.4	65.5
	P_E	[kW]	116.3	355.3	571.4	828.3	1066.0	1286.4	1543.3

(b)	Fn_∇	[–]	1.00	1.50	2.00	2.50	3.00	3.50	4.00
MM	$(R_T/\Delta)_{100000}$	[–]	0.0447	0.0903	0.1087	0.1269	0.1378	0.1454	0.1568
	τ	[deg]	0.5171	2.1505	2.4937	3.1538	3.3977	3.0666	2.6272
	L_M/L_P	[–]	0.9107	0.8183	0.7495	0.6924	0.6433	0.6004	0.5628
	(S)	[–]	7.007	6.517	5.791	5.105	4.502	3.983	3.545
Δ = 100000 lb (45.359t)	v	[m/s]	5.89	8.83	11.78	14.72	17.67	20.61	23.56
	V	[kn]	11.45	17.17	22.90	28.62	34.35	40.07	45.80
	L_M	[m]	20.95	18.82	17.24	15.93	14.80	13.81	12.94
	S	[m²]	24.79	23.06	20.49	18.06	15.93	14.09	12.54
	R_T	[kN]	19.9	40.2	48.4	56.5	61.3	64.7	69.8
	P_E	[kW]	117.1	355.2	570.0	831.9	1083.5	1334.2	1644.1

Table 8.42 Results for the NSS MM

	Fn_∇	[–]	1.00	1.50	2.00	2.50	3.00	3.50	4.00	4.25
MM	$(R_T/\Delta)_{100000}$	[–]	0.0591	0.0784	0.0975	0.1198	0.1424	0.1545	0.1669	0.1744
	τ	[deg]	1.7407	2.5151	2.7365	3.0608	4.0363	4.2438	4.0653	3.9898
	L_M/L_P	[–]	0.9355	0.9334	0.9256	0.9002	0.8443	0.7972	0.7813	0.7798
	(S)	[–]	7.648	7.413	6.966	6.315	5.499	4.771	4.423	4.350
Δ = 100000 lb (45.359t)	v	[m/s]	5.89	8.84	11.78	14.73	17.67	20.62	23.56	25.03
	V	[kn]	11.45	17.18	22.90	28.63	34.35	40.08	45.80	48.67
	L_M	[m]	21.52	21.47	21.29	20.70	19.42	18.33	17.97	17.94
	S	[m²]	27.06	26.23	24.65	22.34	19.46	16.88	15.65	15.39
	R_T	[kN]	26.3	34.9	43.4	53.4	63.4	68.8	74.4	77.7
	P_E	[kW]	155.0	308.5	511.8	785.8	1120.8	1419.1	1752.1	1945.1

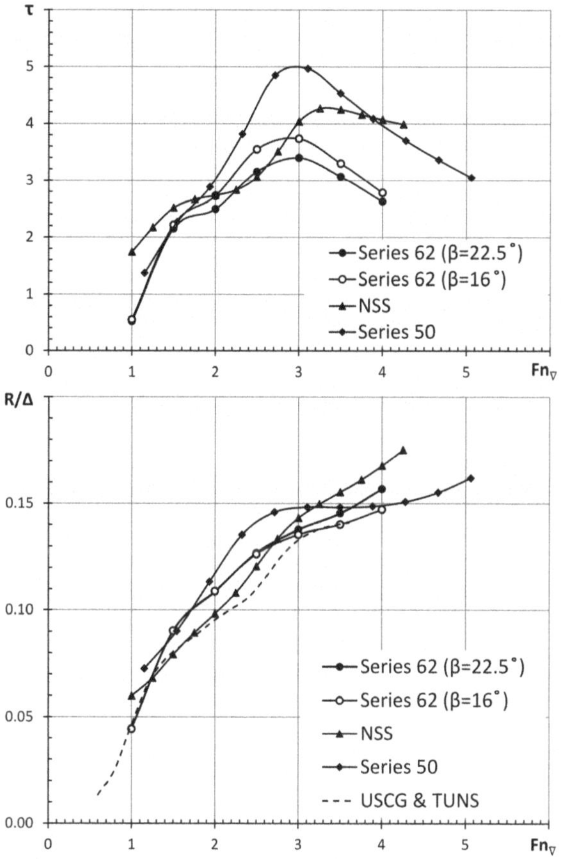

Fig. 8.34 The Predictions of four MMs applicable for a typical 100000 lb (45.4 t) yacht (L_P = 23 m, B_{PX} = 4.1 m, LCG = 8.6 m)

References

Bailey D (1982) A statistical analysis of propulsion data obtained from models of high speed round bilge hulls. RINA Symposium on Small Fast Warships and Security Vessels, London

Blount DL (2014) Performance by design. ISBN 0-978-9890837-1-3

Lahtiharju E, Karppinen T, Hellevaara M, Aitta T (1991) Resistance and seakeeping characteristics of fast transom stern hulls with systematically varied form (trans: SNAME), vol 99

Mercier JA, Savitsky D (1973) Resistance of transom-stern craft in the pre-planing regime. Davidson Laboratory Report 1667

Radojčić D (1985) An approximate method for calculation of resistance and trim of the planing hulls. University of Southampton, Ship Science Report No. 23. Paper presented on SNAME symposium on powerboats, September 1985

Radojčić D, Kalajdžić M (2018) Resistance and trim modeling of naples hard chine systematic series. RINA Trans Int J Small Craft Technol. https://doi.org/10.3940/rina.ijsct.2018.b1.211

Radojčić D, Rodić T, Kostić N (1997) Resistance and trim predictions for the NPL high speed round bilge displacement hull series. RINA conference on power, performance and operability of small craft, Southampton

Radojčić D, Prinčevac M, Rodić T (1999) Resistance and trim predictions for the SKLAD semidisplacement hull series. Oceanic Eng Int 3(1)

Radojčić D, Grigoropoulos GJ, Rodić T, Kuvelić T, Damala DP (2001) The resistance and trim of semi-displacement, double-chine, transom-stern hull series. In: Proceedings of 6th international conference on Fast Sea Transportation (FAST 2001), Southampton

Radojčić D, Zgradić A, Kalajdžić M, Simić A (2014a) Resistance prediction for hard chine hulls in the pre-planing regime. Polish Marit Res 21(2(82)) (Gdansk)

Radojčić D, Morabito M, Simić A, Zgradić A (2014b) Modeling with regression analysis and artificial neural networks the resistance and trim of series 50 experiments with V-bottom motor boats. J Ship Prod Design 30(4)

Radojčić DV, Kalajdžić MD, Zgradić AB, Simić AP (2017) Resistance and trim modeling of systematic planing hull series 62 (with 12.5, 25 and 30 degrees deadrise angles) using artificial neural networks, Part 2: mathematical models. J Ship Prod Design 33(4)

Chapter 9
Propeller's Open-Water Efficiency Prediction

9.1 Programs—VBA Codes for Microsoft Excel

Two examples that explain the application of typical MMs for evaluation of propeller's hydrodynamic performance are given.

The first example, for the regression based MMs for the AEW propeller series for non-cavitating and cavitating conditions, is given in Sect. 9.1.1. All necessary data for the AEW propeller series are given in Sect. 9.2.2. VBA codes for other regression based methods for estimation of hydrodynamic characteristics treated here are similar.

The second example, for the ANN based MMs for the Newton-Rader propeller series (cavitating conditions) is given in Sect. 9.1.2. This example is given for evaluation of K_T only; the same procedure applies for evaluation of K_Q, except that the K_Q coefficients should be applied. All necessary data for the Newton-Rader propeller series are given in the Sect. 9.2.5. VBA codes for other ANN based methods for estimation of hydrodynamic characteristics are similar.

Both examples begin with the flow charts (Figs. 9.1 and 9.6) which describe the structure of the VBA macro code. VBA code follows, and then the tables that contain the required input data, MM's coefficients, and output data, all corresponding to a specific method, are presented. Comments that explain the significant steps in the code are included too; they are marked with an apostrophe " ' " at the beginning of each line. Note that since VBA code reads the data required for calculation from the specified cells, each table containing the input data, MM coefficients, and output data must be placed at the exact position in the Excel Sheet 1, as indicated in Figs. 9.2, 9.3, 9.4, 9.5 and Figs. 9.7, 9.8, 9.9, for the first and second example, respectively.

Electronic supplementary material The online version of this chapter (https://doi.org/10.1007/978-3-030-30607-6_9) contains supplementary material, which is available to authorized users.

D. Radojčić et al., *Power Prediction Modeling of Conventional High-Speed Craft*,
https://doi.org/10.1007/978-3-030-30607-6_9

9.1.1 An Example of Regression Based Mathematical Models

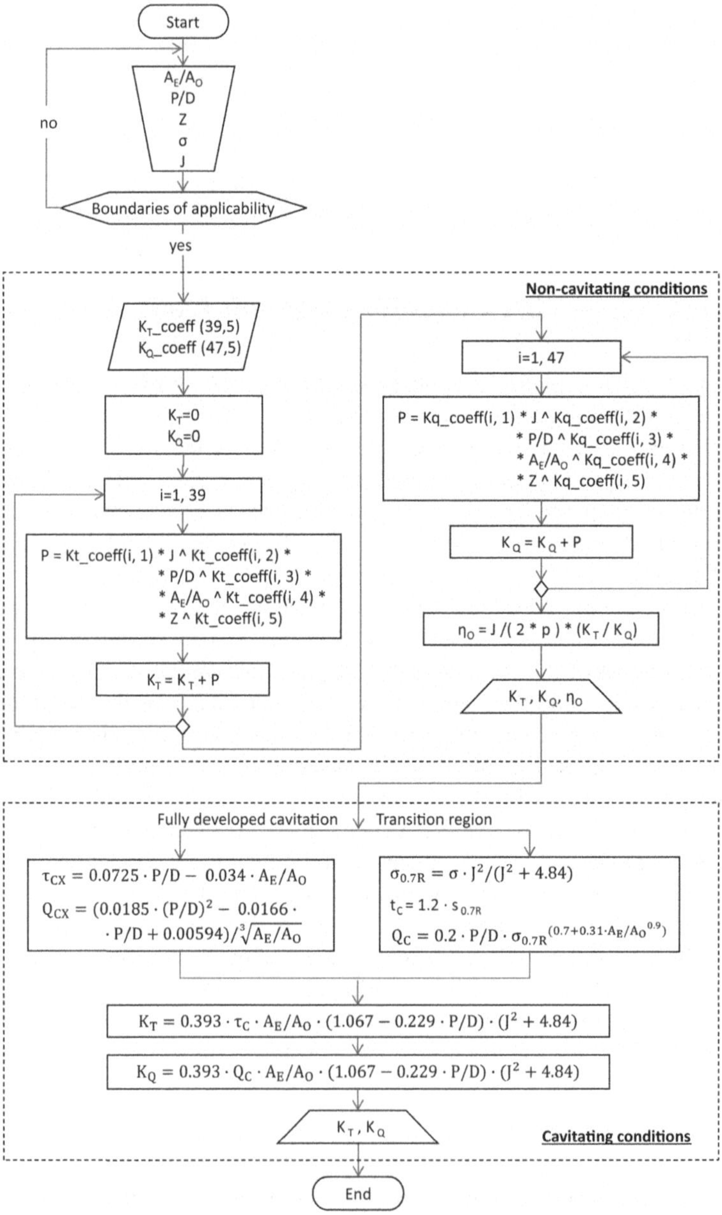

Fig. 9.1 Flow chart for regression based mathematical model (Example for the AEW propeller series for non-cavitating and cavitating conditions)

Example of VBA Macro Code for Regression Based Mathematical Model (AEW Propeller Series, Non-Cavitating and Cavitating Conditions)

```
Sub Program()
Dim Kt_coeff(39, 5), Kq_coeff(47, 5)

'Clear all previous results
Sheet1.Cells.Range("C67", "M81") = ""

'Loading independent variables:
P_D = Sheet1.Cells(6, 3)
Ae_Ao = Sheet1.Cells(7, 3)
Z = Sheet1.Cells(8, 3)
Sigma = Sheet1.Cells(9, 3)

'Loading regression coefficients
For i = 1 To 47
For k = 1 To 5
If i < 40 Then Kt_coeff(i, k) = Sheet1.Cells(i + 14, k + 1)
Kq_coeff(i, k) = Sheet1.Cells(i + 14, k + 7)
Next k
Next i

'CALCULATION OF THE PROPELLER HYDRODYNAMIC CHARACTERISTICS
'(non-cavitating conditions)

'Maximal value of J depend on KT (until KT become negative)
'If the calculation step is decreased, KT curve will become smoother
lok = 3
For J = 0 To 1.6 Step 0.15

'Calculate thrust coefficient
sum_Kt = 0
For i = 1 To 39
P = Kt_coeff(i, 1) * J ^ Kt_coeff(i, 2) * P_D ^ Kt_coeff(i, 3)
P = P * Ae_Ao ^ Kt_coeff(i, 4) * Z ^ Kt_coeff(i, 5)
sum_Kt = sum_Kt + P
Next i

'Calculate torque coefficient
sum_Kq = 0
For i = 1 To 47
P = Kq_coeff(i, 1) * J ^ Kq_coeff(i, 2) * P_D ^ Kq_coeff(i, 3)
P = P * Ae_Ao ^ Kq_coeff(i, 4) * Z ^ Kq_coeff(i, 5)
sum_Kq = sum_Kq + P
Next i
```

```
'Calculate open water efficiency
Eta_O = J * sum_Kt / sum_Kq / 2 / 3.141592

'Write results for non-cavitating conditions
Sheet1.Cells(67, lok) = J
Sheet1.Cells(68, lok) = sum_Kt
Sheet1.Cells(69, lok) = sum_Kq
Sheet1.Cells(70, lok) = Eta_O

'CALCULATION OF THE PROPELLER HYDRODYNAMIC CHARACTERISTICS
'(transition region and fully developed cavitation)

'Calculate K_T and K_Q - transition region
Sigma07R = Sigma * (J ^ 2 / (J ^ 2 + 4.84))
Tau_C = 1.2 * Sigma07R
Q_C = 0.2 * P_D * Sigma07R ^ (0.7 + 0.31 * Ae_Ao ^ 0.9)
Kt = 0.393 * Tau_C * Ae_Ao * (1.067 - 0.229 * P_D) * (J ^ 2 + 4.84)
Kq = 0.393 * Q_C * Ae_Ao * (1.067 - 0.229 * P_D) * (J ^ 2 + 4.84)

'Write results - transition region
Sheet1.Cells(72, lok) = Sigma07R
Sheet1.Cells(73, lok) = Tau_C
Sheet1.Cells(74, lok) = Q_C
Sheet1.Cells(75, lok) = Kt
Sheet1.Cells(76, lok) = Kq

'Calculate K_T and K_Q - fully developed cavitation
Tau_C = 0.0725 * P_D - 0.034 * Ae_Ao
Q_C = (0.0185 * P_D ^ 2 - 0.0166 * P_D + 0.00594) / Ae_Ao ^ (1 / 3)
Kt = 0.393 * Tau_C * Ae_Ao * (1.067 - 0.229 * P_D) * (J ^ 2 + 4.84)
Kq = 0.393 * Q_C * Ae_Ao * (1.067 - 0.229 * P_D) * (J ^ 2 + 4.84)

'Write results - fully developed cavitation
Sheet1.Cells(78, lok) = Tau_C
Sheet1.Cells(79, lok) = Q_C
Sheet1.Cells(80, lok) = Kt
Sheet1.Cells(81, lok) = Kq

lok = lok + 1
Next J

End Sub
```

Note: All data must be in the Sheet1 (Excel), and at the exact position as indicated in Figs. 9.2, 9.3 and 9.4.

	A	B	C	D
1		**INPUT DATA**		
2				
3		**P/D**	*1.4*	
4		**Ae/Ao**	*0.65*	
5		**Z**	*3*	
6		**σ**	*0.7*	
7				

Fig. 9.2 Sample of input data

	A	B	C	D	E	F	G	H	I	J	K	L	M
12		**Regression coefficients**											
13													
14		K_T coefficients						K_Q coefficients					
15		-0.05586363	0	0	0	0		0.00515898	0	0	0	0	
16		-0.21730109	1	0	0	0		0.01606668	2	0	0	0	
17		0.26053140	0	1	0	0		-0.04411530	1	1	0	0	
18		0.15811400	0	2	0	0		0.00682223	0	2	0	0	
19		-0.14758100	2	0	1	0		-0.04088110	0	1	1	0	
20		-0.48149700	1	1	1	0		-0.07732967	1	1	1	0	
21		0.37812278	0	2	1	0		-0.08853810	2	1	1	0	
50		0.00069090	0	0	1	2		-0.00183491	1	1	0	2	
51		0.00421749	0	3	1	2		0.00011245	3	2	0	2	
52		0.00005652	3	6	1	2		-0.00002972	3	6	0	2	
53		-0.00146564	0	3	2	2		0.00026955	1	0	1	2	
54								0.00083265	2	0	1	2	
55								0.00155334	0	2	1	2	
56								0.00030268	0	6	1	2	
57								-0.00018430	0	0	2	2	
58								-0.00042540	0	3	2	2	
59								0.00008692	3	3	2	2	
60								-0.00046590	0	6	2	2	
61								0.00005542	1	6	2	2	

Fig. 9.3 Regression coefficient for K_T and K_Q coefficients (See Table 9.3. for K_T and K_Q)

	A	B	C	D	E	F	G	H	I	J	K	L	M	N
64		**OUTPUT**												
65														
66		**Non-cavitating conditions**												
67		J	0.0000	0.1500	0.3000	0.4500	0.6000	0.7500	0.9000	1.0500	1.2000	1.3500	1.5000	
68		K_T	0.6433	0.5974	0.5430	0.4816	0.4151	0.3451	0.2735	0.2018	0.1320	0.0656	0.0044	
69		K_Q	0.1319	0.1232	0.1130	0.1016	0.0893	0.0763	0.0630	0.0496	0.0365	0.0239	0.0120	
70		η_O	0.0000	0.1158	0.2294	0.3394	0.4439	0.5398	0.6218	0.6797	0.6909	0.5907	0.0876	
71		**Cavitating conditions - transition region**												
72		$\sigma_{0.7R}$	0.0000	0.0032	0.0128	0.0281	0.0485	0.0729	0.1004	0.1299	0.1605	0.1915	0.2221	
73		τ_C	0.0000	0.0039	0.0153	0.0337	0.0582	0.0875	0.1204	0.1558	0.1926	0.2298	0.2666	
74		Q_C	0.0000	0.0015	0.0053	0.0108	0.0178	0.0258	0.0345	0.0437	0.0530	0.0622	0.0712	
75		K_T	0.0000	0.0036	0.0144	0.0324	0.0577	0.0901	0.1297	0.1766	0.2306	0.2919	0.3604	
76		K_Q	0.0000	0.0014	0.0050	0.0104	0.0176	0.0266	0.0372	0.0495	0.0634	0.0790	0.0962	
77		**Cavitating conditions - fully developed cavitation**												
78		τ_C	0.0794	0.0794	0.0794	0.0794	0.0794	0.0794	0.0794	0.0794	0.0794	0.0794	0.0794	
79		Q_C	0.0219	0.0219	0.0219	0.0219	0.0219	0.0219	0.0219	0.0219	0.0219	0.0219	0.0219	
80		K_T	0.0733	0.0736	0.0746	0.0763	0.0787	0.0818	0.0855	0.0900	0.0951	0.1009	0.1073	
81		K_Q	0.0202	0.0203	0.0206	0.0210	0.0217	0.0225	0.0236	0.0248	0.0262	0.0278	0.0296	
82														

Fig. 9.4 Sample of output data

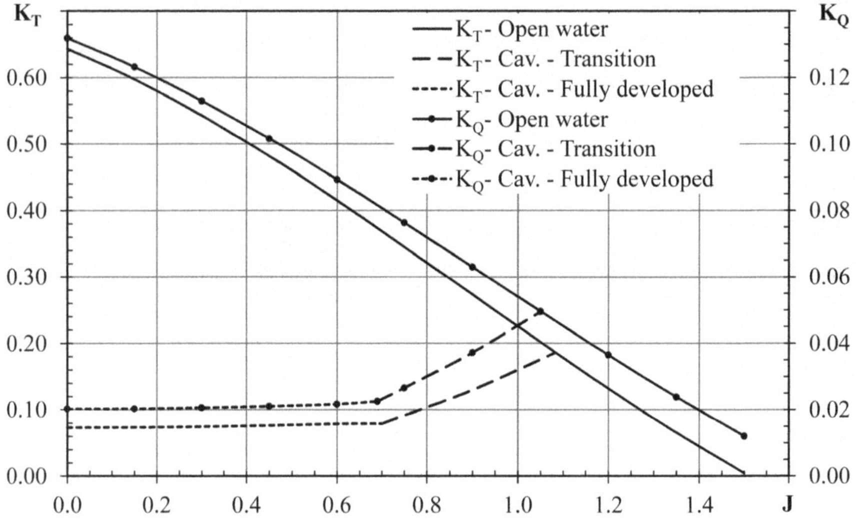

Fig. 9.5 Sample of open water K_T and K_Q chart

9.1.2 An Example of ANN Based Mathematical Models

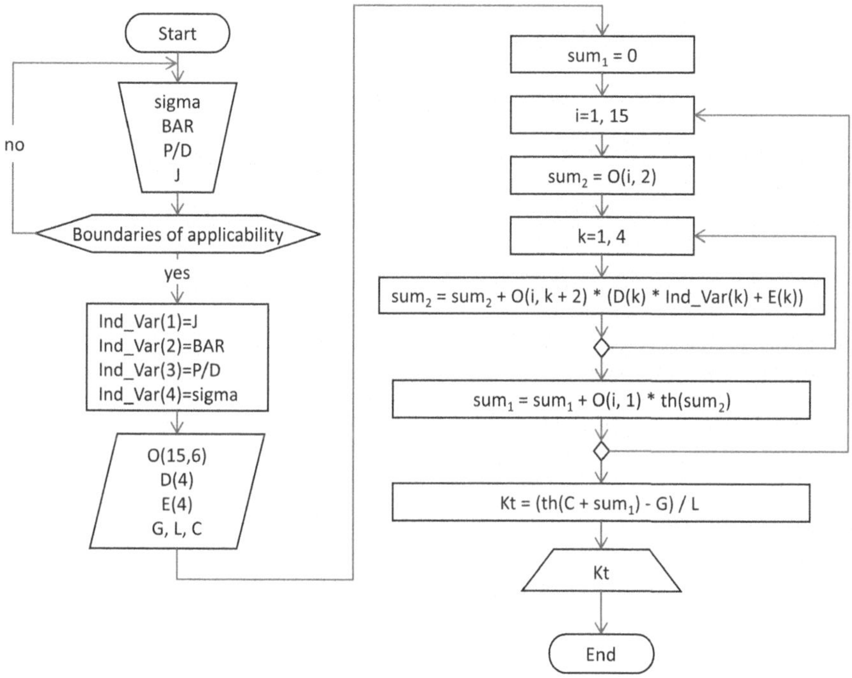

Fig. 9.6 Flow chart of ANN based mathematical model (Example for Newton-Rader propeller series)

Example of VBA Macro Code for ANN Based Mathematical Model (Newton-Rader Series, Cavitating Conditions)

```
Sub Program()
Dim O(15, 6), D(4), E(4), Ind_Var(4)
'Clear all previous results
Sheet1.Cells.Range("C35", "M36") = ""

'Input Independent Variables:
'Nomenclature: Ind_Var(1)=J
'Nomenclature: Ind_Var(2)=BAR
'Nomenclature: Ind_Var(3)=P/D
'Nomenclature: Ind_Var(4)=sigma
Ind_Var(2) = Sheet1.Cells(6, 3)
Ind_Var(3) = Sheet1.Cells(7, 3)
Ind_Var(4) = Sheet1.Cells(8, 3)

'Loading ANN MM coefficients
For i = 1 To 15
For J = 1 To 6
O(i, J) = Sheet1.Cells(i + 15, J + 1)
Next J
Next i
For i = 1 To 4
D(i) = Sheet1.Cells(i + 15, 9)
E(i) = Sheet1.Cells(i + 15, 10)
Next i
G = Sheet1.Cells(16, 11)
L = Sheet1.Cells(16, 12)
C = Sheet1.Cells(16, 13)

'MATHEMATICAL MODEL for KT (cavitating condition)

'Maximal value of J depends on KT (until KT becomes negative)
'If the calculation step is decreased, KT curve will become smoother.
J_min = Ind_Var(3) * (1.08 * Ind_Var(4) ^ 2 - 1.585 * Ind_Var(4) + 1.013)
lok = 3
For J = J_min To 2 Step 0.1
Ind_Var(1) = J

'Calculate thrust coefficient (cavitating conditions)
sum1 = 0
For i = 1 To 15
sum2 = O(i, 2)
  For k = 1 To 4
```

```
        sum2 = sum2 + O(i, k + 2) * (D(k) * Ind_Var(k) + E(k))
        Next k
    sum1 = sum1 + O(i, 1) * th(sum2)
    Next i
    Kt_cav = (th(C + sum1) - G) / L

'Write results
Sheet1.Cells(35, lok) = J
Sheet1.Cells(36, lok) = Kt_cav
If Kt_cav <= 0 Then End
lok = lok + 1
Next J

End Sub

'Calculate hyperbolic tangent function
Function th(arg)
th = (Exp(2 * arg) - 1) / (Exp(2 * arg) + 1)
End Function
```

	A	B	C	D
1		**INPUT DATA**		
2				
3		**BAR**	*0.8*	
4		**P/D**	*1.50*	
5		**σ**	*0.50*	
6				

Fig. 9.7 Sample of input data

	A	B	C	D	E	F	G	H	I	J	K	L	M	N
12		MM Coefficients												
13														
14		Thrust coefficient K_T at cavitating condition												
15		Oi	ai	h1	h2	h3	h4		Dk	Ek	G	H	C	
16		1.3173542	-1.1486914	-0.0877314	0.0930062	0.4013664	0.2672774		1.0158013	-1.4079006	-0.7683440	3.2913985	-0.3777408	
17		-0.7381216	-0.1937162	0.1365919	0.2123235	0.0867300	0.9730156		3.8297873	-2.7382977				
18		-0.9416671	0.6080810	-2.3712103	-0.1302717	1.0149868	-0.8226167		1.7307693	-2.7000000				
19		-0.8044308	-0.1007714	1.3385242	-0.3342977	-0.8950956	0.7887833		2.4000000	-1.5000000				
20		-0.7276427	0.9139260	-4.3714209	-0.0342357	2.6848557	0.9683161							
21		0.9334061	-0.3992116	-5.1717801	-0.5460135	1.7492508	-1.0320357							
22		-0.4368628	-0.0155451	-2.0022478	0.5081350	1.5426202	-0.6679885							
23		-0.7448840	-0.6181244	-0.4951757	0.3894272	-0.6724022	0.1578318							
24		0.7119366	0.8126080	1.9200073	0.2217853	-0.6562962	1.2139996							
25		0.6704036	-0.5812509	-0.1461492	1.3636353	-0.8197115	0.0973382							
26		1.4418043	1.2027683	-4.5123234	0.0197586	2.7100451	0.7360260							
27		0.9745181	0.3642887	-2.8521757	-0.1744290	1.1859676	-0.2120958							
28		1.0017167	0.3914023	4.3152881	0.5462066	-1.3390867	1.0693594							
29		-0.5880054	-0.8103508	-0.7641475	1.5753068	-0.4050508	0.1446425							
30		0.6959038	-0.7087057	-1.6359380	0.4502301	0.1425356	0.2437267							
31														

Fig. 9.8 ANN coefficients for K_T (See Table 9.7)

	A	B	C	D	E	F	G	H	I	J	K	L	M
33		OUTPUT											
34													
35		J	0.7358	0.8358	0.9358	1.0358	1.1358	1.2358	1.3358	1.4358	1.5358	1.6358	
36		K_T	0.1458	0.1581	0.1743	0.1937	0.2100	0.2067	0.1720	0.1132	0.0414	-0.0104	
37													

Fig. 9.9 Sample of output data

9.2 Recommended MMs for High Speed Propellers

This section contains tabulated data of selected MMs necessary for evaluation of propeller's hydrodynamic characteristics, i.e. thrust and torque coefficients from which open water efficiency is derived. Data for each MM is presented as follows:

- **Name** of the Mathematical Model (MM)
- **Reference** (including cross-reference with other sections of this book)
- **Input** (independent variables)—*J*, *P/D*. *EAR/DAR* and/or *z* and $\sigma/\sigma_{0.7R}$
- **Output** (dependent or target variables)—K_T and K_Q
- **Blade outline** of the series that the MM is based on
- **Boundaries of applicability** for a given MM
- **Regression or ANN coefficients** for MMs.

Recommended MMs for high speed propellers are presented as follows:

- **Section 9.2.1—B Series**. Regression based MM for non-cavitating conditions for speeds up to 25–30 kn. Regression coefficients for evaluation of K_T and K_Q are given in Table 9.2.
- **Section 9.2.2—AEW Series**. Regression based MM for 3- and 4-bladed propellers for non-cavitating and cavitating conditions. Regression coefficients for evaluation of K_T and K_Q for non-cavitating conditions are given in Table 9.3. Characteristics for transition region and fully developed cavitation may be evaluated separately from simple expressions, see Sect. 4.3 and Table 9.1.
- **Section 9.2.3—KCA Series (RA)**. Regression based MM for (a) non-cavitating, and (b) cavitating conditions. Regression coefficients for evaluation of K_T and K_Q for non-cavitating conditions are given in Table 9.4. Characteristics for cavitating conditions are evaluated from $K_{Tcav} = K_T - \Delta K_T$ and $K_{Qcav} = K_Q - \Delta K_Q$. Regression coefficients for evaluation of ΔK_T and ΔK_Q are given in Tables 9.5.
- **Section 9.2.4—KCA Series (ANN)**. ANN based, single-function MM for transition and cavitating conditions. ANN coefficients for evaluation of K_T and K_Q in both regimes are given in Table 9.6a and b respectively.
- **Section 9.2.5—Newton-Rader Series**. ANN based MM for non-cavitating and cavitating conditions for speeds of 50+ kn. Two separate MMs are presented; the first one for non-cavitating conditions only, and the second single-function MM for transition and cavitating conditions. Their ANN coefficients for evaluation of K_T and K_Q are given in Tables 9.7, 9.8a and b respectively.

Recommended MMs for B, AEW, KCA, and the Newton-Rader series are summarized in Table 9.1.

Table 9.1 Recommended MMs for evaluation of thrust K_T and torque K_Q coefficients

Name Refer.	Mathematical Model	Boundaries of Applicability	Remarks
B series (see 4.1.1, 9.2.1) Oosterveld & Van Oossanen (1975)	For non-cavitating conditions (Based on B series data, Fig. 9.10) $K_T = \sum_{i=1}^{39} (C_{Ti} \cdot (J)^{s_i} \cdot (P/D)^{t_i} \cdot (EAR)^{u_i} \cdot (z)^{v_i})$ $K_Q = \sum_{i=1}^{47} (C_{Qi} \cdot (J)^{s_i} \cdot (P/D)^{t_i} \cdot (EAR)^{u_i} \cdot (z)^{v_i})$ C_T, C_Q, s, t, u, v - coefficients for non-cavitating conditions Valid for R_n=2·10⁶, for other R_n corrections ΔK_T and ΔK_Q	For non-cavitating conditions z = 3 - 6 $0.5 < P/D < 1.4$ $0.35 < EAR < 1.05$ Extreme values should be avoided	Advantages: By far the largest and best known fixed-pitch, general purpose series. Used for benchmarking purposes. Disadvantages: For speeds up to 20–30 kn (inner blade sections aerofoil, outer segment.). Fig. 9.10.
AEW series (see 4.2.1, 9.2.2, Model 1, and Sect. 4.3) Blount & Hubble (1981)	For non-cavitating conditions (Based on AEW data, Fig. 9.11) $K_T = \sum_{i=1}^{39} (C_{Ti} \cdot (J)^{s_i} \cdot (P/D)^{t_i} \cdot (EAR)^{u_i} \cdot (z)^{v_i})$ $K_Q = \sum_{i=1}^{47} (C_{Qi} \cdot (J)^{s_i} \cdot (P/D)^{t_i} \cdot (EAR)^{u_i} \cdot (z)^{v_i})$ C_T, C_Q, s, t, u, v - coefficients for non-cavitating conditions For cavitating conditions (Based on Blount and Fox 1978) $K_T = 0.393 \cdot \tau_c \cdot EAR \cdot (1.067 - 0.229 \cdot P/D) \cdot (J^2 + 4.84)$ $K_Q = 0.393 \cdot Q_c \cdot EAR \cdot (1.067 - 0.229 \cdot P/D) \cdot (J^2 + 4.84)$ For transition region: $\tau_c = 1.2 \cdot \sigma_{0.7R}$ $Q_c = 0.2 \cdot P/D \cdot \sigma_{0.7R}{}^{(0.7+0.31 \cdot EAR^{0.9})}$ For fully developed cavitation: $\tau_{cx} = 0.0725 \cdot P/D - 0.034 \cdot EAR$ $Q_{cx} = (0.0185 \cdot (P/D)^2 - 0.0166 \cdot P/D + 0.00594)/\sqrt[3]{EAR}$	For non-cavitating conditions z = 3 and 4 0.80 = P/D =1.40 0.50 = EAR =1.10 For cavitating conditions 0.60 = P/D =2.00 $\tau_{CX} \leq \tau_C$ $Q_{CX} \leq Q_C$ $\sigma_{0.7R} \leq 0.4$	Advantages: Applicable for 3- and 4-bladed segmental section propellers. MM valid for heavily cavitating conditions. Disadvantages: Transition zone, non-cav. to cav., not smooth. η_O=f(K_T, K_Q) not good for cavitating regime.

Table 9.1 (continued)

Name Refer.	Mathematical Model	Boundaries of Applicability	Remarks
KCA series (RA) (see 4.2.1, 9.2.3, Model 3) Radojčić (1988)	For non-cavitating conditions (Based on KCA data, Fig. 9.12) $K_T = \sum_{1}^{16}(C_T \cdot 10^e \cdot (DAR)^x \cdot (P/D)^y \cdot (J)^z)$ $K_Q = \sum_{1}^{17}(C_Q \cdot 10^e \cdot (DAR)^x \cdot (P/D)^y \cdot (J)^z)$ C_T, C_Q, e, x, y, z - coefficients for non-cavitating conditions Additional equations for cavitating conditions (Based on KCA data) $\Delta K_T = \sum_{1}^{20}(d_T \cdot 10^e \cdot (DAR)^s \cdot (P/D)^v \cdot (\sigma_{0.7R})^t \cdot (K_T)^u)$ $\Delta K_Q = \sum_{1}^{18}(d_Q \cdot 10^e \cdot (DAR)^s \cdot (P/D)^v \cdot (\sigma_{0.7R})^t \cdot (K_T)^u)$ d_T, d_Q, e, s, v, t, u - coefficients for cavitating conditions	For non-cavitating conditions $0.5 \le DAR \le 1.1$ $0.8 \le P/D \le 2.0$ $J \ge 0.3 \quad J_{min} = 0.492 \cdot P/D + 0.031$ $K_T \le \sqrt{J/2.5 - 0.1}$ For cavitating conditions for $\sigma < 1.00 \quad J_{min} = 0.571 \cdot P/D + 0.021$ for $\sigma \ge 1.00 \quad J_{min} = 0.492 \cdot P/D + 0.031$ $1.25 - 0.3 \cdot (DAR) - 0.2 \cdot \sigma \le P/D \le 1.8$ $K_T \le \sqrt{J/2.5 - 0.1} - (0.07/\sigma)$ $\sigma \ge 0.50$ $\Delta K_T \ge 0, \quad \Delta K_Q \ge 0$ $\eta_o \ge 0.20, \quad \eta_{atm} \ge \eta_{cav}$	Advantages: Good for non-cavitating 3-bladed segmental section propellers. Good representation of $\eta_O = f(K_T, K_Q)$. Disadvantages: Transition zone non-cav. to cav. not good enough.
KCA series (ANN) (see 4.2.1, 9.2.4, Model 4) Koushan (2007)	Equation for transition and cavitating conditions (Based on KCA data) $Y = \frac{tanh\left(C + \sum_{i=1}^{m}\left(O_i \cdot tanh\left(a_i + \sum_{k=1}^{m}\left(h_{k,i} \cdot (D_k \cdot X_k + E_k)\right)\right)\right)\right) - G}{H}$ Y - K_T or K_Q coefficient a, C, D, E, G, h, H, O - constants (depend whether Y is K_T or K_Q and if cavitation is present or not) X - Input variables in following order: J, DAR, P/D, and s 89 and 107 terms for K_T and K_Q respectively for transition and cavitating conditions	For transition and cavitating conditions $0.60 \le P/D \le 2.00$ $0.5 \le DAR \le 1.1$ $0.50 \le \sigma \le 2.00$ $J_{min} = 0.4847 \cdot P/D + 0.0339$ $\left(\frac{P}{D}\right)_{min} = 0.67 + 0.83 \cdot DAR^{-0.32} + 0.61 \cdot \sigma^2 - 1.64 \cdot \sigma$	Advantages: Good for transition zone. Single function for transition and cavitating regimes. Disadvantages: $\eta_O = f(K_T, K_Q)$ inconsistent

Table 9.1 (continued)

Name Refer.	Mathematical Model	Boundaries of Applicability	Remarks
Newton-Rader series (see 4.2.2, 9.2.5) Koushan (2005)	Separate equations for non-cavitating, and transition and cavitating conditions (Based on Newton-Rader data, Fig. 9.14) $Y = \frac{tanh\left(C + \sum_{i=1}^{m}\left(O_i \cdot tanh\left(a_i + \sum_{k=1}^{m}\left(h_{k,i} \cdot (D_k \cdot X_k + E_k)\right)\right)\right)\right) - G}{H}$ Y - K_T or K_Q coefficient a, C, D, E, G, h, H, O - constants (depend whether Y is K_T or K_Q and if cavitation is present or not) X - Input variables in following order: J, EAR, P/D and s 34 equation terms for both K_T and K_Q for non-cavitating conditions 101 equation terms for both K_T and K_Q for transition and cavitating conditions .	For both regimes $1.04 \leq P/D \leq 2.08$ $0.48 \leq EAR \leq 0.95$ $0.25 \leq \sigma \leq 1.00$ For non-cavitating conditions only $J_{min} = 0.538 \cdot P/D + 0.129$ For transition and cavitating conditions only $J_{min} = \frac{P}{D}(1.08 \cdot \sigma^2 - 1.585 \cdot \sigma + 1.013)$	Advantages: The only MM for Newton-Rader propeller series. Disadvantages: η_O=f(K_T, K_Q) inconsistent for both regimes.

9.2.1 B Series

Reference Oosterveld and Van Oossanen (1975); see Sect. 4.1.1 (Fig. 9.10).

INPUT
$X_1 = J$
$X_2 = P/D$
$X_3 = EAR$
$X_4 = z$

OUTPUT
K_T
K_Q

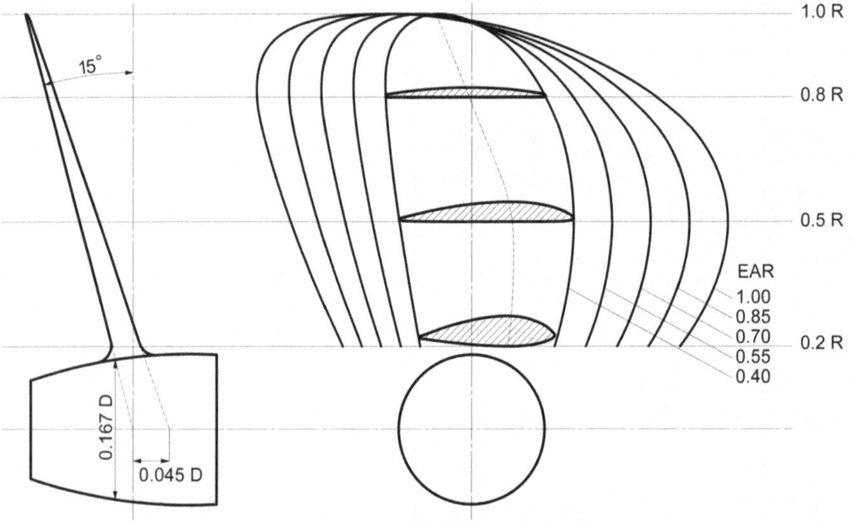

Fig. 9.10 Blade outline of the B series. Blade sections: inner aerofoil- and outer segmental-type

Boundaries of Applicability

$$z = 3\text{--}6$$
$$0.5 < P/D < 1.4$$
$$0.35 < EAR < 1.05$$

Note: Input parameters which are out of the bounds of data distribution should be avoided, see Fig. 4.1.

Table 9.2 B Series regression coefficients for K_T and K_Q (non-cavitating conditions)

$$K_T = \sum_{i=1}^{39} \left(C_{Ti} \cdot (X_1)^{s_i} \cdot (X_2)^{t_i} \cdot (X_3)^{u_i} \cdot (X_4)^{v_i} \right)$$

n	C_{Tn}	s	t	u	v
1	0.00880496	0	0	0	0
2	−0.20455400	1	0	0	0
3	0.16635100	0	1	0	0
4	0.15811400	0	2	0	0
5	−0.14758100	2	0	1	0
6	−0.48149700	1	1	1	0
7	0.41543700	0	2	1	0
8	0.01440430	0	0	0	1
9	−0.05300540	2	0	0	1
10	0.01434810	0	1	0	1
11	0.06068260	1	1	0	1
12	−0.01258940	0	0	1	1
13	0.01096890	1	0	1	1
14	−0.13369800	0	3	0	0
15	0.00638407	0	6	0	0
16	−0.00132718	2	6	0	0
17	0.16849600	3	0	1	0
18	−0.05072140	0	0	2	0
19	0.08545590	2	0	2	0
20	−0.05044750	3	0	2	0
21	0.01046500	1	6	2	0
22	−0.00648272	2	6	2	0
23	−0.00841728	0	3	0	1
24	0.01684240	1	3	0	1
25	−0.00102296	3	3	0	1
26	−0.03177910	0	3	1	1
27	0.01860400	1	0	2	1
28	−0.00410798	0	2	2	1
29	−0.00060685	0	0	0	2
30	−0.00498190	1	0	0	2
31	0.00259830	2	0	0	2
32	−0.00056053	3	0	0	2
33	−0.00163652	1	2	0	2
34	−0.00032879	1	6	0	2
35	0.00011650	2	6	0	2
36	0.00069090	0	0	1	2
37	0.00421749	0	3	1	2
38	0.00005652	3	6	1	2
39	−0.00146564	0	3	2	2

$$K_Q = \sum_{i=1}^{47} \left(C_{Qi} \cdot (X_1)^{s_i} \cdot (X_2)^{t_i} \cdot (X_3)^{u_i} \cdot (X_4)^{v_i} \right)$$

n	C_{Qn}	s	t	u	v
1	0.00379368	0	0	0	0
2	0.00886523	2	0	0	0
3	−0.03224100	1	1	0	0
4	0.00344778	0	2	0	0
5	−0.04088110	0	1	1	0
6	−0.10800900	1	1	1	0
7	−0.08853810	2	1	1	0
8	0.18856100	0	2	1	0
9	−0.00370871	1	0	0	1
10	0.00513691	0	1	0	1
11	0.02094490	1	1	0	1
12	0.00474319	2	1	0	1
13	−0.00723408	2	0	1	1
14	0.00438388	1	1	1	1
15	−0.02694030	0	2	1	1
16	0.05580820	3	0	1	0
17	0.01618860	0	3	1	0
18	0.00318086	1	3	1	0
19	0.01589600	0	0	2	0
20	0.04717290	1	0	2	0
21	0.01962830	3	0	2	0
22	−0.05027820	0	1	2	0
23	−0.03005500	3	1	2	0
24	0.04171220	2	2	2	0
25	−0.03977220	0	3	2	0
26	−0.00350024	0	6	2	0
27	−0.01068540	3	0	0	1
28	0.00110903	3	3	0	1
29	−0.00031391	0	6	0	1
30	0.00359850	3	0	1	1
31	−0.00142121	0	6	1	1
32	−0.00383637	1	0	2	1
33	0.01268030	0	2	2	1
34	−0.00318278	2	3	2	1
35	0.00334268	0	6	2	1
36	−0.00183491	1	1	0	2
37	0.00011245	3	2	0	2
38	−0.00002972	3	6	0	2
39	0.00026955	1	0	1	2
40	0.00083265	2	0	1	2
41	0.00155334	0	2	1	2
42	0.00030268	0	6	1	2
43	−0.00018430	0	0	2	2
44	−0.00042540	0	3	2	2
45	0.00008692	3	3	2	2
46	−0.00046590	0	6	2	2
47	0.00005542	1	6	2	2

9.2.2 AEW Series

Reference Blount and Hubble (1981); see Sect. 4.2.1 (Model 1) and Sect. 4.3 (Fig. 9.11).

INPUT	OUTPUT
$X_1 = J$	K_T
$X_2 = P/D$	K_Q
$X_3 = EAR$	
$X_4 = z$	

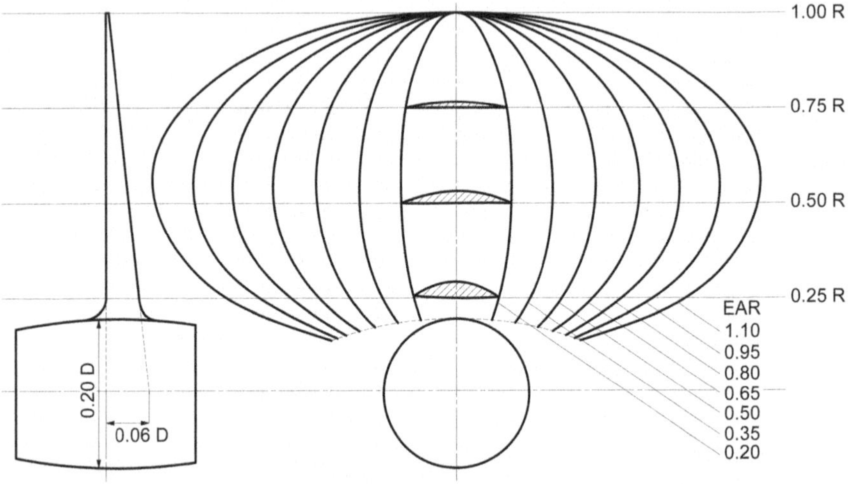

Fig. 9.11 Blade outline of the AEW series. Blade sections: segmental

Boundaries of Applicability

Non-cavitating conditions	Cavitating conditions
$z = 3$ and 4	$0.60 \leq P/D \leq 2.00$
$0.80 \leq P/D \leq 1.40$	$\tau_{CX} \leq \tau_C$
$0.50 \leq EAR \leq 1.10$	$Q_{CX} \leq Q_C$
	$\sigma_{0.7R} \leq 0.4$

Note: For cavitating conditions (transition region and fully developed cavitation) see Table 9.1.

Table 9.3 AEW Series regression coefficients for K_T and K_Q (non-cavitating conditions)

$$K_T = \sum_{i=1}^{39}\left(C_{Ti} \cdot (X_1)^{s_i} \cdot (X_2)^{t_i} \cdot (X_3)^{u_i} \cdot (X_4)^{v_i}\right) \quad K_Q = \sum_{i=1}^{47}\left(C_{Qi} \cdot (X_1)^{s_i} \cdot (X_2)^{t_i} \cdot (X_3)^{u_i} \cdot (X_4)^{v_i}\right)$$

n	C_{Tn}	s	t	u	v
1	−0.0558636300	0	0	0	0
2	−0.2173010900	1	0	0	0
3	0.2605314000	0	1	0	0
4	0.1581140004	0	2	0	0
5	−0.1475810000	2	0	1	0
6	−0.4814970000	1	1	1	0
7	0.3781227800	0	2	1	0
8	0.0144043000	0	0	0	1
9	−0.0530054000	2	0	0	1
10	0.0143481000	0	1	0	1
11	0.0606826000	1	1	0	1
12	−0.0125894000	0	0	1	1
13	0.0109689000	1	0	1	1
14	−0.1336980000	0	3	0	0
15	0.0024115700	0	6	0	0
16	−0.0005300200	2	6	0	0
17	0.1684960000	3	0	1	0
18	0.0263454200	0	0	2	0
19	0.0436013600	2	0	2	0
20	−0.0311849300	3	0	2	0
21	0.0124921500	1	6	2	0
22	−0.0064827200	2	6	2	0
23	−0.0084172800	0	3	0	1
24	0.0168424000	1	3	0	1
25	−0.0010229600	3	3	0	1
26	−0.0317791000	0	3	1	1
27	0.0186040000	1	0	2	1
28	−0.0041079800	0	2	2	1
29	−0.0006068480	0	0	0	2
30	−0.0049819000	1	0	0	2
31	0.0025983000	2	0	0	2
32	−0.0005605280	3	0	0	2
33	−0.0016365200	1	2	0	2
34	−0.0003287870	1	6	0	2
35	0.0001165020	2	6	0	2
36	0.0006909040	0	0	1	2
37	0.0042174900	0	3	1	2
38	0.0000565229	3	6	1	2
39	−0.0014656400	0	3	2	2

n	C_{Qn}	s	t	u	v
1	0.0051589800	0	0	0	0
2	0.0160666800	2	0	0	0
3	−0.0441153000	1	1	0	0
4	0.0068222300	0	2	0	0
5	−0.0408811000	0	1	1	0
6	−0.0773296700	1	1	1	0
7	−0.0885381000	2	1	1	0
8	0.1693750200	0	2	1	0
9	−0.0037087100	1	0	0	1
10	0.0051369600	0	1	0	1
11	0.0209449000	1	1	0	1
12	0.0047431900	2	1	0	1
13	−0.0072340800	2	0	1	1
14	0.0043838800	1	1	1	1
15	−0.0269403000	0	2	1	1
16	0.0558082000	3	0	1	0
17	0.0161886000	0	3	1	0
18	0.0031808600	1	3	1	0
19	0.0129043500	0	0	2	0
20	0.0244508400	1	0	2	0
21	0.0070064300	3	0	2	0
22	−0.0271904600	0	1	2	0
23	−0.0166458600	3	1	2	0
24	0.0300449000	2	2	2	0
25	−0.0336974900	0	3	2	0
26	−0.0035002400	0	6	2	0
27	−0.0106854000	3	0	0	1
28	0.0011090300	3	3	0	1
29	−0.0003139120	0	6	0	1
30	0.0035895000	3	0	1	1
31	−0.0014212100	0	6	1	1
32	−0.0038363700	1	0	2	1
33	0.0126803000	0	2	2	1
34	−0.0031827800	2	3	2	1
35	0.0033426800	0	6	2	1
36	−0.0018349100	1	1	0	2
37	0.0001124510	3	2	0	2
38	−0.0000297228	3	6	0	2
39	0.0002695510	1	0	1	2
40	0.0008326500	2	0	1	2
41	0.0015533400	0	2	1	2
42	0.0003026830	0	6	1	2
43	−0.0001843000	0	0	2	2
44	−0.0004253990	0	3	2	2
45	0.0000869243	3	3	2	2
46	−0.0004659000	0	6	2	2
47	0.0000554194	1	6	2	2

9.2.3 KCA Series (RA)

Reference Radojčić (1988); see Sect. 4.2.1 (Model 3) (Fig. 9.12).

INPUT	OUTPUT
$X_1 = DAR$	K_T
$X_2 = P/D$	K_Q
$X_3 = J$	ΔK_T
$X_4 = \sigma_{0.7R}$	ΔK_Q

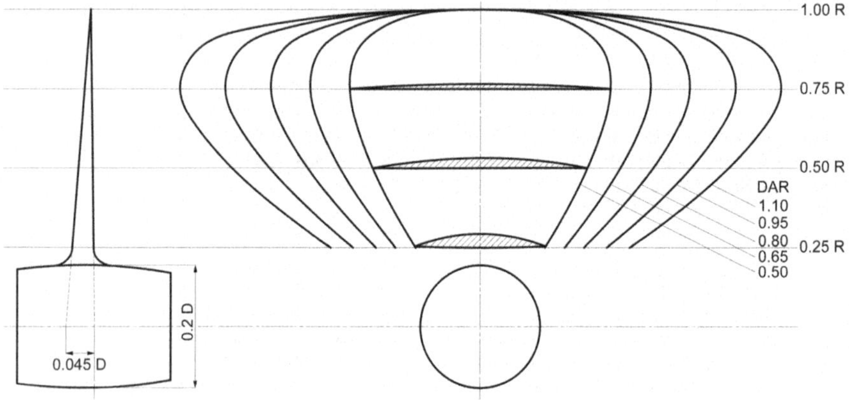

Fig. 9.12 Blade outline of the KCA series. Blade sections: segmental

Boundaries of Applicability

Non-cavitating conditions	Cavitating conditions
$0.5 \leq DAR \leq 1.1$	for $\sigma < 1.00$
$0.8 \leq P/D \leq 2.0$	$J_{\min} = 0.571 \cdot P/D + 0.021$
$J \geq 0.3$	for $\sigma \geq 1.00$
$J_{\min} = 0.492 \cdot P/D + 0.031$	$J_{\min} = 0.492 \cdot P/D + 0.031$
$K_T \leq \sqrt{J/2.5 - 0.1}$	$1.25 - 0.3 \cdot DAR - 0.2 \cdot \sigma \leq P/D \leq 1.8$
	$K_T \leq \sqrt{J/2.5 - 0.1} - (0.07/\sigma)$
	$\sigma \geq 0.50$
	$\Delta K_T \geq 0,\ \Delta K_Q \geq 0$
	$\eta_o \geq 0.20,\ \eta_{atm} \geq \eta_{cav}$

Note: $K_{Tcav} = K_T - \Delta K_T$ and $K_{Qcav} = K_Q - \Delta K_Q$.
ΔK_T, ΔK_Q = f (DAR, P/D, J, $\sigma_{0.7R}$, K_T), see Table 9.5.

Cavitation criteria

$$A_d/A_0 \geq 2.55 \cdot (1 - w)^2 K_T \left[K \sigma_0 J^2 (1.067 - 0.229P/D)\right]^{-1},$$

where $K \approx 0.70$, 0.55 and 0.36 for, respectively, $K_{T\text{ break down}}$, 10% and 2.5% back cavitation

Table 9.4 KCA Series (RA) regression coefficients for K_T and K_Q (non-cavitating conditions)

$$K_T = \sum_{1}^{16} (C_T\, 10^e\, (X_1)^x\, (X_2)^y\, (X_3)^z) \qquad K_Q = \sum_{1}^{17} (C_Q\, 10^e\, (X_1)^x\, (X_2)^y\, (X_3)^z)$$

n	C_{Tn}	e	x	y	z
1	0.1193852	0	0	0	0
2	−0.6574682	0	0	0	1
3	0.3493294	0	0	1	0
4	0.4119366	0	0	1	1
5	−0.1991927	0	0	2	1
6	5.8630510	−2	0	2	2
7	−1.1077350	−2	0	2	3
8	−0.1341679	0	1	0	0
9	0.2628839	0	1	0	1
10	−0.5217023	0	1	1	1
11	0.2970728	0	1	2	0
12	6.1525800	−2	2	1	3
13	−2.4708400	−2	2	2	3
14	−4.0801660	−3	1	6	0
15	4.1542010	−3	1	6	1
16	−1.1364520	−3	2	6	0

n	C_{Qn}	e	x	y	z
1	1.5411660	−3	0	0	0
2	0.1091688	0	0	0	1
3	−0.3102420	0	0	0	2
4	0.1547428	0	0	0	3
5	−4.3706150	−2	0	1	0
6	0.2490295	0	0	1	2
7	−0.1594602	0	0	1	3
8	8.5367470	−2	0	2	0
9	−9.5121630	−2	0	2	1
10	−9.3203070	−3	0	2	2
11	3.2878050	−2	0	2	3
12	5.4960340	−2	1	0	1
13	−4.8650630	−2	1	1	0
14	−0.1062500	0	1	1	1
15	8.5299550	−2	1	2	0
16	1.1010230	−2	2	0	3
17	−3.1517560	−3	2	2	2

Table 9.5 KCA Series (RA) regression coefficients for ΔK_T and ΔK_Q (needed for cavitating conditions)

$$\Delta K_T = \sum_{1}^{20} (d_T\, 10^e (X_1)^s (X_2)^v (X_4)^t (K_T)^u) \quad \Delta K_Q = \sum_{1}^{18} (d_Q\, 10^e (X_1)^s (X_2)^v (X_4)^t (K_T)^u)$$

n	d_{Tn}	e	s	v	t	u
1	6.688144	−2	0	0	0	0
2	3.579195	0	0	0	2	0
3	−5.70035	0	0	0	3	0
4	−1.359994	0	1	1	0	0
5	−8.111903	0	1	1	1	0
6	4.770548	1	1	1	3	0
7	−2.313208	−1	2	1	0	0
8	−1.387858	1	2	1	2	0
9	4.992201	1	2	1	3	0
10	−7.161204	1	2	1	4	0
11	1.721436	1	2	2	0	0
12	2.322218	1	2	2	1	0
13	−1.156897	2	2	2	2	0
14	5.014178	−2	0	0	0	2
15	−6.555364	−2	0	0	1	2
16	2.852867	−1	1	0	1	1
17	−8.081759	−1	1	1	0	1
18	8.671852	1	3	2	5	0
19	−3.727835	1	3	3	0	0
20	8.04397	1	3	3	1	0

n	d_{Qn}	e	s	v	t	u
1	4.024475	−3	0	0	0	0
2	1.202447	−1	0	0	2	0
3	−9.83607	−2	1	1	0	0
4	−8.31884	−1	1	1	1	0
5	5.098177	0	1	1	3	0
6	−5.192839	−1	2	1	1	0
7	2.641109	0	2	2	0	0
8	−1.688934	1	2	2	3	0
9	4.928417	−2	0	0	1	1
10	1.024274	−2	0	0	0	2
11	−1.194521	−1	0	1	1	1
12	5.498736	−2	1	0	1	1
13	−2.488235	−1	1	1	0	1
14	−5.832879	−1	0	0	5	0
15	1.503955	−1	0	3	0	0
16	−3.316121	0	3	3	0	0
17	3.890792	0	3	3	1	0
18	1.682032	1	3	3	3	0

9.2.4 KCA Series (ANN)

Reference Koushan (2007); see Sect. 4.2.1 (Model 4) (Fig. 9.13).

INPUT	OUTPUT
$X_1 = J$	K_T
$X_2 = DAR$	K_Q
$X_3 = P/D$	
$X_4 = \sigma$	

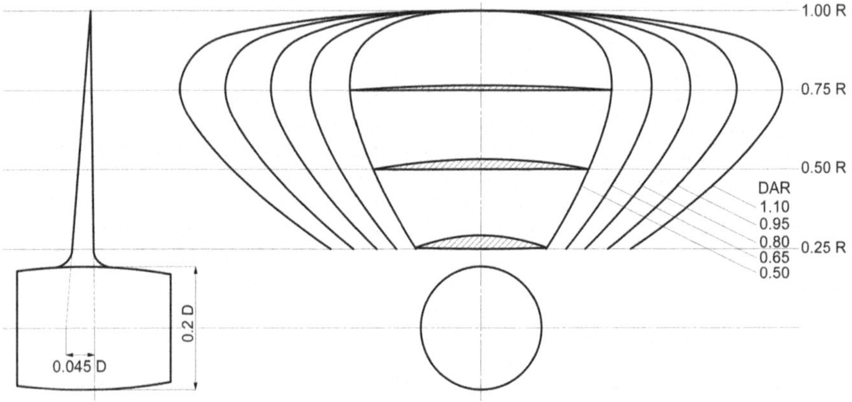

Fig. 9.13 Blade outline of the KCA series. Blade sections: segmental

Boundaries of Applicability

Non-cavitating and Cavitating conditions
$0.60 \leq P/D \leq 2.00$
$0.5 \leq DAR \leq 1.1$
$0.50 \leq \sigma \leq 2.00$
$J_{\min} = 0.4847 \cdot P/D + 0.0339$
$\left(\frac{P}{D}\right)_{\min} = 0.67 + 0.83 \cdot DAR^{-0.32} + 0.61 \cdot \sigma^2 - 1.64 \cdot \sigma$

Table 9.6 (a) KCA series (ANN) coefficients of the mathematical model for K_T for transition and cavitating conditions

(a)
$$K_T = \frac{tanh\left(C + \sum_{i=1}^{13}\left(O_i \cdot tanh\left(a_i + \sum_{k=1}^{4}\left(h_{k,i} \cdot (D_k \cdot X_k + E_k)\right)\right)\right)\right) - G}{H}$$

k	D_k	E_k
1	0.995973706245422	−1.22352087497711
2	3.00	−2.40
3	1.285714268684380	−1.67142856121063
4	1.20	−1.50

coeff.	$K_T - cav$
H	3.140624761581420
G	−0.859633564949035
C	−0.885586142539978

i	O_i	a_i	h_{1i}	h_{2i}	h_{3i}	h_{4i}
1	0.664608895778656	−2.193300008773800	−0.726629436016082	0.669164776802062	0.594114363193511	0.452542901039123
2	0.721204936504364	−6.024066925048820	−2.597256183624260	0.676047623157501	4.161122798919670	0.654244840145111
3	−0.151763767004013	−0.696689069271087	−1.391208052635190	0.003160835243762	−2.212373733520500	0.222368687391281
4	−0.191058859229087	2.055507898330680	−0.251667559146881	−0.321265518665313	−2.592654466629020	0.202032059431076
5	0.567672729492187	3.509091377258300	6.716553688049310	0.497569680213928	−3.072792053222650	2.371959447860710
6	0.421842813491821	2.553286075592040	2.385244846343990	−0.185903996229171	0.432899504899978	−0.338320553302764
7	0.321790337562561	−0.383505642414093	−2.364685535430900	0.436348795890808	1.089763522148130	−0.435716688632965
8	0.229824930429458	3.388423681259150	1.019828915596000	0.402370303869247	−1.506125688552850	2.763916254043570
9	0.632828950881958	1.804630279541010	5.007183074951170	0.561147570610046	−1.819651365280150	0.961106181144714
10	0.427665591239929	−2.073302984237670	−6.628965854644770	−0.602317452430725	2.469989299774160	−0.835814237594604
11	−0.396704137325286	3.194185256958000	6.446861267089840	0.275236368179321	−2.984740972518920	2.455309391021720
12	−0.950892448425292	−0.859452128410339	4.109339237213130	−0.023789746686816	−3.386325359344480	−0.135593101382255
13	−0.552885591983795	0.044888626784086	1.941959857940670	0.439951628446579	−0.886807858943939	−0.087464213371277

Table 9.6 (b) KCA series (ANN) coefficients of the mathematical model for K_Q for transition and cavitating conditions

(b)

$$K_Q = \frac{tanh\left(C + \sum_{i=1}^{16}\left(O_i \cdot tanh\left(a_i + \sum_{k=1}^{4}\left(h_{k,i} \cdot (D_k \cdot X_k + E_k)\right)\right)\right)\right) - G}{H}$$

k	D_k	E_k
1	1.00594687461853	−1.24491155147552
2	3.00	−2.40
3	1.28571426868438	−1.67142856121063
4	1.20	−1.50

coeff.	$K_Q - cav$
H	10.571378469467100
G	−0.907461941242218
C	−0.577850341796875

i	O_i	a_i	h_{1i}	h_{2i}	h_{3i}	h_{4i}
1	0.442078769207000	1.768611073493950	2.497873067855830	0.297473937273025	−0.718023419380187	−0.338622987270355
2	−0.421647399663925	0.556160688400268	−1.340572357177730	−0.536944448947906	0.026032123714685	1.099946141242980
3	−0.886713325977325	0.497339546680450	−0.610023617744445	−0.223969161510467	−0.482632517814636	−0.487264901399612
4	−1.867102861404410	−1.616421222686760	−4.256658554077140	−0.553267300128936	1.646532058715820	−1.156178951263420
5	0.227311909198761	−1.463332295417780	−3.104576826095580	−0.634656071662902	1.578319191932670	−1.895716309547420
6	0.243787378072738	1.097079634666440	0.997569739818572	−0.004510338418186	0.048544548451900	1.191902160644530
7	0.525087118148803	0.746584773063659	−1.354389071464530	−0.745467662811279	0.125915661454200	0.952797472476959
8	0.164162874221801	−1.385915160179130	−0.368810653686523	0.847794473171234	1.370921611785880	−0.681915104389190
9	−0.590143740177154	0.897126734256744	1.256091117858880	0.385022848844528	−0.775024771690368	−0.261870503425598
10	0.486191481351852	0.352565705776214	−0.895469605922698	−0.374975472688674	0.425471633672714	−0.931823551654815
11	−0.852708101272583	−0.913129091262817	3.201136112213130	−0.005423143506050	−2.411940813064570	−0.047307174652815
12	0.168456405401229	1.328734755516050	−0.571855306625366	−0.023150026798248	−2.945253372192380	−0.074185602366924
13	−0.760893166065216	1.990051507949820	5.714661121368400	0.656298816204071	−2.378326892852780	1.671640872955320
14	0.475182175636291	−0.504588842391967	−2.376817941665640	0.273024111986160	0.981634378433227	−0.401308417320251
15	−0.554895162582397	1.609721302986140	4.739315986633300	0.629641473293304	−1.693942904472350	0.747780144214630
16	1.064848780632010	−3.546957015991210	−1.059366464614860	0.509716451168060	2.275857210159300	0.615991234779357

9.2.5 Newton-Rader Series

Reference Koushan (2005); see Sect. 4.2.2 (Fig. 9.14).

INPUT	OUTPUT
$X_1 = J$	K_T
$X_2 = EAR$	K_Q
$X_3 = P/D$	
$X_4 = \sigma$	

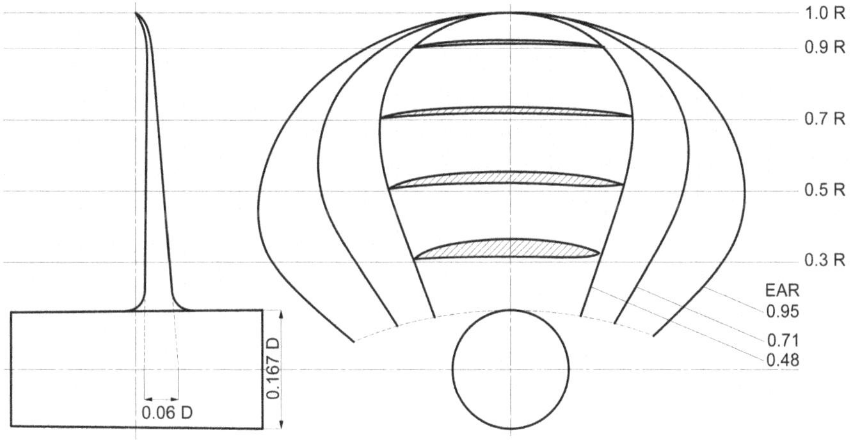

Fig. 9.14 Blade outline of the Newton-Rader series. Blade sections: concave back, hollow face, modified leading and trailing edge

Boundaries of Applicability

Non-cavitating conditions	Cavitating conditions
$1.04 \leq P/D \leq 2.08$	$1.04 \leq P/D \leq 2.08$
$0.48 \leq EAR \leq 0.95$	$0.48 \leq EAR \leq 0.95$
$0.25 \leq \sigma \leq 1.00$	$0.25 \leq \sigma \leq 1.00$
$J_{\min} = 0.538 \cdot \frac{P}{D} + 0.129$	$J_{\min} = \frac{P}{D}(1.08 \cdot \sigma^2 - 1.585 \cdot \sigma + 1.013)$

Table 9.7 Newton-Rader Series ANN coefficients for K_T and K_Q (non-cavitating conditions only)

$$K_T \text{ or } K_Q = \frac{tanh\left(C + \sum_{i=1}^{5}\left(O_i \cdot tanh\left(a_i + \sum_{k=1}^{3}\left(h_{k,i} \cdot (D_k \cdot X_k + E_k)\right)\right)\right)\right) - G}{H}$$

k	D_k	E_k	coeff.	$K_T - non\text{–}cav$
1	1.13278782367706	−1.69295144081115	H	3.224303960800170
2	3.82978725433349	−2.73829770088195	G	−0.738784790039062
3	1.73076927661895	−2.70	C	−0.376195490360260

i	O_i	a_i	h_{1i}	h_{2i}	h_{3i}
1	0.940852165222167	−0.270164817571640	−0.124824509024620	0.103370726108551	0.865557014942169
2	1.826129317283630	1.490583539009090	−2.384441137313840	0.085133850574493	1.728574037551880
3	−1.893094539642330	2.630162954330440	2.234081983566280	−0.431566774845123	−1.169071674346920
4	0.675479471683502	0.368309527635574	−0.579336047172546	0.385407656431198	−0.127583935856819
5	0.971990406513214	0.286899149417877	−0.223387002944946	−0.373791098594665	−0.299132347106933

k	D_k	E_k	coeff.	$K_Q - non\text{–}cav$
1	1.13278782367706	−1.69295144081115	H	10.597585439682000
2	3.82978725433349	−2.73829770088195	G	−0.90
3	1.73076927661895	−2.70	C	0.226830631494522

i	O_i	a_i	h_{1i}	h_{2i}	h_{3i}
1	−3.202843666076660	2.700680255889890	1.662291646003720	−0.318319201469421	−1.399070382118220
2	−0.370148748159408	−0.347793906927108	0.045066494494677	0.684060215950012	0.614231705665588
3	−0.624823510646820	0.170036911964416	0.412147045135498	−0.227014541625976	−1.119494080543510
4	0.904307782649993	1.116573095321650	0.164921686053276	0.199502408504486	−0.272221803665161
5	−1.995479822158810	−1.274008274078360	1.738352537155150	−0.122752793133258	−1.131857395172110

Table 9.8 (a) Newton-Rader Series ANN coefficients for K_T (transition and cavitating condition) coefficients of the mathematical models for K_T and K_Q coefficients for cavitating conditions

(a)

$$K_T = \frac{tanh\left(C + \sum_{i=1}^{15}\left(O_i \cdot tanh\left(a_i + \sum_{k=1}^{4}\left(h_{k,i} \cdot (D_k \cdot X_k + E_k)\right)\right)\right)\right) - G}{H}$$

k	D_k	E_k
1	1.015801310539240	–1.407900571823120
2	3.829787254333490	–2.738297700881950
3	1.730769276618950	–2.70
4	2.40	–1.50

coeff.	$K_T - cav$
H	3.291398525238030
G	–0.768344044685363
C	–0.377740830183029

i	O_i	a_i	h_{1i}	h_{2i}	h_{3i}	h_{4i}
1	1.317354202270500	–1.148691415786740	–0.087731383740902	0.093006208539009	0.401366353034973	0.267277389764785
2	–0.738121628761291	–0.193716168403625	0.136591866612434	0.212323457002639	0.086729958653450	0.973015606403350
3	–0.941667079925537	0.608081042766571	–2.371210336685180	–0.130271703004837	1.014986753463740	–0.822616696357727
4	–0.804430842399597	–0.100771367549896	1.338524222373960	–0.334297746419906	–0.895095586776733	0.788783311843872
5	–0.727642714977264	0.913926005363464	–4.371420860290520	–0.034235712140799	2.684855699539180	0.968316078186035
6	0.933406054973602	–0.399211555719375	–5.171780109405510	–0.546013474464416	1.749250769615170	–1.032035708427420
7	–0.436862766742706	–0.015545075759292	–2.002247810363770	0.508135020732879	1.542620182037350	–0.667988538742065
8	–0.744883954524993	–0.618124425411224	–0.495175689458847	0.389427244663238	–0.672402203083038	0.157831832766532
9	0.711936593055725	0.812608003616333	1.920007348060600	0.221785292029380	–0.656296193599700	1.213999629020690
10	0.670403599739074	–0.581250905990600	–0.146149247884750	1.363635301589960	–0.819711506366729	0.097338154911995
11	1.441804289817810	1.202768325805660	–4.512323379516600	0.019758621230721	2.710045099258420	0.736025989055633
12	0.974518120288848	0.364288687705993	–2.852175712585440	–0.174428954720497	1.185967564582820	–0.212095811963081
13	1.001716732978820	0.391402333974838	4.315288066864010	0.546206593513488	–1.339086651802060	1.069359421730040
14	–0.588005423545837	–0.810350835323333	–0.764147520065307	1.575306773185720	–0.405050814151763	0.144642546772956
15	0.695903778076171	–0.708705723285675	–1.635938048362730	0.450230121612548	0.142535582184791	0.243726670742034

Table 9.8 (b) Newton-Rader Series ANN coefficients for K_Q (transition and cavitating condition) coefficients of the mathematical models for K_T and K_Q coefficients for cavitating conditions

(b)

$$K_Q = \frac{tanh\left(C + \sum_{i=1}^{15}\left(O_i \cdot tanh\left(a_i + \sum_{k=1}^{4}\left(h_{k,i} \cdot (D_k \cdot X_k + E_k)\right)\right)\right)\right) - G}{H}$$

k	D_k	E_k
1	1.01580131053924	−1.40790057182312
2	3.82978725433349	−2.73829770088195
3	1.73076927661895	−2.70
4	2.40	−1.50

coeff.	K_Q - *cav*
H	10.520163774490300
G	−0.847399175167083
C	−1.040541768074030

i	O_i	a_i	h_{1i}	h_{2i}	h_{3i}	h_{4i}
1	0.969991624355316	0.053129803389311	−0.240043446421623	−0.168751984834671	−0.135895714163780	−1.043172240257260
2	0.231118693947792	−1.052649378776550	4.257379055023190	0.002107465174049	−2.515236377716060	−1.653260469436640
3	1.088238000869750	0.451427340507507	4.661829471588130	0.559997856616973	−1.431214809417720	1.101806521415710
4	0.505151510238647	−1.368094325065610	−2.136204957962030	−0.108278773725032	1.294241428375240	−0.081455551087856
5	−1.071812391281120	0.452628254890441	5.355250358581540	0.566431343555450	−1.808804035186760	1.026858925819390
6	0.992358982563018	0.340347319841384	0.191872522234916	0.402995556592941	0.512346208095550	−0.521808624267578
7	1.214749336242670	0.638038694858551	1.526024580001830	0.188212499022483	−0.427222281694412	1.122764825820920
8	0.371759355068206	−0.326672583818435	0.534627437591552	−0.875299453735351	−1.236307501792900	0.604217648506164
9	−0.970895588397979	−1.393670439720150	4.323750019073480	−0.092468962073326	−2.450195789337150	−0.723869979381561
10	0.547687768936157	0.640973389148712	0.989770770072937	−0.878276824951171	0.170171350240707	−0.296523988246917
11	−0.162236914038658	0.774095118045806	4.051302909851070	−0.044920612126589	−1.689987540245050	1.831228494644160
12	−0.666321873664855	0.585490763187408	1.492992401123040	−0.288998872041702	−0.202796399593353	−0.292661637067794
13	−0.647844552993774	−0.583688259124755	3.405870437622070	0.317586600780487	−1.321955084800720	0.191683411598205
14	−1.081090211868280	0.614605069160461	−1.642314791679380	−0.230708897113800	0.398518145084381	−0.706239044666290
15	0.155757471919059	−0.351223945617675	−0.764876306056976	1.591475725173950	−0.645161092281341	0.231215476989746

9.3 Some Typical Examples

A comparison of the MMs presented in Sect. 9.2 with the series' and commercial propellers is presented in Sects. 9.3.1 and 9.3.2 respectively. The experimental data for the KCA 112 propeller is given in Table 9.9, while comparable results obtained with the regression and ANN based MMs are given in Tables 9.10 and 9.11 respectively. Results for the same propeller obtained by the AEW MM are given in Table 9.12. Comparison of the results of all three MMs with the experimental data for non-cavitating and cavitating conditions is shown in Fig. 9.15.

The same procedure is used for the parent Newton-Rader propeller, i.e. experimental data, results of ANN based MM, and comparison of results are given in Tables 9.13, 9.14 and Fig. 9.16, respectively.

A comparison of the results obtained by the KCA (RA), KCA (ANN) and AEW MMs is performed for two commercial propellers. Experimental results are given in Tables 9.15 and 9.19, while the MM's results are given in Tables 9.16, 9.17, 9.18 and 9.20, 9.21, 9.22, respectively. Comparison of the results of all three MMs with the experimental data for non-cavitating and cavitating conditions is shown in Figs. 9.17 and 9.18.

Note: For the KCA series, *DAR* (Developed Area Ratio) is used. Other series use *EAR* (Expanded Area Ratio). Consequently, *BAR* (Blade Area Ratio) notation is used in Figs. 9.15, 9.17 and 9.18, denoting that *BAR* value is used as *DAR* and EAR values for KCA and AEW series, respectively. It is, however assumed that differences between *DAR* and *EAR* are negligible. Namely, $EAR \approx 0.34 \cdot DAR \cdot [2.75 + DAR/z]$; see O'Brien (1969).

9.3.1 Comparison of MMs with the Series' Propellers

Table 9.9 Experimental Data for KCA 112 propeller

σ = atm											
J	0.6507	0.7068	0.8092	0.9084	1.0077	1.1037	1.1998				
K_T	0.3344	0.2972	0.2360	0.1781	0.1243	0.0738	0.0250				
J	0.6515	0.7036	0.8028	0.9037	1.0029	1.1038	1.2014	1.2468			
$10\,K_Q$	0.6172	0.5608	0.4612	0.3667	0.2771	0.1941	0.1112	0.0747			
J	0.7079	0.8067	0.9087	0.9856	1.0090	1.1106	1.1785	1.2047	1.2418		
η_O	0.5863	0.6497	0.6981	0.7149	0.7150	0.6586	0.5058	0.3711	0.0069		
σ = 0.75											
J	0.6522	0.7008	0.8031	0.9054	1.0056	1.1058	1.2026	1.2210			
K_T	0.0961	0.1110	0.1375	0.1606	0.1203	0.0691	0.0121	0.0020			
J	0.6534	0.7056	0.8048	0.8468	0.9040	1.0046	1.1033	1.1987	1.2171		
$10\,K_Q$	0.2163	0.2445	0.2861	0.3028	0.3294	0.2745	0.1880	0.0965	0.0782		
J	0.6503	0.7055	0.8077	0.9081	0.9783	1.0083	1.1100	1.1613	1.2039	1.2184	
η_O	0.4638	0.5168	0.6161	0.6957	0.7139	0.7122	0.6493	0.5019	0.2021	0.0033	
σ = 0.50											
J	0.6517	0.7054	0.7508	0.8063	0.9021	0.9610	1.0046	1.1034	1.1803		
K_T	0.0829	0.0772	0.0756	0.0815	0.0958	0.1026	0.0960	0.0514	0.0068		
J	0.6520	0.7026	0.8008	0.9025	0.9855	1.0024	1.0996	1.1832			
$10\,K_Q$	0.1847	0.1776	0.1918	0.2208	0.2384	0.2333	0.1611	0.0808			
J	0.6583	0.7069	0.7606	0.8076	0.8631	0.9101	0.9705	1.0074	1.1026	1.1539	1.1849
η_O	0.4584	0.4766	0.4981	0.5229	0.5676	0.6189	0.6470	0.6421	0.5229	0.3028	0.0050

Table 9.10 Regression MM's data for KCA 112 propeller

σ = atm													
J	0.650	0.700	0.750	0.800	0.850	0.900	0.950	1.000	1.050	1.100	1.150	1.200	1.250
K_T	0.325	0.294	0.265	0.236	0.207	0.179	0.151	0.124	0.097	0.071	0.046	0.021	−0.003
$10\,K_Q$	0.610	0.560	0.511	0.463	0.415	0.368	0.323	0.278	0.235	0.193	0.152	0.114	0.077
η_O	0.551	0.586	0.619	0.648	0.674	0.695	0.708	0.710	0.694	0.649	0.553	0.356	−0.080
σ = 0.75													
J	0.650	0.700	0.750	0.800	0.850	0.900	0.950	1.000	1.050	1.100	1.150	1.200	1.250
K_T	0.107	0.121	0.133	0.142	0.145	0.143	0.134	0.118	0.097	0.071	0.046	0.021	−0.003
$10\,K_Q$	0.236	0.266	0.288	0.303	0.309	0.305	0.290	0.265	0.235	0.193	0.152	0.114	0.077
η_O	0.469	0.507	0.550	0.595	0.636	0.672	0.699	0.712	0.694	0.649	0.553	0.356	−0.080
σ = 0.50													
J	0.650	0.700	0.750	0.800	0.850	0.900	0.950	1.000	1.050	1.100	1.150	1.200	1.250
K_T	0.059	0.069	0.078	0.087	0.092	0.094	0.092	0.084	0.070	0.052	0.029	0.003	−0.027
$10\,K_Q$	0.152	0.175	0.194	0.209	0.218	0.221	0.217	0.205	0.187	0.162	0.131	0.094	0.052
η_O	0.404	0.437	0.480	0.527	0.573	0.612	0.639	0.649	0.630	0.564	0.409	0.050	−1.030

Table 9.11 ANN MM's data for KCA 112 propeller

$\sigma = 0.75$

J	0.650	0.700	0.750	0.800	0.850	0.900	0.950	1.000	1.050	1.100	1.150	1.200	1.250
K_T	0.102	0.110	0.121	0.137	0.151	0.155	0.144	0.121	0.094	0.065	0.038	0.013	−0.007
$10 K_Q$	0.214	0.234	0.256	0.281	0.303	0.314	0.306	0.276	0.232	0.184	0.139	0.098	0.059
η_O	0.495	0.523	0.564	0.620	0.676	0.709	0.711	0.698	0.675	0.621	0.495	0.247	−0.249

$\sigma = 0.50$

J	0.650	0.700	0.750	0.800	0.850	0.900	0.950	1.000	1.050	1.100	1.150	1.200	1.250
K_T	0.077	0.078	0.080	0.083	0.088	0.093	0.096	0.091	0.075	0.053	0.029	0.007	−0.011
$10 K_Q$	0.175	0.182	0.189	0.197	0.205	0.212	0.217	0.213	0.195	0.163	0.124	0.084	0.047
η_O	0.453	0.476	0.503	0.536	0.580	0.631	0.672	0.679	0.645	0.567	0.425	0.149	−0.482

Table 9.12 AEW MM's data for KCA 112 propeller

σ = atm

J	0.650	0.700	0.750	0.800	0.850	0.900	0.950	1.000	1.050	1.100	1.150	1.200	1.250
K_T	0.322	0.295	0.268	0.241	0.214	0.187	0.161	0.135	0.109	0.084	0.059	0.035	0.012
$10 K_Q$	0.632	0.588	0.543	0.498	0.453	0.409	0.365	0.321	0.279	0.237	0.196	0.157	0.118
η_O	0.527	0.559	0.589	0.616	0.639	0.657	0.667	0.667	0.654	0.619	0.551	0.427	0.196

$\sigma = 0.75$ (left side: fully developed cavitation, right side: transition region)

J	0.300	0.350	0.400	0.450	0.500	0.585	0.600	0.650	0.700	0.750	0.800	0.850	0.900
K_T	0.073	0.074	0.074	0.075	0.076	0.077	0.081	0.095	0.110	0.126	0.143	0.162	0.181
$10 K_Q$	0.167	0.168	0.170	0.171	0.173	0.176	0.185	0.216	0.249	0.284	0.321	0.361	0.402

$\sigma = 0.50$ (left side: fully developed cavitation, right side: transition region)

J	0.400	0.450	0.500	0.550	0.600	0.650	0.730	0.750	0.800	0.850	0.900	0.950	0.990
K_T	0.074	0.075	0.076	0.077	0.077	0.078	0.080	0.084	0.096	0.108	0.121	0.135	0.146
$10 K_Q$	0.170	0.171	0.173	0.175	0.176	0.179	0.183	0.193	0.218	0.245	0.273	0.303	0.328

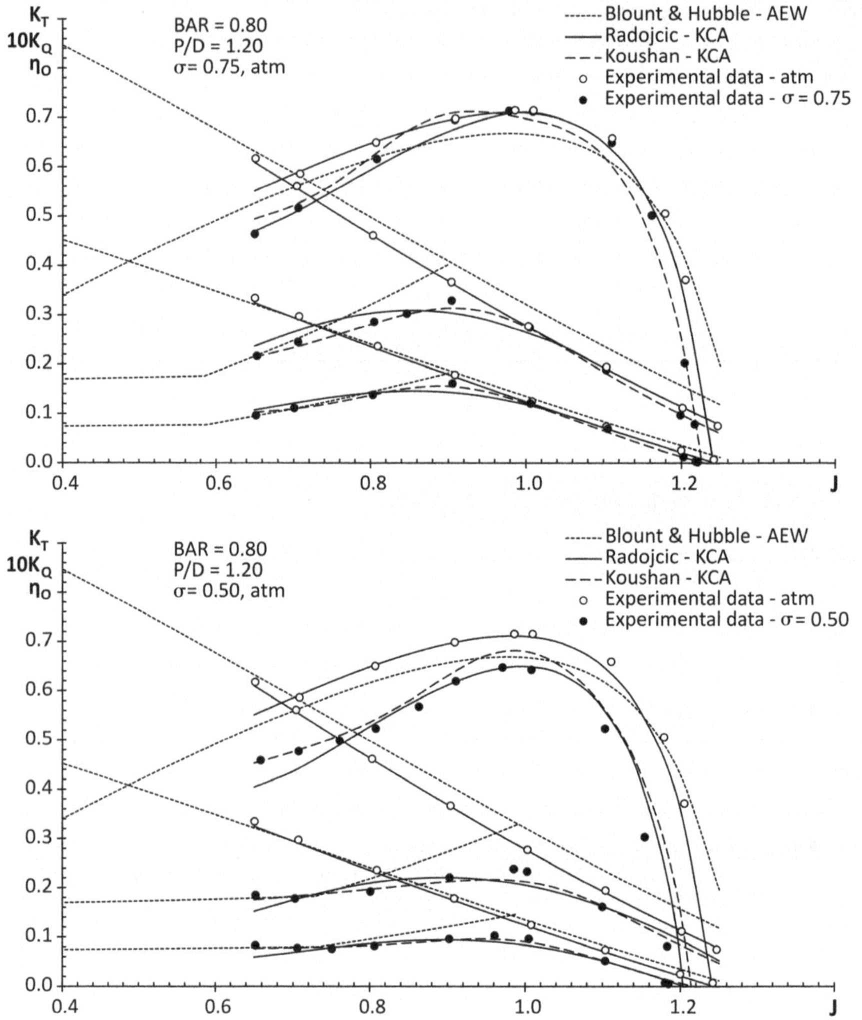

Fig. 9.15 Comparison of three MMs with KCA 112 propeller (DAR = 0.8, P/D = 1.2, σ = atm, 0.5, and 0.75)

Table 9.13 Experimental Data for a parent Newton-Rader propeller

σ = atm														
J	0.900	0.950	1.000	1.050	1.100	1.150	1.200	1.250	1.300	1.350	1.400	1.428		
K_T	0.253	0.232	0.211	0.190	0.168	0.145	0.122	0.097	0.072	0.044	0.016	0.000		
10 K_Q	0.539	0.508	0.476	0.441	0.405	0.368	0.330	0.289	0.246	0.198	0.150	–		
η_O	0.672	0.690	0.706	0.718	0.724	0.721	0.703	0.668	0.602	0.477	0.238	0.000		
σ = 0.75														
J	0.750	0.800	0.850	0.900	0.950	1.000	1.050	1.100	1.150	1.200	1.250	1.300	1.350	1.367
K_T	0.174	0.184	0.196	0.207	0.215	0.205	0.186	0.163	0.138	0.110	0.079	0.046	0.012	0.000
10 K_Q	0.362	0.387	0.415	0.446	0.474	0.463	0.434	0.396	0.353	0.304	0.249	0.189	0.126	0.000
η_O	0.574	0.607	0.639	0.665	0.686	0.703	0.715	0.722	0.715	0.690	0.632	0.502	0.205	0.000
σ = 0.50														
J	0.650	0.700	0.750	0.800	0.850	0.900	0.950	1.000	1.050	1.100	1.150	1.200	1.250	1.300
K_T	0.128	0.128	0.130	0.136	0.148	0.161	0.171	0.174	0.167	0.150	0.124	0.089	0.048	0.012
10 K_Q	0.272	0.273	0.278	0.291	0.319	0.353	0.384	0.402	0.399	0.376	0.332	0.252	0.186	0.083
η_O	0.484	0.523	0.559	0.595	0.627	0.653	0.674	0.690	0.699	0.699	0.684	0.640	0.513	0.030

Table 9.14 ANN MM's data for a parent Newton-Rader propeller

σ = atm										
J	0.813	0.913	1.013	1.113	1.213	1.263	1.313	1.377	1.440	
K_T	0.284	0.243	0.203	0.161	0.114	0.087	0.059	0.024	−0.009	
10 K_Q	0.598	0.531	0.463	0.388	0.305	0.261	0.215	0.156	0.100	
η_O	0.615	0.664	0.708	0.736	0.718	0.672	0.577	0.334	−0.206	
σ = 0.75										
J	0.540	0.640	0.740	0.840	0.940	1.040	1.140	1.240	1.340	1.440
K_T	0.101	0.130	0.161	0.190	0.206	0.192	0.146	0.086	0.022	−0.026
10 K_Q	0.230	0.284	0.344	0.407	0.457	0.451	0.377	0.275	0.148	0.028
η_O	0.377	0.465	0.549	0.623	0.675	0.706	0.701	0.613	0.309	−2.065
σ = 0.50										
J	0.613	0.713	0.813	0.913	1.013	1.113	1.213	1.263	1.313	
K_T	0.114	0.126	0.141	0.158	0.165	0.140	0.069	0.029	−0.003	
10 K_Q	0.249	0.268	0.306	0.356	0.389	0.348	0.219	0.144	0.077	
η_O	0.449	0.535	0.598	0.643	0.685	0.715	0.608	0.403	−0.077	

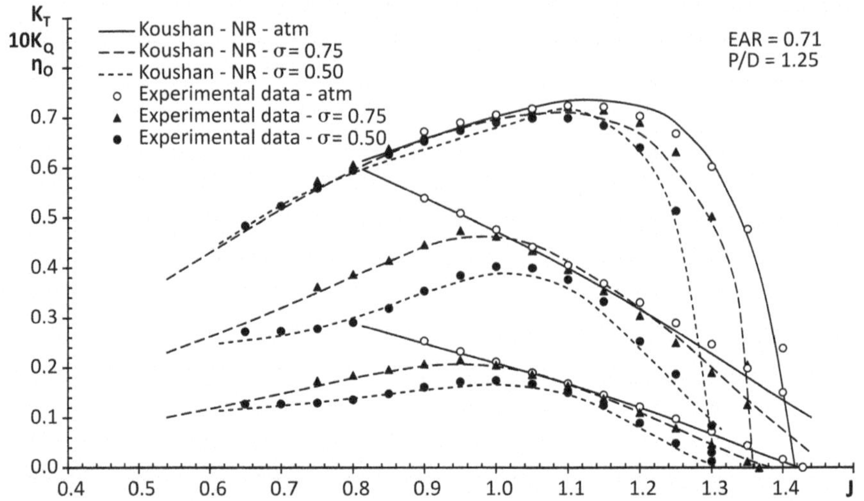

Fig. 9.16 Comparison of Koushan's (2005) MM with a parent Newton-Rader propeller A3/71/125 (EAR = 0.71, P/D = 1.25, σ = atm, 0.75, and 0.50)

9.3.2 Comparison of MMs with the Commercial Propellers

A comparison with the full scale tested commercial propellers is shown in Figs. 9.17 and 9.18. Experimental data is from Denny et al. (1988); some propellers have subsequently been retested and were used in discussion of Radojčić (1988). Note that the MMs are based on the laboratory propellers, while the commercially available propellers are not manufactured to the same tolerances. Moreover, experiments are not always repeatable; see Fig. 9.17.

Table 9.15 Experimental Data for commercial propellers (BAR = 0.54, P/D = 1.004)

BAR = 0.54	**P/D = 1.004**	**σ = 2.3**							
Measured in 1987									
J	0.5943	0.6292	0.6362	0.6564	0.6912	0.7093	0.7344	0.7714	0.8397
K_T	0.1748	0.1740	0.1839	0.1853	0.1825	0.1832	0.1838	0.1718	0.1448
$10\ K_Q$	0.3064	0.3113	0.3177	0.3205	0.3148	0.3119	0.3091	0.2956	0.2608
η_O	0.5290	0.5638	0.5850	0.6119	0.6360	0.6651	0.6913	0.7119	0.7467
Measured in 1981									
J	0.7950	0.8281	0.8572	0.8943					
K_T	0.1518	0.1415	0.1248	0.1074					
$10\ K_Q$	0.2921	0.2734	0.2528	0.2316					
η_O	0.6546	0.6746	0.6677	0.6478					

Table 9.16 KCA-RA MM's data for commercial propellers (BAR = 0.54, P/D = 1.004)

σ = atm												
J	0.550	0.600	0.650	0.700	0.750	0.800	0.850	0.900	0.950	1.000	1.050	1.100
K_T	0.254	0.228	0.203	0.177	0.152	0.128	0.103	0.079	0.056	0.032	0.009	–0.014
10 K_Q	0.395	0.363	0.330	0.296	0.262	0.228	0.194	0.161	0.128	0.096	0.065	0.036
η_O	0.562	0.600	0.636	0.668	0.695	0.715	0.722	0.709	0.659	0.535	0.232	–0.681
σ = 2.3												
J	0.550	0.600	0.650	0.700	0.750	0.800	0.850	0.900	0.950	1.000	1.050	1.100
K_T	0.175	0.185	0.186	0.177	0.152	0.128	0.103	0.079	0.056	0.032	0.009	–0.014
10 K_Q	0.298	0.316	0.318	0.296	0.262	0.228	0.194	0.161	0.128	0.096	0.065	0.036
η_O	0.515	0.558	0.604	0.668	0.695	0.715	0.722	0.709	0.659	0.535	0.232	–0.681

Table 9.17 KCA-ANN MM's data for commercial propellers (BAR = 0.54, P/D = 1.004)

σ = 2.3												
J	0.550	0.600	0.650	0.700	0.750	0.800	0.850	0.900	0.950	1.000	1.050	1.100
K_T	0.168	0.183	0.183	0.168	0.145	0.122	0.099	0.076	0.053	0.030	0.009	–0.009
10 K_Q	0.299	0.317	0.316	0.293	0.258	0.219	0.181	0.146	0.113	0.081	0.051	0.023
η_O	0.493	0.551	0.600	0.639	0.673	0.707	0.738	0.749	0.714	0.593	0.291	–0.665

Table 9.18 AEW MM's data for commercial propellers (BAR = 0.54, P/D = 1.004)

σ = atm													
J	0.000	0.100	0.200	0.300	0.400	0.500	0.600	0.700	0.800	0.900	1.000	1.050	1.100
K_T	0.445	0.416	0.384	0.348	0.310	0.269	0.226	0.182	0.137	0.092	0.046	0.023	0.001
10 K_Q	0.643	0.605	0.562	0.517	0.469	0.418	0.365	0.310	0.253	0.196	0.137	0.107	0.077
η_O	0.000	0.110	0.217	0.322	0.421	0.512	0.592	0.655	0.689	0.671	0.534	0.362	0.014
σ = 2.23 (left side: fully developed cavitation, right side: transition region)													
J	0.050	0.100	0.150	0.200	0.260	0.310	0.350	0.400	0.450	0.500	0.550	0.600	0.650
K_T	0.047	0.047	0.047	0.047	0.047	0.047	0.060	0.078	0.099	0.122	0.148	0.176	0.207
J	0.000	0.050	0.100	0.150	0.200	0.260	0.310	0.350	0.400	0.450	0.500	0.550	0.600
10 K_Q	0.084	0.084	0.084	0.084	0.084	0.084	0.115	0.142	0.180	0.222	0.267	0.317	0.369

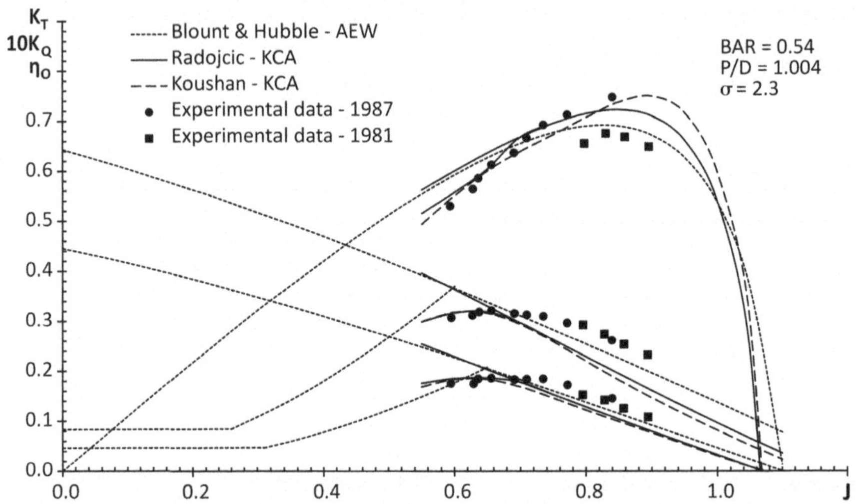

Fig. 9.17 Comparison of three MMs with a commercial propeller EAR = 0.54 ($EAR_{nominal}$ = 0.50), P/D = 1.004 ($P/D_{nominal}$ = 1.00), σ = 2.3

Table 9.19 Experimental Data for commercial propellers (BAR = 0.53, P/D = 1.238)

BAR = 0.53		**P/D = 1.238**		**σ = 2.3**	
Measured					
J	0.7946	0.8350	0.8859	0.9094	0.9720
K_T	0.2054	0.2064	0.1921	0.1824	0.1584
10 K_Q	0.4572	0.4467	0.4241	0.4091	0.3646
η_O	0.5723	0.6215	0.6436	0.6505	0.6787

Table 9.20 KCA (RA) MM's data for commercial propellers (BAR = 0.53, P/D = 1.238)

σ = atm													
J	0.750	0.800	0.850	0.900	0.950	1.000	1.050	1.100	1.150	1.200	1.250	1.300	1.350
K_T	0.275	0.249	0.224	0.198	0.173	0.149	0.125	0.101	0.078	0.054	0.032	0.009	−0.013
10 K_Q	0.524	0.482	0.440	0.398	0.357	0.317	0.277	0.238	0.200	0.164	0.128	0.093	0.060
η_O	0.627	0.659	0.689	0.714	0.735	0.749	0.753	0.742	0.708	0.635	0.492	0.203	−0.468
σ = 2.3													
J	0.650	0.700	0.750	0.800	0.850	0.900	0.950	1.000	1.050	1.100	1.150	1.200	1.250
K_T	0.230	0.242	0.247	0.242	0.224	0.198	0.173	0.149	0.125	0.101	0.078	0.054	0.032
10 K_Q	0.462	0.485	0.492	0.480	0.440	0.398	0.357	0.317	0.277	0.238	0.200	0.164	0.128
η_O	0.515	0.556	0.598	0.642	0.689	0.714	0.735	0.749	0.753	0.742	0.708	0.635	0.492

Table 9.21 KCA (ANN) MM's data for commercial propellers (BAR = 0.53, P/D = 1.238)

σ = 2.3													
J	0.650	0.700	0.750	0.800	0.850	0.900	0.950	1.000	1.050	1.100	1.150	1.200	1.250
K_T	0.224	0.243	0.247	0.235	0.213	0.189	0.168	0.149	0.130	0.109	0.086	0.062	0.036
$10\ K_Q$	0.468	0.498	0.506	0.489	0.455	0.412	0.369	0.328	0.289	0.250	0.212	0.172	0.131
η_O	0.495	0.543	0.583	0.612	0.635	0.659	0.690	0.723	0.751	0.763	0.747	0.685	0.549

Table 9.22 AEW MM's data for commercial propellers (BAR = 0.53, P/D = 1.238)

σ = atm													
J	0.100	0.200	0.300	0.400	0.500	0.600	0.700	0.800	0.900	1.000	1.100	1.200	1.300
K_T	0.509	0.478	0.445	0.409	0.370	0.329	0.287	0.244	0.200	0.155	0.111	0.067	0.024
$10\ K_Q$	0.891	0.843	0.790	0.734	0.674	0.612	0.547	0.481	0.412	0.343	0.273	0.202	0.132
η_O	0.091	0.181	0.269	0.355	0.437	0.514	0.585	0.646	0.693	0.720	0.710	0.631	0.376
σ = 2.23 (left side: fully developed cavitation, right side: transition region)													
J	0.150	0.200	0.250	0.320	0.360	0.400	0.450	0.500	0.550	0.600	0.650	0.700	0.760
K_T	0.057	0.057	0.057	0.058	0.058	0.072	0.091	0.113	0.136	0.162	0.190	0.221	0.260
J	0.100	0.150	0.200	0.250	0.320	0.360	0.400	0.450	0.500	0.550	0.600	0.650	0.700
$10\ K_Q$	0.134	0.135	0.135	0.136	0.139	0.171	0.206	0.253	0.305	0.361	0.421	0.485	0.553

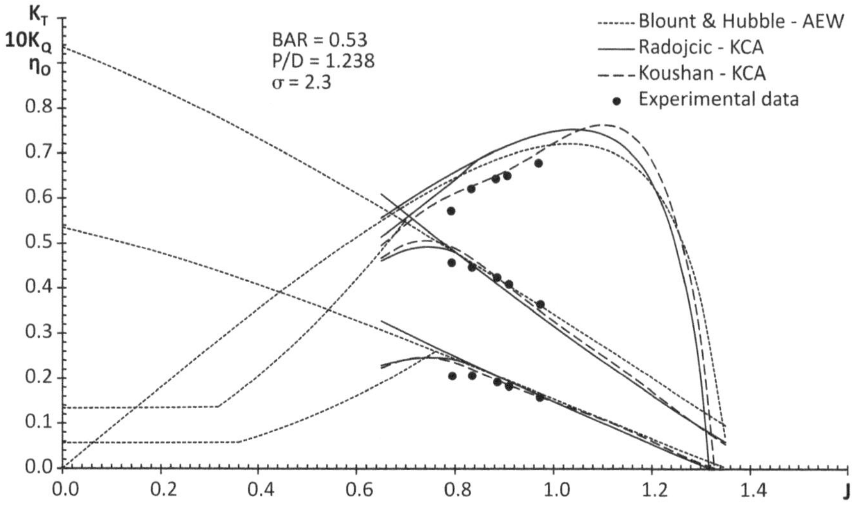

Fig. 9.18 Comparison of three MMs with a commercial propeller EAR = 0.53 ($EAR_{nominal}$ = 0.50), P/D = 1.238 ($P/D_{nominal}$ = 1.25), σ = 2.3

References

Blount DL, Hubble EN (1981) Sizing segmental section commercially available propellers for small craft. In: Propellers '81 symposium, SNAME, Virginia Beach

Denny SB, Puckette LT, Hubble EN, Smith SK, Najarian RF (1988) A new usable propeller series. SNAME, Hampton Road Section

Koushan K (2005) Mathematical expressions of thrust and torque of Newton-Rader propeller series for high speed crafts using artificial neural networks. In: Proceedings of 8th international conference on fast sea transportation (FAST 2005), St. Petersburg

Koushan K (2007) Mathematical Expressions of thrust and torque of Gawn-Burrill propeller series for High speed crafts using artificial neural networks. In: Proceedings of 9th international conference on fast sea transportation (FAST 2007), Shanghai

Oosterveld MWC, van Oossanen P (1975) Further computer-analyzed data of the Wageningen B-screw series. Int Shipbuilding Progr 22(251)

O'Brien TP (1969) The design of marine screw propellers. Hutchinson and Co. Publishers Ltd., London

Radojčić D (1988) Mathematical model of segmental section propeller series for open-water and cavitating conditions applicable in CAD. In: Propellers '88 Symposium, SNAME, Virginia Beach

Chapter 10
Additional Topics on Resistance, Propulsion and Powering

10.1 Prediction of Additional Resistance Components

Mathematical models for prediction of bare hull resistance and propeller's open water efficiency have been outlined in Chaps. 8 and 9, respectively. These parameters are order of magnitude more significant than other HSC powering components and as such are the focus of this book. Nevertheless, accurate power prediction involves other resistance and propulsion components, and hence a few approximate methods, as well as worked examples, for the prediction of these factors are also given in this chapter.

The current section consists of four parts, following similar logic as was used in Chap. 5:

- Section 10.1.1 presents the empirical equations for evaluation of appendage resistance (Note: Appendage lift is assumed to be negligible)
  - For each of the appendage component (skeg, shaftline, rudder)
  - For forward appendages without rudder (i.e. shaftline only)
  - For total appendage drag (valid for twin-screw, hard-chine hulls)
  - For appendage drag per shaftline (hard-chine, shaftline and rudder).
- Section 10.1.2 presents an equation for evaluation of air and wind resistance.
- Section 10.1.3 presents an empirically based equation for prediction of added resistance due to seaway, valid for planing craft only.
- Section 10.1.4 provides the prediction of resistance in shallow water. Since this is not a straight forward procedure, a worked example is given later in Sect. 10.3.2.

Electronic supplementary material The online version of this chapter (https://doi.org/10.1007/978-3-030-30607-6_10) contains supplementary material, which is available to authorized users.

D. Radojčić et al., *Power Prediction Modeling of Conventional High-Speed Craft*,
https://doi.org/10.1007/978-3-030-30607-6_10

10.1.1 Prediction of Appendage Resistance

Empirically Based Equations for Each of the Appendage Components (Hadler 1966)

- **Skeg and Centreline Keels**

$$\text{Resistance} - R = \frac{1}{2}\rho S v^2 C_F$$

where

ρ mass density of water
S wetted surface of skeg or keel
v speed of advance, i.e. speed of a craft
C_F ATTC or ITTC skin friction coefficient (at the appropriate Rn based on length of appendage in direction of flow).

- **Shafts and Strut Bossings**

$$\text{Resistance} - R = \frac{1}{2}\rho l d v^2\left(1.1\sin^3\varepsilon + \pi C_F\right)$$

where

l length of shaft or bossing
d diameter of shaft or bossing
ε angle of flow
Other symbols as above.
Equation valid for $Rn_d < 5.5 \times 10^5$, where $Rn_d = \frac{vd}{\upsilon}$ (based on shaft diameter).

- **Rudder and Struts (Streamline Bodies)**

$$\text{Resistance} - R = \frac{1}{2}\rho S_R v^2\left(2C_F\left(1 + 2t/c + 60(t/c)^4\right)\right)$$

where

S_R profile area of rudder or strut as seen from one side
t/c section thickness—chord ratio
Other symbols as above.

Forward Appendages Without Rudder (Radojčić 1991)

Radojčić (1991) makes the following assumptions: (a) propeller trust increases when the rudder is in propeller race, hence the rudder drag may be neglected, and (b) appendage's wetted surface is proportional to propeller disc area. If required, rudder drag may be added separately according to the equation provided above.

Rudder design is dictated by the maneuvering demands (approximately $A_{rudder} =$ 4% of lateral underwater area at rest, or for planing craft $0.0016L_P^2$ as per Hadler and Hubble 1971).

$$\text{Resistance}-R_{AP} = \frac{1}{2}\rho\frac{\pi D^2}{4}v^2 C_{DP} \approx 0.4\rho D^2 v^2 C_{DP}$$

where

$C_{DP} = \left[0.224\psi^3 - 8.8\psi^2 + 115.4\psi - (171 - 14S_t)\right] \times 10^{-4}$
$\psi = 7.5 - 15°$ – shaft angle
$v < 33\,\text{kn}$ – speed of advance, i.e. speed of a craft
$S_t = \begin{cases} 0, \textit{for I strut} \\ 1, \textit{for V strut} \end{cases}$ – strut type

Appendage Allowance (Semi-Displacement SKLAD Series, Gamulin 1996)

Appendage allowance is 3.5% R_{TBH} per shaft line, where appendages consists of a propeller shaft, bracket with single strut and a rudder.

Total Appendage Drag (Twin-Screw, Hard-Chine, Blount and Fox 1976)

$$\text{Resistance}-R_{AP} = R_{BH}\left(\frac{1}{\eta_a} - 1\right)$$

where $\eta_a = \frac{1}{0.005F_{n\nabla}^2 + 1.05}$
Note: Slightly higher values are obtained.

Appendage Drag per Shaftline (Hard-Chine, Blount 2014)

$$\text{Resistance}-R_{AP} = R\left(0.003F_{n\nabla}^2 + 0.025\right)$$

10.1.2 Prediction of Air and Wind Resistance

$$\text{Resistance} - R_{AA} = \frac{1}{2}\rho_A A_w v_R^2 C_{AA}$$

where

ρ_A	mass density of air (1.23 kg/m^3 at 15 °C and 1 atm)
A_w	area exposed to wind
$v_R = v_{SHIP} \pm v_{WIND}$	speed of advance
$C_{AA} = 0.5 - 0.8$	wind resistance coefficient (lower values for streamlined superstructure)
Note	A_w might be substituted with B^2
	R_{AA} should not be accounted for if incorporated in C_A

10.1.3 Prediction of Resistance in a Seaway

Empirically based equations for planing craft (Hoggart 1979; Hoggart and Jones 1980) follow:

$$\text{Added Resistance} \; -R_A/\Delta = 1.3\left(H_{1/3}/B_{PX}\right)^{0.5}\left(L_P/\nabla^{1/3}\right)^{-2.5} Fn_\nabla$$

$$\text{Vertical acceleration} - \eta_{1/10CG} = 7\left(H_{1/3}/B_{PX}\right)(1+\tau/2)^{0.25}\left(L_P/B_{PX}\right)^{-1.25} Fn_\nabla$$

where $H_{1/3}$—significant wave height in irregular sea

Boundaries of applicability (in parentheses for vertical acceleration) are:

$$C_\Delta = 0.151(0.174) - 1.266$$
$$L_P/\nabla^{1/3} = 4.54 - 7.54$$
$$L_P/B_{PX} = 2.824(2.664) - 6.43$$
$$\beta = 0(10) - 24^\circ$$
$$H_{1/3}/B_{PX} = 0.133(0.122) - 0.769(0.8)$$
$$V/\sqrt{L_P} = 1 - 6$$

Note: Acceleration $\eta_{1/10CG}$ has dimensions in gs. Same reference gives equations for acceleration at bow as $\eta_{1/10bow}$. The values in brackets refer to acceleration.

10.1.4 Prediction of Resistance in Shallow Water

Ratio of shallow- to deep-water wave resistance (R_{Rh}/R_{Rd}) can be evaluated from the relatively simple MM given in (Radojčić and Bowles 2010). This approach is based on the fact that L/h is the dominant factor and that other parameters (L/T, C_B, hull cross-section, etc.) are of secondary importance. The algorithm shown in Fig. 10.1 explains this procedure, and a worked numerical example is given in Sect. 10.3.2. Note that equations evaluate resistance increments in the subcritical region only since the primary problem for a vessel is to overcome the large resistance hump around the critical speed. Resistance decrements in the supercritical regime are neglected since are less important, see Fig. 10.2.

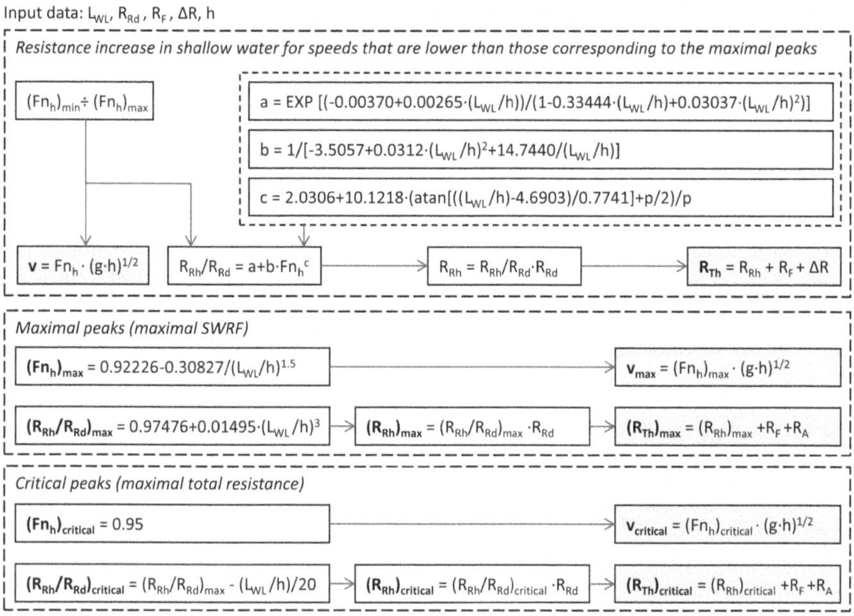

Fig. 10.1 Flow chart for computation of resistance in shallow water

Fig. 10.2 Resistance in various water depths showing *maximal* and *critical* peaks

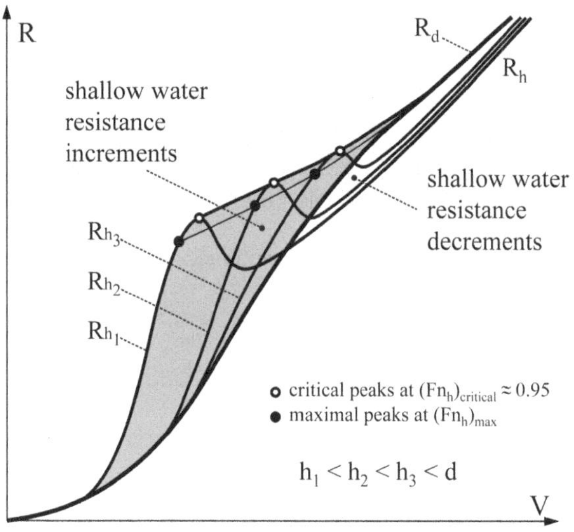

10.2 Prediction of Propulsive Coefficients

The purpose of this section is to highlight a few approximate methods for determination of propulsive coefficients to be used for preliminary power predictions. For high-speed round-bilge hulls with propellers on inclined shafts, two methods are given. The first one is based on the model experiments of several similar hulls. The second is for SKLAD series already treated in Sect. 8.2.4. Tabulated data for HSC with various propulsor types follow. Note that the results obtained are approximate mean values, and are only indicative, hence should be used with care.

High-Speed Round-Bilge Hulls (Bailey 1982)

This reference gives approximate MMs for evaluation of η_D, η_R, η_H, w and t. Only first one is replicated here.

- **Propulsive Efficiency η_D**

$\eta_D = \eta_0 \eta_R \eta_H$	η_D or QPC (quasi propulsive coefficient)
η_0	propeller open-water efficiency
η_R	relative rotative efficiency
$\eta_H = \frac{1-t}{1-w}$	hull efficiency.

$$\eta_D = a_0 + a_1(M) + a_2 LCB + a_3\psi + a_4\frac{P}{D} + a_5 BAR + a_6\frac{t_a}{D} + a_7\beta + a_8 J$$

where

ψ	shaft inclination relative to buttock
LCB	%L positive aft of amidship
$\frac{t_a}{D}$	propeller blade thickness ratio at shaft axis
β	transom wedge angle (if fitted). $\beta = \delta w$
a_i	coefficients (Table 10.1)

Table 10.1 Regression coefficient for η_D

$C_B < 0.45$

Coeff.	Multipl.	a_i	Variation in parameter
a_0	1	0.45005	–
a_1	(M)	–0.00185	4.98–7.44
a_2	LCB	–0.00926	2.55–6.73
a_3	ψ	–0.00424	7–16°
a_4	P/D	–0.11153	0.903–1.35
a_5	BAR	–0.08421	0.491–0.978
a_6	t_a/D	1.14984	0.038–0.068
a_7	β	0.00246	0–10°
a_8	J	0.37675	0.6–1.15

$C_B > 0.45$

Coeff.	Multipl.	a_i	Variation in parameter
a_0	1	0.09232	–
a_1	(M)	–0.00443	5.16–8.01
a_2	LCB	0.01743	2.0–6.5
a_3	ψ	–0.01887	10–13°
a_4	P/D	0.03292	0.96–1.34
a_5	BAR	–0.26592	0.487–0.8
a_6	t_a/D	–5.47884	0.045–0.07
a_7	β	0.00000	–
a_8	J	0.33943	0.651–1.21

Semi-Displacement SKLAD Series (Gamulin 1996)

- **Propulsive Efficiencies**

For rough evaluations $$\eta_D \approx \begin{cases} 0.50\, for \quad P_D = P_{EBH}/\eta_D \\ 0.55\, for \quad P_D = P_{EAP}/\eta_D \end{cases}$$

For more accurate power evaluation $\eta_D = \eta_0 \eta_R \eta_H$. Unfortunately, available are w, t and η_R coefficients only for the hulls having $C_B \leq 0.45$

- **Wake Fraction**

$$w = \begin{cases} 0.056 - 0.066 \cdot Fn_\nabla & for \quad Fn_\nabla = 0.6 - 1.45 \\ 0.04 \cdot Fn_\nabla - 0.1 & for \quad Fn_\nabla = 1.45 - 3.0 \end{cases}$$

- **Thrust Deduction Fraction**

$$t = \begin{cases} 0.15 - 0.08 \cdot Fn_\nabla & for \quad Fn_\nabla = 0.6 - 1.45 \\ 0.02 \cdot Fn_\nabla & for \quad Fn_\nabla = 1.45 - 3.0 \end{cases}$$

- **Relative Rotative Efficiency**

$$\eta_R = 0.035 \cdot Fn_\nabla + 0.9 \quad for \quad Fn_\nabla = 0.6 - 3.0$$

High-Speed Craft (Blount and Bjarne 1989)

See Table 10.2.

Table 10.2 Typical propulsive coefficients for various HSC types

Propulsion concept		Speed range								
		$Fn_L \leq 0.4$			$0.4 < Fn_L \leq 2.5$			$Fn_L > 2.5$		
		w	t	η_R	w	t	η_R	w	t	η_R
Propeller on inclined shaft	6° shaft	0.01 to −0.02	0.01	0.97 to 1.01	0 to 0.04	0.01 to 0.02	0.97 to 1.01	0 to −0.10	0.03	0.97 to 1.01
	12° shaft	0.02 to −0.02	0.05	0.97 to 1.01	0.04 to −0.05	0.05 to 0.07	0.97 to 1.01	0.03 to −0.05	0.07 to 0.11	0.97 to 1.01
Propeller in Tunnel	40% D	−0.03	0.10	0.92	0.02 to −0.03	0.07 to 0.10	0.90 to 0.93	0.03	0.03 to 0.07	0.88 to 0.90
	65% D	−0.03	0.12	0.92	0 to 0.05	0.10 to 0.12	0.90 to 0.93	0.04 to 0.05	0.08 to 0.10	0.88 to 0.90
Propeller under hull	Pull (tractor) type	0	0 to 0.05	1.00	0	0 to 0.05	1.00	0	0 to 0.05	1.00
	Push type	0.05 to 0.07	0.05 to 0.07	0.97 to 1.01	0.05 to 0.07	0.05 to 0.07	0.97 to 1.01	0.05 to 0.07	0.05	0.97 to 1.01
Outboard & Inboard (Outdrive Propeller)		0.03	0	0.97 to 1.01	0.03	0	0.97 to 1.01	0.03	0	0.97 to 1.01
Surface Pearcing Propeller (SPP)		0	0	0.97 to 1.01	0	0	0.97 to 0.98	0	0	0.97 to 1.01

Note For propeller on inclined shaft, t includes all effects of shaft inclination and dynamic trim. For contra-rotating propellers $\eta_R = 1.06 - 1.08$

Propulsive Coefficients in Shallow Water

All propulsive coefficients change in shallow water, i.e. w decreases, t increases and η_D decreases. Open water efficiency η_0 decreases mostly due to increased propeller loading, which is essentially a result of increased shallow water resistance. The approximate multiplying correction factor for extremely shallow water relative to the deep water propulsive efficiency η_D is of the order of 0.8, 0.6 and even 0.4 for L/h of around 5, 10 and 20 respectively. This is a very rough estimate based on Helm (1961, 1963), also Hofman and Radojčić (1997). These peaks/hollows are between $Fn_h = 0.82$ and 0.90 (lower values correspond to extremely shallow water, i.e. to L/h values given above). Note that given correction factors are oversimplified due to (a) general lack of reliable data, and (b) h/T ratio, LCG and other consequential parameters are neglected in spite of the fact that they may be influential. Hence these estimated correction factors should be used with caution.

10.3 An Example of Powering Process for High-Speed Craft

This section contains a complete worked example for a typical semi-displacement mega yacht of 500 t. The same numerical example was already used in Sect. 8.4.1 with the purpose of comparing the recommended MMs for the semi-displacement hull forms. A comprehensive calculation procedure for evaluation of bare hull resistance for the NPL hull form is given here. Evaluation of the additional resistance components, i.e. appendage resistance and air resistance is presented in Sect. 10.3.1, to enable the prediction of total calm water resistance in deep water. Evaluation of the shallow water resistance is presented in Sect. 10.3.2. Deep and shallow water power estimation using approximate equations (MMs) for evaluation of propulsive coefficients (Sect. 10.2) is given in Sect. 10.3.3. The procedures presented correspond roughly to the powering process typically performed during the preliminary design stage.

10.3.1 Deep Water Resistance

Prediction of Bare Hull Resistance

Input data for semi-displacement hull form is given in Table 10.3.

Table 10.3 Input data

Hull characteristics		
L_{WL}	[m]	52
B_{WL}	[m]	9.3
T	[m]	2.55
Δ	[t]	500

Other data		
ρ	[t/m^3]	1.025
ν	[m^2/s]	1.1907E–06
g	[m/s^2]	9.81

Throughout this example C_F according to ITTC–1957

Bare hull resistance is calculated according to the NPL speed-dependent MM (see Sects. 8.1.1 and 8.2.3, see Table 10.4). The correlation allowance is $C_A = 0$.

Table 10.4 Evaluation of the bare hull resistance (R_T)

No.		Fn_∇	[–]	0.8	1.0	1.2	1.4	1.6	1.8
1	Δ = 100000 lb (45.359 t)	$(R_T/\Delta)_m$	[–]	0.0137	0.0250	0.0452	0.0588	0.0678	0.0756
2		R_{Tm}	[kN]	6.11	11.14	20.12	26.15	30.16	33.65
3		v_m	[m/s]	4.71	5.89	7.07	8.25	9.42	10.60
4		Rn_m 10^{-8}	[–]	0.9232	1.1540	1.3847	1.6155	1.8463	2.0771
5		C_{Fm} 10^3	[–]	2.1077	2.0408	1.9885	1.9459	1.9100	1.8792
6		C_{Tm} 10^3	[–]	5.8393	6.8164	8.5489	8.1643	7.2076	6.3533
7		C_{Rm} 10^3	[–]	3.7317	4.7756	6.5603	6.2185	5.2976	4.4741
8	Δ = 500 t	v	[m/s]	7.034	8.793	10.552	12.310	14.069	15.827
9		V	[kn]	13.675	17.094	20.512	23.931	27.350	30.769
10		Fn_L	[–]	0.311	0.389	0.467	0.545	0.623	0.701
11		Rn 10^{-8}	[–]	3.0721	3.8401	4.6081	5.3761	6.1441	6.9121
12		C_F 10^3	[–]	1.7820	1.7300	1.6891	1.6557	1.6275	1.6032
13		C_T 10^3	[–]	5.5137	6.5055	8.2494	7.8741	6.9251	6.0774
14		R_F	[kN]	20.64	31.31	44.02	58.74	75.41	94.02
15		R_R	[kN]	43.23	86.44	170.99	220.60	245.47	262.38
16		R_T	[kN]	63.87	117.75	215.01	279.34	320.88	356.40
17		P_E	[kW]	449	1035	2269	3439	4514	5641

Values in rows 1–7 refer to the geometrically similar hull form (model) with $\Delta = 100{,}000$ lb. Ship-model scale is $\lambda = \sqrt[3]{\Delta/45.359} = 2.22895$, where Δ is displacement of the new design. Therefore, the length of the model is: $L_m = L/\lambda = 23.329\,\mathrm{m}$.

1. The model resistance to weight ratio is calculated according to the mathematical model given in Table 8.6. Note: Here $R_T = R_{BH}$.
2. Total resistance of the model: $R_{Tm} = (\mathrm{R_T}/\Delta) \cdot 45.359 \cdot g$, where g = 9.81 m/s^2.
3. The model speed: $v_m = Fn_\nabla \cdot \sqrt{g \cdot \nabla_m^{1/3}} = 5.8896 \cdot Fn_\nabla$.
4. Reynolds number for the model: $Rn_m = L_m \cdot v_m/\nu$.
5. Frictional resistance coefficient for the model: $C_{Fm} = 0.075/(\log Rn_m - 2)^2$.
6. Total resistance coefficient for the model: $C_{Tm} = 2 \cdot R_{Tm}/(\rho \cdot S_m \cdot v_m^2)$; wetted surface of the model is $S_m = 91.94\,\mathrm{m}^2$, according to the MM given in Table 8.8.
7. Residuary resistance coefficient of the model: $C_{Rm} = C_{Tm} - C_{Fm}$.

Values in rows from 8 to 17 refer to the new design (500 t). Volumetric displacement of the new design is: $\nabla = L \cdot B \cdot T \cdot 0.397 = 489.57\,\mathrm{m}^3$.

8. Speed of the new design in m/s: $v = Fn_\nabla \cdot \sqrt{g \cdot \nabla^{1/3}} = 8.792 \cdot Fn_\nabla$.
9. Speed of the new design in kn.
10. Length Froude number: $Fn_L = v/\sqrt{g \cdot L}$.
11. Reynolds number for the new design: $Rn = L \cdot v/\nu$.
12. Frictional resistance coefficient for the new design: $C_F = 0.075/(\log Rn - 2)^2$.
13. Total resistance coefficient for the new design: $C_T = C_F + C_R + C_A$, where $C_R = C_{Rm}$ and correlation allowance is assumed to be $C_A = 0$.
14. Frictional resistance of the new design: $R_F = 0.5 \cdot \rho \cdot S \cdot v^2 \cdot C_F$, where S is wetted surface of the new design $S = S_m \cdot \lambda^2 = 456.78m^2$.
15. Residual resistance of the new design: $R_R = 0.5 \cdot \rho \cdot S \cdot v^2 \cdot C_R$
16. Total bare hull resistance in deep water of the new design: $R_T = 0.5 \cdot \rho \cdot S \cdot v^2 \cdot C_T$
17. Effective power of the new design: $P_E = R_T \cdot v$.

Prediction of Appendage Resistance

The quadruple-screw propulsion is adopted. Consequently, appendages consist of four sets of propeller shafts, struts (I type), strut bossings and rudders. The characteristics of appendages are assumed as follows:

Skeg:	$l_{Skeg} = 6.9$ m, $S_{Skeg} = 14$ m^2
Shaft:	$d_S = 0.13$ m, $l_S = 4.4$ m (for propeller diameter $D = 1.4$ m)
Strut Bossing:	$l_{SB} = 0.56$ m, $d_{SB} = 0.22$ m
Strut:	0.8 m × 0.4 m, $c = 0.45$ m, $t/c = 0.16$ m
Rudder:	$h = 1.35$ m, $c = 0.75$ m, $t/c = 0.15$

Equations for Each Appendage (Hadler 1966)

- *Skeg*

Appendage resistance due to skeg can be estimated from (Table 10.5):

$$R_{Skeg} = \frac{1}{2}\rho S v^2 C_F$$

where: $l_{Skeg} = 6.9\,\mathrm{m}, S_{Skeg} = 14\,\mathrm{m}^2$

Table 10.5 Evaluation of appendage resistance due to skeg

No.	Fn_∇	[–]	0.8	1.0	1.2	1.4	1.6	1.8
1	v	[m/s]	7.034	8.793	10.552	12.310	14.069	15.827
2	$Rn_{Skeg}\ 10^{-8}$	[–]	0.4076	0.5095	0.6115	0.7134	0.8153	0.9172
3	$C_{F\ Skeg}\ 10^3$	[–]	2.3828	2.3026	2.2400	2.1891	2.1463	2.1097
4	R_{Skeg}	[kN]	0.8460	1.2774	1.7894	2.3802	3.0481	3.7919
5	R_{skeg}/R_T	[%]	1.325	1.085	0.832	0.852	0.950	1.064

- *Propeller Shafts*

Appendage resistance due to propeller shaft can be estimated from (Table 10.6):

$$R_{Sh} = \frac{1}{2}\rho l d v^2 \left(1.1 \sin^3 \varepsilon + \pi C_{FShaft}\right)$$

where: $l_S = 4.4\,\mathrm{m}, d_S = 0.13\,\mathrm{m}, \varepsilon = 9^\circ$

Table 10.6 Evaluation of appendage resistance due to propeller shafts

No.	Fn_∇	[–]	0.8	1.0	1.2	1.4	1.6	1.8
1	v	[m/s]	7.034	8.793	10.552	12.310	14.069	15.827
2	$Rn_d\ 10^{-5}$	[–]	7.680	9.600	11.520	13.440	15.360	17.280
3	$Rn_{Sh}\ 10^{-6}$	[–]	25.9943	32.4928	38.9914	45.4900	51.9885	58.4871
4	$C_{F\ Sh}\ 10^3$	[–]	2.5579	2.4687	2.3993	2.3429	2.2956	2.2550
5	$R_{Sh\ per\ Shaft}$	[kN]	0.1776	0.2711	0.3833	0.5138	0.6624	0.8290
6	R_{Sh}	[kN]	0.7102	1.0844	1.5330	2.0551	2.6497	3.3161
7	R_{Sh}/R_T	[%]	1.112	0.921	0.713	0.736	0.826	0.930

Note $Rn_d > 5.5 \times 10^5$, hence this method is not applicable

- *Strut Bossings*

Appendage resistance due to strut bossing can be estimated from (Table 10.7):

$$R_{SB} = \frac{1}{2}\rho l_{SB} d_{SB} v^2 \left(1.1 \sin^3 \varepsilon + \pi C_{FSB}\right)$$

where: $l_{SB} = 0.56\,\mathrm{m}, d_{SB} = 0.22\,\mathrm{m}, \varepsilon = 9°$.

Table 10.7 Evaluation of appendage resistance due to strut bossings

No.	Fn_∇	[–]	0.8	1.0	1.2	1.4	1.6	1.8
1	v	[m/s]	7.034	8.793	10.552	12.310	14.069	15.827
2	$Rn_d\ 10^{-5}$	[–]	12.997	16.246	19.496	22.745	25.994	29.244
3	$Rn_{SB}\ 10^{-6}$	[–]	3.3084	4.1355	4.9625	5.7896	6.6167	7.4438
4	$C_{F\ SB}\ 10^3$	[–]	3.6716	3.5191	3.4014	3.3065	3.2274	3.1600
5	$R_{SB\ per\ Shaft}$	[kN]	0.0492	0.0745	0.1047	0.1396	0.1793	0.2235
6	R_{SB}	[kN]	0.1967	0.2980	0.4187	0.5585	0.7170	0.8941
7	R_{SB}/R_T	[%]	0.308	0.253	0.195	0.200	0.223	0.251

Note $Rn_d > 5.5 \times 10^5$, hence this method is not applicable

- *Struts*

Appendage resistance due to strut can be estimated from (Table 10.8):

$$R_{St} = \frac{1}{2}\rho S_R v^2 \left(2C_{FSt}\left(1 + 2t/c + 60(t/c)^4\right)\right)$$

where: $S_R = 0.32\,\mathrm{m}^2, c = 0.45\,\mathrm{m}, t/c = 0.16$.

Table 10.8 Evaluation of appendage resistance due to struts

No.	Fn_∇	[–]	0.8	1.0	1.2	1.4	1.6	1.8
1	v	[m/s]	7.034	8.793	10.552	12.310	14.069	15.827
2	$Rn_{St}\ 10^{-6}$	[–]	2.6585	3.3231	3.9878	4.6524	5.3170	5.9816
3	$C_{F\ St}\ 10^3$	[–]	3.8309	3.6685	3.5433	3.4424	3.3584	3.2869
4	$R_{St\ per\ Strut}$	[kN]	0.0845	0.1265	0.1759	0.2326	0.2964	0.3671
5	R_{St}	[kN]	0.3381	0.5058	0.7036	0.9304	1.1855	1.4685
6	R_{St}/R_T	[%]	0.529	0.430	0.327	0.333	0.369	0.412

Note Values given in row 5 correspond to four struts

- *Rudders*

Appendage resistance due to rudder can be estimated from (Table 10.9):

$$R_{Rudder} = \frac{1}{2}\rho S_R v^2 \left(2C_{FR}\left(1 + 2t/c + 60(t/c)^4\right)\right)$$

where: $S_R = 1.0125\,\text{m}^2, c = 0.75\text{m}, t/c = 0.15.$

Table 10.9 Evaluation of appendage resistance due to rudders

No.	Fn_∇	[–]	0.8	1.0	1.2	1.4	1.6	1.8
1	Va	[m/s]	9.145	11.431	13.717	16.003	18.289	20.576
2	$Rn_R\ 10^{-6}$	[–]	5.7601	7.2001	8.6401	10.0802	11.5202	12.9602
3	$C_{FR}\ 10^3$	[–]	3.3096	3.1788	3.0777	2.9958	2.9276	2.8693
4	$R_{Rudder\ single}$	[kN]	0.3821	0.5735	0.7995	1.0593	1.3521	1.6772
5	R_{Rudder}	[kN]	1.5285	2.2939	3.1981	4.2373	5.4083	6.7086
6	R_{Rudder}/R_T	[%]	2.393	1.948	1.487	1.517	1.685	1.882

Note Rudder inflow is assumed to be $Va = 1.3 \cdot v$ (accounted for water acceleration due to the propeller). Values given in row 5 correspond to four rudders

Forward Appendages Without Rudder (Radojčić 1991)

Appendage resistance due to the shaft, strut and strut bossing can be estimated from (Table 10.10):

$$R_{App} = \frac{1}{2}\rho\frac{\pi D^2}{4}v^2 C_{DP},$$

where: $C_{DP} = \left[0.224\psi^3 - 8.8\psi^2 + 115.4\psi - (171 - 14S_t)\right] \cdot 10^{-4}$

$$D = 1.4\,\text{m}, S_t = 0(\text{strut} - \text{I type}), \psi = 9^\circ$$

Table 10.10 Evaluation of appendage resistance due to forward appendages without rudders

No.	Fn_∇	[–]	0.8	1.0	1.2	1.4	1.6	1.8
1	v	[m/s]	7.034	8.793	10.552	12.310	14.069	15.827
2	$R_{App\ per\ Shaft}$	[kN]	1.2418	1.9403	2.7941	3.8030	4.9672	6.2866
3	R_{App}	[kN]	4.9672	7.7613	11.1762	15.2121	19.8688	25.1465
4	R_{App}/R_T	[%]	7.777	6.591	5.198	5.446	6.192	7.056

Note Values given in row 3 correspond to four shafts configuration

Appendage Allowance According to the SKLAD Series (Gamulin 1996)

Resistance of the appendages which consists of a propeller shaft, bracket with single strut, and the rudder is 3.5% R_T per shaft line, i.e. is 14.0% R_T for quadruple-screw installation (Table 10.11).

Table 10.11 Evaluation of appendage resistance according to the SKLAD series

No.	Fn_∇	[–]	0.8	1.0	1.2	1.4	1.6	1.8
1	v	[m/s]	7.034	8.793	10.552	12.310	14.069	15.827
2	R_T	[kN]	63.87	117.75	215.01	279.34	320.88	356.40
3	R_{App}	[kN]	8.9418	16.4848	30.1014	39.1075	44.9229	49.8954
4	R_{App}/R_T	[%]	14.0	14.0	14.0	14.0	14.0	14.0

Total Appendage Drag (Blount and Fox 1976)

Total appendage resistance for twin-screw hard-chine planing crafts can be estimated from (Table 10.12):

$$R_{App} = R_T\left(\frac{1}{\eta_a} - 1\right), \eta_a = \frac{1}{0.005Fn_\nabla^2 + 1.05}$$

Table 10.12 Evaluation of the total appendage resistance

No.	Fn_∇	[–]	0.8	1.0	1.2	1.4	1.6	1.8
1	v	[m/s]	7.034	8.793	10.552	12.310	14.069	15.827
2	R_T	[kN]	63.87	117.75	215.01	279.34	320.88	356.40
3	η_a	[–]	0.9495	0.9479	0.9459	0.9436	0.9409	0.9379
4	R_{App}	[kN]	6.7958	12.9523	24.5972	33.4089	40.3022	47.1868
5	R_{App}/R_T	[%]	10.640	11.000	11.440	11.960	12.560	13.240

Note Values given in row 4 correspond to four shafts configuration

Appendage Resistance per Shaftline (Blount 2014)

Total appendage resistance per shaftline, for hard-chine planing crafts, can be estimated from (Table 10.13):

$$R_{App} = R_T\left(0.003Fn_\nabla^2 + 0.025\right)$$

Table 10.13 Evaluation of total appendage resistance

No.	Fn_∇	[–]	0.8	1.0	1.2	1.4	1.6	1.8
1	v	[m/s]	7.034	8.793	10.552	12.310	14.069	15.827
2	R_T	[kN]	63.87	117.75	215.01	279.34	320.88	356.40
3	R_{App}	[kN]	6.8775	13.1878	25.2164	34.5040	41.9451	49.4962
4	R_{App}/R_T	[%]	10.768	11.200	11.728	12.352	13.072	13.888

Note Values given in row 3 correspond to four shafts configuration

Comparison of Different Appendage Resistance Methods

Comparison of results for the appendage resistance is given in Table 10.16. The following appendages are considered: propeller shafts, struts, strut bossings and rudders. Note that skeg resistance is not accounted for. For a method based on the forward appendages only (Radojčić 1991), the rudder resistance evaluated according to Hadler (1966) is added to enable comparison of results on equal basis (Table 10.14).

Table 10.14 Appendage resistance presented as R_{APP}/R_T (here $R_T = R_{BH}$)

No.	Method	Fn_∇	[–]	0.8	1.0	1.2	1.4	1.6	1.8
1	Hadler (1966)	R_{App}/R_T	[%]	4.34	3.55	2.72	2.79	3.10	3.48
2	Radojčić (1991)	R_{App}/R_T	[%]	10.17	8.54	6.69	6.96	7.88	8.94
3	SKLAD	R_{App}/R_T	[%]	14.00	14.00	14.00	14.00	14.00	14.00
4	Blount and Fox (1976)	R_{App}/R_T	[%]	10.64	11.00	11.44	11.96	12.56	13.24
5	Blount 2014	R_{App}/R_T	[%]	10.77	11.20	11.73	12.35	13.07	13.89

It should be noted that Hadler's method is not applicable (shaft and strut bossing), hence results are, as expected, not comparable with other methods.

Prediction of Air Resistance

Air resistance can be estimated according to recommendations given in 10.1.2.

$$R_{AA} = \frac{1}{2}\rho_A A_w v_R^2 C_{AA}$$

where: $\rho_A = 1.23\,\text{kg/m}^3$, $A_W = 86.5\,\text{m}^2$ and C_{AA} is assumed to be 0.65

Table 10.15 Air resistance

No.	Fn_∇	[–]	0.8	1.0	1.2	1.4	1.6	1.8
1	v	[m/s]	7.034	8.793	10.552	12.310	14.069	15.827
2	R_{AA}	[kN]	1.711	2.673	3.849	5.239	6.843	8.661
3	R_{AA}/R_T	[%]	2.585	2.203	1.748	1.829	2.073	2.353

Prediction of Total In-Service Resistance

For the prediction of total in-service resistance, but in calm and deep water, which approximately corresponds to trial conditions, the following is assumed:

- Model-ship correlation allowance of $C_A = 0.2 \times 10^{-3}$
- The appendage resistance of 10% R_T.

Other data are evaluated, see Table 10.4 for bare hull resistance, Table 10.5 for skeg resistance, and Table 10.15 for air resistance. Total in-service resistance in calm and deep water is given in Table 10.16.

Table 10.16 Components of in-service resistance (calm and deep water)

No.	Fn_∇	[–]	0.8	1.0	1.2	1.4	1.6	1.8
1	V	[kn]	13.67	17.09	20.51	23.93	27.35	30.77
2	v	[m/s]	7.03	8.79	10.55	12.31	14.07	15.83
3	R_F	[kN]	20.64	31.31	44.02	58.74	75.41	94.02
4	R_R	[kN]	43.23	86.44	170.99	220.60	245.47	262.38
5	$R_{T\ (CA=0)}$	[kN]	63.87	117.75	215.01	279.34	320.88	356.40
6	R_A	[kN]	2.32	3.62	5.21	7.10	9.27	11.73
7	$R_{T\ (CA=0.0002)}$	[kN]	66.19	121.37	220.22	286.43	330.14	368.12
8	R_{App}	[kN]	6.62	12.14	22.02	28.64	33.01	36.81
9	R_{Skeg}	[kN]	0.85	1.28	1.79	2.38	3.05	3.79
10	R_{AA}	[kN]	1.71	2.67	3.85	5.24	6.84	8.66
11	ΔR	[kN]	11.49	19.71	32.87	43.36	52.17	60.99
12	R_T*	[kN]	75.36	137.46	247.88	322.70	373.05	417.39
13	P_E*	[kW]	530	1209	2616	3972	5248	6606
14	$P_E*/P_{E\ BH}$	[%]	114	113	113	113	113	113

Note
Row 6 $R_A = 0.5 \cdot \rho \cdot S \cdot v^2 \cdot C_A$

Row 11 $\Delta R = R_A + R_{App} + R_{Skeg} + R_{AA}$
Row 12 $R_T* = R_T + \Delta R$

The final deep water results for a 500 t semi-displacement mega yacht with quadruple-screw propulsion are shown in Fig. 10.3. Presented are total in-service resistance (R_T*), bare hull resistance (R_T), residuary resistance (R_R) and frictional resistance (R_F) for a speed range between 13 and 32 kn.

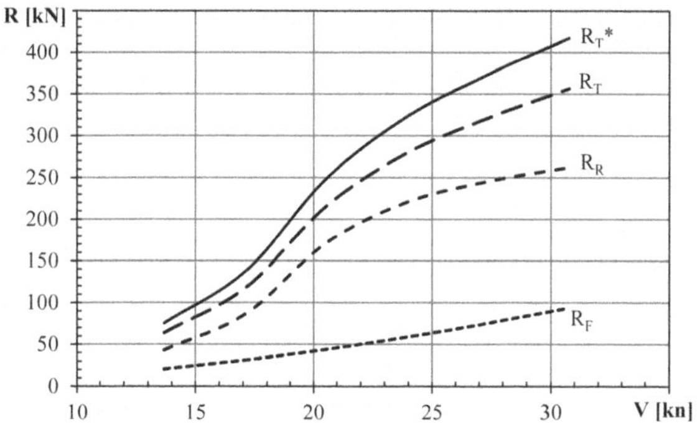

Fig. 10.3 Components of in-service resistance (calm and deep water)

10.3.2 Shallow Water Resistance

In this section an example of evaluation of shallow water resistance is given. The vessel used in the example is again a 52 m, 500 t NPL hull examined in Sect. 10.3.1. This time, however, it is assumed that she will sail from deep into shallow waters, with the depth of h = 10.4 m (corresponding to L/h = 5) being the shallowest. The calculation procedure is same as the one presented in Sect. 10.1.4. Four water depths are examined: L/h = 2, 3, 4 and 5. Other (deep water) data is from Table 10.16 and Fig. 10.3. Calculation is performed in three segments, see Figs. 10.1 and 10.2:

- Total resistance increase in shallow water for speeds that are lower than those corresponding to the maximal peaks (Table 10.17)
- Maximal peaks (Table 10.18) and
- Critical peaks (Table 10.19).

In-service resistance in shallow water (R_{Th}*, see Table 10.17) is calculated from:

$$R^*_{Th} = R_F + R_{Rh} + \Delta R,$$

where: $R_{Rh} = (R_{Rh}/R_{Rd}) \cdot R_{Rd}$, $R_{Rd} = R_R$ (see row 4 in Table 10.16).

The final results for shallow water resistance increments are presented in Fig. 10.4 (results for L/h = 2 are not shown because resistance increment with respect to the deep water is negligible). The shaded area in Fig. 10.4 corresponds to resistance increment due to considered water depths.

Table 10.17 Evaluation of resistance increase in shallow water (valid for speeds that are lower than those corresponding to the maximal peaks)

h=26.0 m (L/h)=2		a = 1.003541		b = 0.250557		c = 2.933268	
Fn_h	[–]	0.44	0.55	0.66	0.77	0.88	0.99
R_{Rh}/R_{Rd}	[–]	1.026	1.047	1.078	1.120	1.176	1.248
R_{Rh}	[kN]	44.36	90.50	184.29	247.14	288.74	327.33
R_{Th}*	[kN]	76.49	141.52	261.19	349.24	416.32	482.34
h=17.3 m (L/h)=3		a = 1.015876		b = 0.592091		c = 3.414676	
Fn_h	[–]	0.54	0.67	0.81	0.94	1.08	1.21
R_{Rh}/R_{Rd}	[–]	1.088	1.170	1.303	1.502	1.784	2.164
R_{Rh}	[kN]	47.02	101.14	222.85	331.44	437.79	567.67
R_{Th}*	[kN]	79.16	152.16	299.74	433.53	565.38	722.68
h=13.0 m (L/h)=4		a = 1.047673		b = 1.471670		c = 4.745220	
Fn_h	[–]	0.62	0.78	0.93	1.09	1.25	1.40
R_{Rh}/R_{Rd}	[–]	1.203	1.497	2.114	3.264	5.224	8.350
R_{Rh}	[kN]	52.02	129.36	361.46	719.97	1282.21	2190.91
R_{Th}*	[kN]	84.15	180.38	438.36	822.06	1409.79	2345.92
h=10.4 m (L/h)=5		a = 1.115951		b = 4.482295		c = 8.317657	
Fn_h	[–]	0.70	0.87	1.04	1.22	1.39	1.57
R_{Rh}/R_{Rd}	[–]	1.337	2.531	7.562	24.349	71.660	189.018
R_{Rh}	[kN]	57.80	218.74	1292.92	5371.47	17590.26	49593.83
R_{Th}*	[kN]	89.93	269.76	1369.82	5473.56	17717.84	49748.85

Table 10.18 Evaluation of maximal peaks

No.	L/h	[m]	2	3	4	5
1	h	[m]	26.0	17.3	13.0	10.4
2	$(Fn_h)_{max}$	[–]	0.81	0.86	0.88	0.89
3	v_{max}	[m/s]	12.99	11.25	9.98	9.04
4	V_{max}	[kn]	25.25	21.87	19.40	17.57
5	R_F	[kN]	64.91	49.64	39.69	32.98
6	R_R	[kN]	231.38	195.37	146.10	98.80
7	ΔR	[kN]	47.12	37.34	28.41	21.55
8	$(R_{Rh}/R_{Rd})_{max}$	[–]	1.09	1.38	1.93	2.84
9	$(R_{Rh})_{max}$	[kN]	253.21	269.34	282.21	280.93
10	$(R_{Th})_{max}$	[kN]	365.25	356.32	350.31	335.46

Table 10.19 Evaluation of critical peaks

No.	L/h	[m]	2	3	4	5
1	h	[m]	26.0	17.3	13.0	10.4
2	$(Fn_h)_{critical}$	[–]	0.95	0.95	0.95	0.95
3	$v_{critical}$	[m/s]	15.17	12.39	10.73	9.60
4	$V_{critical}$	[kn]	29.49	24.08	20.86	18.65
5	R_F	[kN]	86.85	59.40	45.42	36.89
6	R_R	[kN]	257.79	221.88	177.74	127.41
7	ΔR	[kN]	57.16	44.08	33.80	25.58
8	$(R_{Rh}/R_{Rd})_{critical}$	[–]	0.99	1.23	1.73	2.59
9	$(R_{Rh})_{critical}$	[kN]	256.34	272.61	307.78	330.45
10	$(R_{Th})_{critical}$	[kN]	400.35	376.09	386.99	392.92

Note In Tables 10.18 and 10.19 friction resistance, residuary resistance and additional resistance components, given in rows 5, 6 and 7 respectively, correspond to deep water speeds given in row 3. These values are obtained from Fig. 10.3

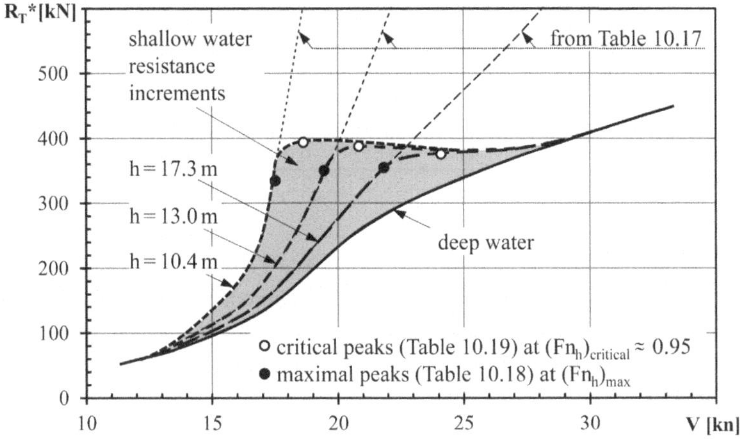

Fig. 10.4 Estimated in-service resistance increase in shallow water. *Note* Deep water data are from Table 10.16, row 12

10.3.3 Power Prediction

In real-life, power estimation must take into account the engine (with a gearbox), the propeller and the hull (see Chap. 6). The focus of this example however, is the propeller performance, while the characteristics of the engine with a gearbox are ignored. The hull performance in both deep and shallow water was already addressed in Sects. 10.3.1 and 10.3.2 respectively. Note that in this example an assumption is made that each of the four propellers operates under uniform conditions, and hence generate the same thrust. It is also assumed that propellers are optimized for a speed of 28 kn in deep water, and that operation in shallow water is occasional.

Two methods that explain the power prediction are elaborated here:

1. Conventional method
2. Short method.

(1) **Conventional Method**

This method is performed in three steps and applies for deep water only:

- Evaluation of propulsive coefficients (i.e. hull-propeller interaction)
- Propeller selection
- Estimation of off-design propeller characteristics

Estimation of Propulsive Coefficients (Gamulin 1996)

Wake fraction (w), thrust deduction fraction (t) and relative rotative efficiency depend on Fn_∇ only (see Sect. 10.2). The results obtained for the speed range considered, including the design speed (V = 28 kn), are given in Table 10.20.

Table 10.20 Propulsive coefficients

V [kn]	Fn_∇ [–	] w [–	]	η_R [–]
13	0.761	0.0058	0.0892	0.9266
18	1.053	–0.0135	0.0658	0.9369
23	1.346	–0.0328	0.0424	0.9471
28	1.638	–0.0345	0.0328	0.9573
33	1.931	–0.0228	0.0386	0.9676

Propeller Selection

Propeller selection techniques per se are not the subject of this book, though propeller optimization procedures and loading criteria are mentioned in Sects. 4.1.1 and 4.3. Propeller optimization is treated, for instance, in Molland et al. (2011) and Blount (2014).

In the present example, the propeller is optimized for V = 28 kn through few iterations applying the KCA (RA) MM (Sect. 9.2.3, Table 9.4). Cavitation and loading criteria are neglected in this example. Influence of shaft inclination on propeller thrust is assumed to be negligible. The procedure applied will be partly explained later (see *procedure for selecting an optimal propeller*).

Total in-service resistance in deep and calm water for the speed range 13–33 kn is given in Table 10.21 based on Fig. 10.3.

Table 10.21 Total in-service resistance

No.	V [kn]	13	18	23	28	33
1	R_T* [kN]	63.2	168.0	305.9	381.8	418.5
2	R_T*/4 [kN]	15.80	41.99	76.47	95.45	104.63

For quadruple-screw propulsion and for $V = 28$ kn total in-service resistance is $R_T* = 381.8$ kN, or $R_T*/4 = 95.45$ kN per propeller. Input and output data for selected optimal propeller (three bladed KCA propeller, D = 1.4 m, P/D = 1.35, DAR = 0.8) is given in Table 10.22. Other hydrodynamic characteristics are obtained from the open-water data of a KCA propeller with the abovementioned characteristics, i.e. from the KCA (RA) MM. Thrust and torque of the propeller are determined from $T = K_T \cdot \rho \cdot n^2 \cdot D^4$ and $Q_0 = K_Q \cdot \rho \cdot n^2 \cdot D^5$.

Table 10.22 Input and output data of a selected propeller

INPUT			OUTPUT		
Resistance per propeller	95.45	kN	Propeller rotational speed	648	rpm
Propeller diameter	1.4	m	Pitch–diameter ratio	1.35	–
Speed	28	kn	Propeller open–water efficiency	0.6976	–
Number of blades	3	–	Advance coefficient	0.9851	–
Propeller area ratio (DAR)	0.8	–	Thrust ceofficient	0.2147	–
Thrust deduction fraction	0.0328	–	Torque coeffcient	0.0483	–
Wake fraction	–0.0345	–	Propeller thrust	98.69	kN
Relative rotative efficiency	0.9573	–	Propeller torque	31.05	kNm

Off-Design Conditions

The total resistance which should be overcome by the thrust of the four propellers is given in Table 10.21. Comprehensive calculation procedure for the design speed $V = 28$ kn only, which is identical for all other of-design speeds, follows.

The speed of advance for $V = 28$ kn is $v_A = V \cdot 0.5144 \cdot (1 - w) = 14.9\,\text{m/s}$. The propeller thrust required is $(R_T{}^*/4)/(1 - t) = 98.69\,\text{kN}$. The evaluation of the wake fraction and thrust deduction fraction is given in Table 10.20.

In this example the propeller's rotational speed (n) is an unknown quantity which should be evaluated from the know input values: $R_T{}^*, D$ and v_A. The usual procedure is to eliminate the unknown values, for instance by superimposing a parabola $K_T{}^* = CJ^2$ (where C is a constant) on a standard K_T-K_Q-J open water propeller chart. This is readily implemented by customized program subroutines.

In particular, for $V = 28$ kn

$$K_T{}^* = J^2 \cdot T/\left(\rho \cdot v_A^2 \cdot D^2\right) = 0.2212 \cdot J^2.$$

Accordingly, the parabola $K_T{}^* = 0.2212 \cdot J^2$ is superimposed on the selected KCA open water propeller chart as shown in Fig. 10.5. The intersection between $K_T{}^* = 0.2212 \cdot J^2$ and thrust coefficient K_T for P/D = 1.35 defines an advance coefficient J and torque coefficient K_Q (see Fig. 10.5), i.e.

$$J = 0.9851 \text{ and } K_Q = 0.0483.$$

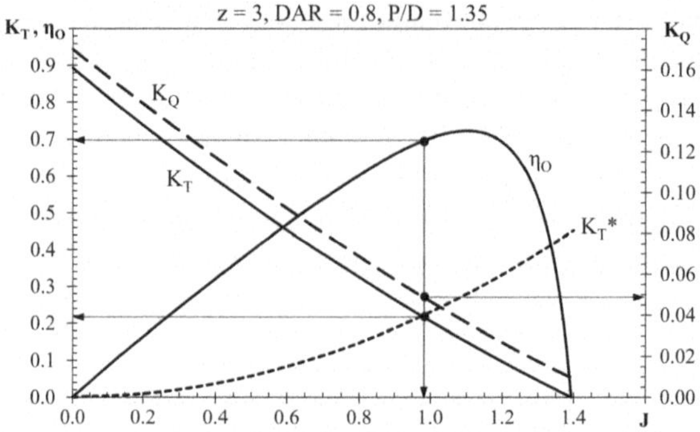

Fig. 10.5 Determination of required advance and torque coefficients

The required rotational speed follows from:

$$n = \frac{v_A}{J \cdot D} = 10.8034\,\mathrm{s}^{-1} = 648.2\ \mathrm{rpm}$$

and propeller torque:

$$Q_O = K_Q \cdot \rho \cdot n^2 \cdot D^5 = 31.05\ \mathrm{kNm}, Q = Q_O/\eta_R = 32.44\,\mathrm{kNm}.$$

Therefore the required power is:

$$P_D = 2 \cdot \pi \cdot n \cdot Q = 2201\,\mathrm{kW}, P_B = P_D/\eta_S = 2269\,\mathrm{kW}.$$

The same procedure should be repeated for other considered speeds, essentially resulting in different parabolas with new slopes. These results are given in Table 10.23 and Fig. 10.6, showing required propeller curve $P_B = f(n)$.

Table 10.23 Propeller curve in deep and calm water

V	[kn]	13	18	23	28	33
n	[rpm]	282	421	554	648	724
P_B	[kW]	182	677	1559	2269	2830

Note that *procedure for selecting an optimal propeller* is basically similar. The difference is that particular propeller is not yet known (i.e. P/D is not defined) while the thrust for one design point is constant. This means that a single parabola $K_T^* = CJ^2$ intersects various K_T curves, for each P/D ratio considered. Each intersection defines the corresponding open water efficiency η_o, see Fig. 10.5. The intersection point corresponding to the highest η_o value, determines an optimal propeller, i.e. P/D ratio. This is also easily executed by custom program routines.

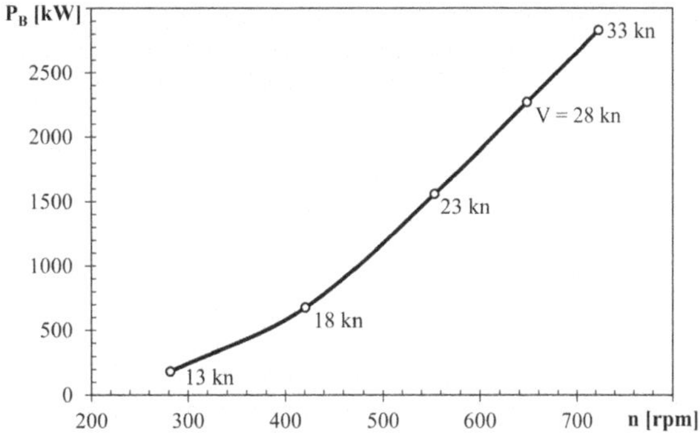

Fig. 10.6 Propeller curve in deep and calm water

(2) Short Method

Short method is less accurate, but is quicker than the conventional method, and is shown here for:

- Deep water performance
- Shallow water performance

Deep Water Performance

- According to the Conventional method the propulsive efficiency for $V = 28$ kn is $\eta_D = \eta_O \eta_R \eta_H = \frac{\eta_O \eta_R (1-t)}{1-w} = 0.6243$ (see Table 10.22).
- According to Gamulin (1996), the estimated value of propulsive efficiency for the SKLAD Series is 0.55.
- The propulsive efficiency estimated from Bailey (1982) (short method, see Sect. 10.2, Table 10.1) is 0.577.

The last MM (Bailey 1982) depends on hull and propeller geometry. Corresponding evaluation procedure is given in Table 10.24 with LCG = 5% L_{WL}, $\psi = 9°$ configuration without stern wedge ($\beta = 0$), ta/D = 0.063 and main hull and propeller parameters as in Tables 10.3 and 10.22.

Note that since the cavitation criteria were neglected, the selected propeller in this example has relatively high value of η_0. If cavitation was to be considered, it may be expected that the DAR value would be higher, resulting in lower η_0, i.e. η_D too.

Table 10.24 Evaluation of propulsive efficiency

Coeff	Multipl.	a_i	Multipl.	$ai \cdot$ Mul.
a_0	1	0.45005	1	0.450
a_1	(M)	–0.00185	6.598	–0.012
a_2	LCB	–0.00926	5	–0.046
a_3	ψ	–0.00424	9	–0.038
a_4	P/D	–0.11153	1.35	–0.151
a_5	BAR	–0.08421	0.8	–0.067
a_6	ta/D	1.14984	0.063	0.072
a_7	β	0.00246	0	0.000
a_8	J	0.37675	0.98	0.369
			Σ	**0.577**

Propulsion Performances in Shallow Water

Propulsion coefficient in shallow water is expressed as a percentage reduction of deep water propulsion coefficient, see Sect. 10.2. Therefore, deep water η_D should be evaluated first for the speed range of interest, see Table 10.25.

Table 10.25 Evaluation of η_D in deep water

No.	V	[kn]	13	18	23	28	33
1	n	[rpm]	282	421	554	648	724
2	η_R	[–]	0.9266	0.9369	0.9471	0.9573	0.9676
3	w	[–]	0.0058	–0.0135	–0.0328	–0.0345	–0.0228
4	t	[–]	0.0892	0.0658	0.0424	0.0328	0.0386
5	η_H	[–]	0.9162	0.9218	0.9272	0.9350	0.9400
6	Q_O	[kNm]	5.5566	13.9532	24.6864	31.0503	35.0505
7	K_Q	[–]	0.0457	0.0514	0.0525	0.0483	0.0437
8	η_O	[–]	0.7060	0.6859	0.6814	0.6975	0.7116
9	η_D	[–]	0.5993	0.5923	0.5984	0.6243	0.6472

Values given in rows 1 and 2 to 4 are from Tables 10.23 and 10.20 respectively.

5. Hull efficiency: $\eta_H = (1 - t)/(1 - w)$.
6. Propeller torque: $Q_O = P_B \cdot \eta_S \cdot \eta_R/(2 \cdot \pi \cdot n)$. Here η_s refers to shaft and gearbox efficiency (assumed to be 3%).
7. Torque coefficient: $K_Q = Q_o/\left(\rho \cdot n^2 \cdot D^5\right)$.
8. Open-water efficiency is evaluated as indicated in Fig. 10.5.
9. Propulsive efficiency in deep water: $\eta_D = \eta_R \cdot \eta_H \cdot \eta_O$.

The distribution of shallow water η_D reduction is assumed to be linear (between minimal and maximal values), see Fig. 10.7, as:

- Below $Fn_h < 0.65$ there is no influence of a shallow water
- Estimated shallow water η_D reduction for L/h = 5 is around 20% (i.e. multiplying factor is 0.8, see Sect. 10.2) at $Fn_h \approx 0.85$
- Above $Fn_h \approx 1.2$ shallow water resistance starts to decrease compared to deep water resistance (this is neglected here, see Fig. 10.2).

This evaluation procedure is given in Table 10.26. The engine power for operation in shallow water is given in Fig. 10.8. Note that a curve $P_B = f(V)$ obtained with a multiplying factor 1 means that a shallow water propulsive coefficient is practically the same as in a deep water. In reality this cannot happen. Applying a multiplying factor of less than 1 is more realistic and exhibits the influence of shallow water on a propulsive coefficient.

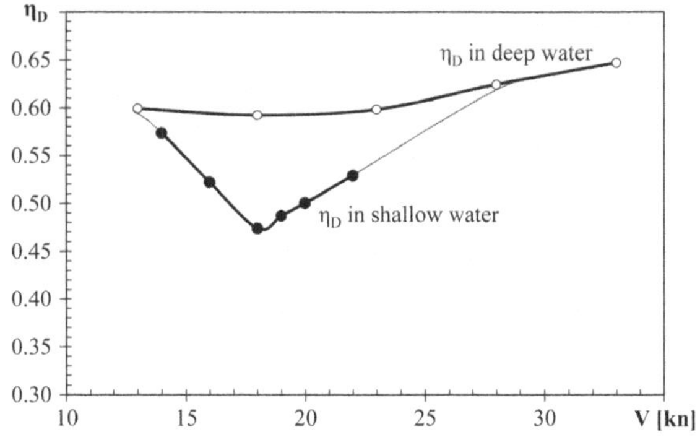

Fig. 10.7 η_D distribution in deep and shallow water for L/h = 5, i.e. h = 10.4 m

Table 10.26 Evaluation of required brake power in shallow water (L/h = 5)

No.	Fn_h	[–]	0.71	0.81	0.92	0.97	1.02	1.12
1	V	[kn]	14	16	18	19	20	22
2	η_D	[–]	0.5970	0.5930	0.5920	0.5920	0.5925	0.5950
3	$\eta_{D\ red.,\ 20\%}$	[%]	4.00	12.00	20.00	17.78	15.56	11.11
4	η_{Dh}	[–]	0.57	0.52	0.47	0.49	0.50	0.53
5	$R_{Th}*/4$	[kN]	25.57	44.02	97.13	99.75	99.34	98.11
6	P_{Eh}	[kW]	184.2	362.3	899.4	974.9	1022.1	1110.3
7	$P_{Dh,\ 0\%}$	[kW]	308	611	1519	1647	1725	1866
8	$P_{Bh,\ 0\%}$	[kW]	318	630	1566	1698	1778	1924
9	$P_{Dh,\ 20\%}$	[kW]	321	694	1899	2003	2043	2099
10	$P_{Bh,\ 20\%}$	[kW]	331	716	1958	2065	2106	2164

1. Hull speed in deep water.
2. η_D in deep water, determined from Fig. 10.7.
3. The reduction of η_D in %.
4. η_{Dh} in shallow water.
5. Resistance in shallow water (L/h = 5, h = 10.4 m) per propeller, determined from Fig. 10.4.
6. The effective power in shallow water: $P_{Eh} = \left(R^*_{Th}/4\right) \cdot V \cdot 0.5144$.
7. The delivered power in shallow water: $P_{Dh} = P_{Eh}/\eta_D$, without correction of propulsive efficiency (multiplying factor = 1).
8. The brake power per propeller in shallow water: $P_{Bh} = P_{Dh}/\eta_S$, where η_S refers to total shaft transmission efficiency, including gearbox losses (multiplying factor = 1).
9. The delivered power in shallow water: $P_{Dh} = P_{Eh}/\eta_{Dh}$, with correction of propulsive efficiency (multiplying factor less than 1, Fig. 10.7).
10. The brake power per propeller in shallow water: $P_{Bh} = P_{Dh}/\eta_S$, where η_S refers to total shaft transmission efficiency, including gearbox losses (multiplying factor less than 1, Fig. 10.7).

The engineering estimates for this example for the deep and shallow water curves P_B = f(V) per propeller are presented in Fig. 10.8. However, two facts concerning this example, should be noted:

(a) Engine characteristics are ignored; analysis concerns a FPP only. In addition, cavitation and loading criteria are neglected, as well as the influence of shaft inclination on propeller thrust. For extreme cases as above a CPP should be considered.
(b) Extra thrust margin, usually necessary to account for speeds corresponding to the resistance hump, was not used (see Blount 2014).

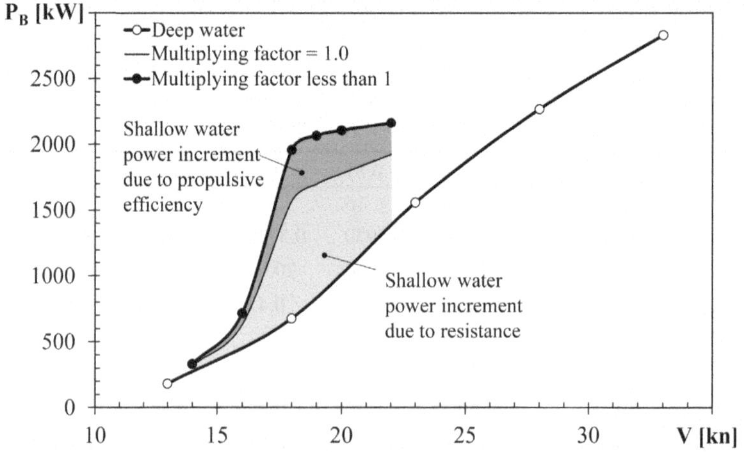

Fig. 10.8 The brake power per propeller in deep and shallow water (h = 10.4 m, L/h = 5, multiplying factor 0.8)

10.4 Catamaran Series '89

MM's for estimation of various hydrodynamic characteristics of hard chine catamarans are presented here. These models are based on extensive Series '89 resistance and self-propulsion tests with relatively large models, intended for high-speed passenger and cargo vessels, and hence the results obtained are reliable. Three groups of MMs developed by regression analysis (see Müller-Graf et al. 2002 and Sect. 6.3) are presented in Sect. 10.4.1:

- MM's for estimation of wetted surface coefficient $WSC = S/(\nabla/2)^{2/3}$ and length-to-displacement ratio $L_{WL}/(\nabla/2)^{1/3}$ as $\mathrm{f}(\beta_M, L_{WL}/B_{XDH})$—Table 10.28.
- Speed-independent MM for estimation of residual resistance-weight ratio $\varepsilon_R = R_R/(g \cdot \Delta)$ as $\mathrm{f}(Fn_{\nabla/2}, \beta_M, \delta_W, L_{WL}/B_{XDH})$—Table 10.29.
- Deadrise-independent MM's for estimation of dynamic trim (θ), propulsive coefficients (η_D, η_H, η_O, $1 - t_X$, $1 - w_T$) and specific brake power (ε_{BTR}) as $\mathrm{f}(Fn_{\nabla/2}, L_{WL}/B_{XDH})$—Tables 10.30, 10.31 and 10.32.

Note that unlike the MM's presented up to now, this specific MM accounts for the appendage resistance. Therefore, there is no need for addition of appendage resistance when power is estimated, regardless if the Conventional or Short method is used. Results obtained are valid for towing tank conditions (R_{TTK}). Namely:

$$R_{TTK} = \varepsilon_R \cdot \rho \cdot \nabla \cdot g + 0.5 \cdot \rho \cdot v_S^2 \cdot (C_F + \Delta C_F) \cdot 2 \cdot S.$$

Included in R_{TTK} is an appendage resistance consisting of a typical appendage arrangement consisting of (per demihull, see Fig. 10.9):

- A deadwood carrying the propeller shaft (Area = (0.18 − 0.20) $L_{WL} \cdot T$)
- A spade rudder (Area (0.014 − 0.017) $L_{WL} \cdot T$, with NACA 0015 profile).

The characteristics of a stock propeller used for the propulsion experiments (from now on just *stock propeller*), are as follows: P/D = 1.4, A_E/A_O = 1.10, z = 3, with modified NACA 16 profiles and cupped trailing edge. Consequently, the propulsion coefficients derived are based on this propeller.

Two flowcharts, showing the evaluation procedure of brake power are presented in Figs. 10.10 and 10.11. They describe brake power evaluation for a subject hull (trial conditions) equipped with both, the propellers whose characteristics correspond to those of the stock propellers and to some other propellers. Worked example of a hard-chine hull form catamaran is given in Sect. 10.4.2.

10.4.1 Mathematical Models (MMs) for Power Evaluation

Reference Müller-Graf et al. (2002); see Sect. 6.3 (Table 10.27).

INPUT	OUTPUT
$X_1 = Fn_{\nabla/2}$	ε_R
$X_2 = \beta_M$	θ
$X_3 = \delta_W$	$\eta_D, \eta_H, \eta_O, 1 - t_X, 1 - w_T$
$X_4 = L_{WL}/B_{XDH}$	ε_{BTR}

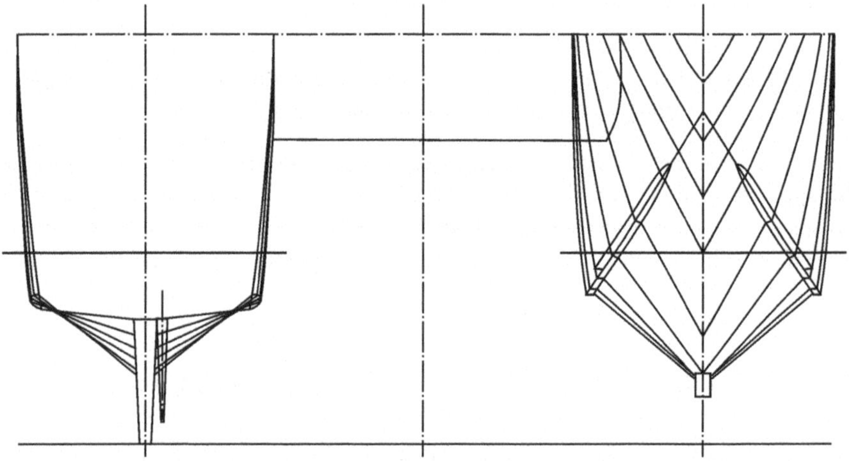

Fig. 10.9 Body plan of the Catamaran series '89

Table 10.27 Boundaries of applicability for three math model groups

Mathematical model	Continuous boundaries	Discrete boundaries
WSC, $L_{WL}/(\nabla/2)^{1/3}$	$L_{WL}/B_{XDH} = (7.55 \div 13.55)$	$\beta_M = 16°, 27°, 38°$
ε_R	$L_{WL}/B_{XDH} = (7.55 \div 13.55)$ $\beta_M = (16° \div 38°)$	$Fn_{\nabla/2} = 0.50, 0.75, 1.00, 1.25, 1.50, 2.00, 2.50, 3.00, 3.50, 4.00$ $\delta_W = 0°, 4°, 8°, 12°$
θ, $1 - t_X$, $1 - w_T$, η_H, η_O, η_D, ε_{BTR}	$L_{WL}/B_{XDH} = (7.55 \div 13.55)$ $Fn_{\nabla/2} = (0.75 \div 4.00)$	$\beta_M = 16°, 27°, 38°$ $\delta_W = 8°$

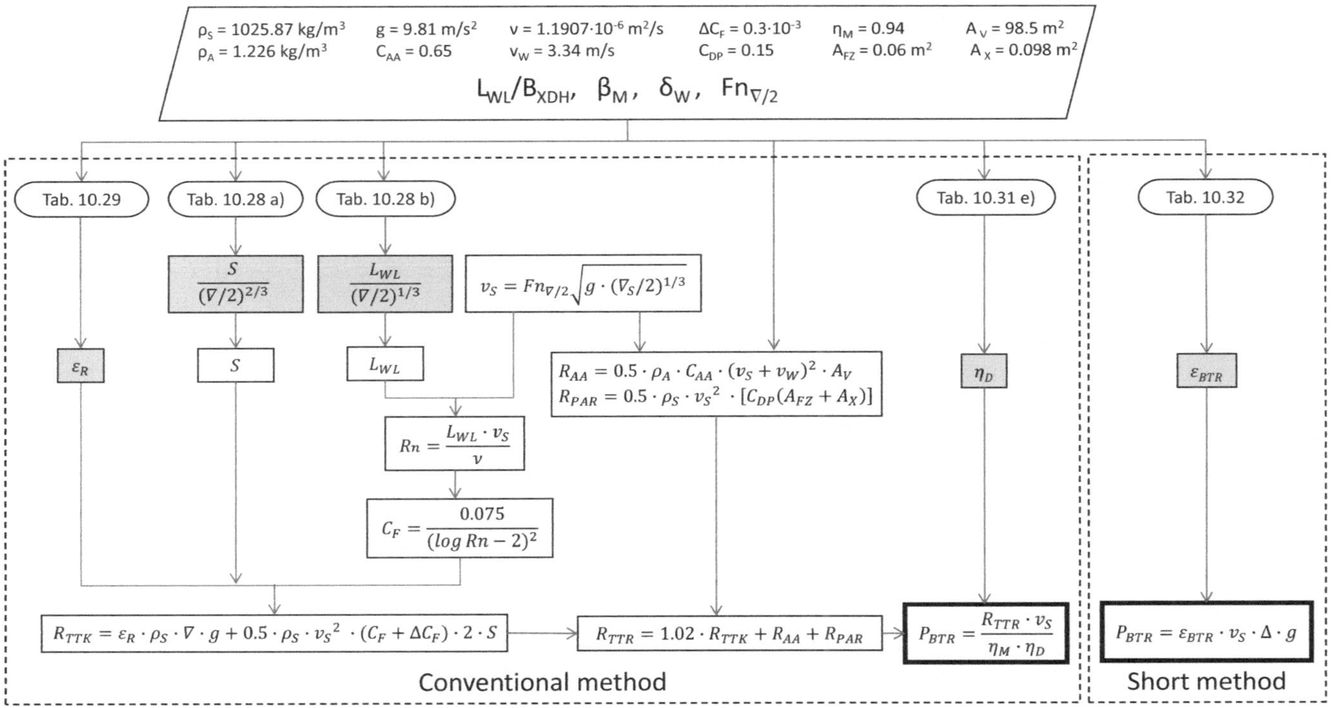

Fig. 10.10 Flow chart showing evaluation procedure for obtaining P_{BTR} (shaded blocks indicate values which are calculated by MMs)

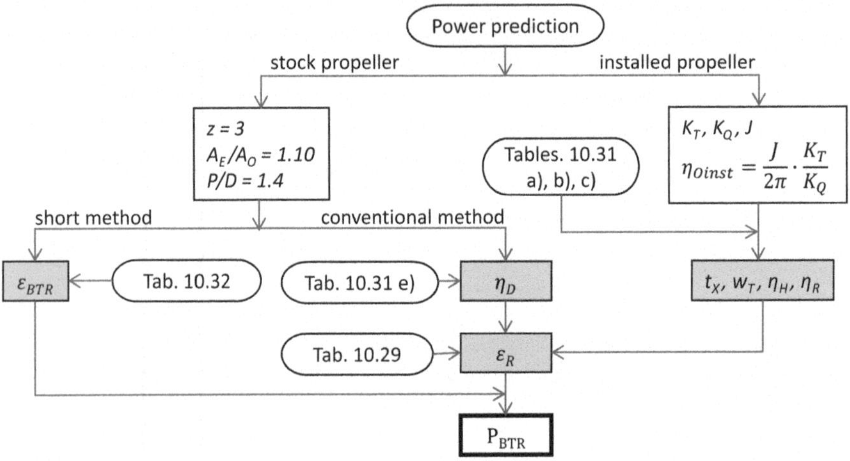

Fig. 10.11 Flow chart showing the evaluation of power prediction (shaded blocks indicate values which are calculated by MM). *Note* The relative rotative coefficient has an almost constant value $\eta_R = 0.96$

Wetted Surface and Length-to-Displacement Ratio

See Table 10.28.

Table 10.28 Polynomial terms and regression coefficients for (a) WSC and (b) $L/(\nabla/2)^{1/3}$

$$WSC = \sum_{i=1}^{11} coeff_i \cdot (X_2)^{b_i} \cdot (X_4)^{d_i} \qquad L_{WL}/(\nabla/2)^{1/3} = \sum_{i=1}^{11} coeff_i \cdot (X_2)^{b_i} \cdot (X_4)^{d_i}$$

i	b	d	$coeff_i$	i	b	d	$coeff_i$
1	0	0	–963.4005565	1	0	0	29.96173317
2	0	1/2	173.5484274	2	0	1/4	–22.23455773
3	1/4	4	3.39391E–05	3	0	1	2.07593765
4	1/4	1/4	1870.596868	4	0	2	–0.053750199
5	1/4	1/3	–1419.106142	5	0	4	0.00028278
6	2/3	4	9.90050E–06	6	1/4	4	–0.000125747
7	1	1/3	–16.12995062	7	1	2	0.001814593
8	1	2/3	5.942104476	8	2	0	0.000212214
9	3	3	–7.49963E–07	9	2	3/4	–0.001845546
10	4	0	2.58292E–05	10	3	1/4	0.000169785
11	4	4	6.71128E–10	11	4	0	–3.32164E–06

Note Wetted surface S with deadwood and rudder

Residuary Resistance

See Table 10.29.

Table 10.29 Polynomial terms and regression coefficients for ε_R

$$\varepsilon_R = \sum_{i=1}^{61} coeff_i \cdot (X_2)^{b_i} \cdot (X_3)^{c} \cdot (X_4)^{d_i}$$

				$Fn_{\nabla/2}$									
i	**b**	**c**	**d**	**0.50**	**0.75**	**1.00**	**1.25**	**1.50**	**2.00**	**2.50**	**3.00**	**3.50**	**4.00**
1	0	0	0	–0.173909006	–112.2651783	–509.5614157	–31.32927612	–6.483827938	–213.9264914	–5.538630171	580.1108040	54.27603261	–184.6978916
2	0	0	1/3	0.098273967	294.2994508	1408.766278	19.14025823	–	539.9987966	–	–1605.025331	–66.19286297	914.8184421
3	0	0	1/2	–	–199.8506111	–994.5100187	–	–	–370.4512859	–9.917725821	1114.673987	–	–699.4893401
4	0	0	1	–	11.2865500	68.12267465	–	1.787214421	25.31768871	7.541644475	–66.93925993	14.77337705	32.83829292
5	0	0	2	–0.000929905	–	–0.650371987	–0.150712610	–0.157478409	–0.375827222	–0.545922051	0.078326823	–0.780955509	1.632325762
6	0	0	3	4.79565E–05	–0.002683374	0.003227611	0.005329248	0.004460691	0.006446521	0.016109245	0.014394597	0.018708262	–0.067066879
7	0	1/3	0	1.401975010	–0.111023232	28.89895758	–	–	5.031697200	–	8.206185884	–	–34.43201745
8	0	1/3	1	–0.373589592	–0.044838426	–	–0.424619179	–	–0.753512558	–	–1.275464296	0.932429098	3.542040660
9	0	1/3	2	0.020557836	–0.001122606	0.004875129	0.031895407	0.013447629	0.046551082	–	0.158227557	–	–0.050695694
10	0	1/3	3	–0.000773178	0.000110826	0.000289405	–0.000640324	0.000773728	–0.000413110	–	–	0.007031444	0.010607503
11	0	1/2	0	–	–	–29.85438839	0.770268435	0.077603426	–3.532728525	3.855052774	–	0.148577600	–
12	0	1/2	1	–	0.111640079	–	0.236867941	–	0.480056793	–1.079438402	–0.602500625	–1.053511336	5.919370038
13	0	1/2	2	0.012565673	–0.004349477	–0.004101338	–0.018451146	–0.011978316	–0.027533194	0.098886531	–	–0.011266240	–0.814803057
14	0	1/2	3	–0.000272806	–	–0.000259073	0.000356916	–0.000684932	–	–0.003000144	–0.003750377	–0.004510684	0.018031341
15	0	1	0	–0.052890922	–0.114161550	3.482000005	–0.154514430	–0.034293080	–	–0.002368950	–	1.064647330	–1.133737713
16	0	2	2	–7.32486E–05	1.06422E–05	0.000135917	–	0.000333708	0.000199097	6.05472E–05	0.001351622	0.001297060	0.002401113
17	0	3	0	–	–	–	–0.003437854	0.011157241	–	–0.003847711	0.061737442	0.060719114	–0.063958353
18	0	3	1/3	5.54611E–06	0.000221070	–0.009114445	0.006701751	–0.030525933	–0.002236196	0.003141962	–0.170204772	–0.182770750	0.019759093
19	0	3	1/2	–	–	0.007117478	–0.003990541	0.021737137	0.003218849	–	0.119972110	0.128987686	0.027020249
20	0	3	1	4.01738E–05	–3.69420E–05	–0.000632067	0.000162902	–0.001584395	–0.000631249	–0.000333281	–0.008100977	–0.008262745	–0.007566737
21	1/3	0	0	–	96.433396322	169.5730735	32.94677916	–273.8738375	23.91614886	16.37966409	–	–50.03192156	1065.910968
22	1/3	0	1/3	0.009026319	–248.0043541	–454.6815211	–19.99056577	782.3842726	–	–	–	64.55774071	–3400.216618
23	1/3	0	1/2	–	166.1281198	311.7049875	–	–557.5559477	–17.87113665	–	26.14833521	–	2480.012934
24	1/3	0	1	0.003117269	–8.698809939	–18.03232553	–	38.40772215	3.459033128	–4.847018415	–15.05328000	–15.21465888	–163.0347301
25	1/3	0	2	–0.000367979	–0.035862798		0.154586141	–0.337995278	–	0.469421912	0.941812585	0.836166796	–
26	1/3	0	3	–	0.003138442	0.004637667	–0.005429348	0.001123062	–0.001945232	–0.014796664	–0.026172290	–0.020484234	0.048761198
27	1/3	1/3	0	–	–1.800692237	–1.054333112		–0.953287538	–9.988112947	–1.426699542	–21.43073246	–	–47.73068432
28	1/3	1/3	1/2	1.067690131	–	–	0.315020592	–	–	0.754873832	7.050646806	2.813607718	–155.6303109
29	1/3	1/3	2	0.002477064	0.001171068	–	–0.004374674	–	–0.006515007	–0.010719862	–0.070689132	–0.026902607	0.392691486
30	1/3	1/3	3	0.000592035	9.70116E–06	1.05017E–05	–	4.35928E–05	0.000243565	0.002221158	0.003126801	–0.001061452	–0.027985432

(continued)

Table 10.29 (continued)

i	b	c	d	$Fn_{\nabla/2}$ 0.50	0.75	1.00	1.25	1.50	2.00	2.50	3.00	3.50	4.00
31	1/3	1/2	0	-1.887287619	2.02892835	0.923845326	-0.651101585	0.835446570	6.759523003	-9.537350769	5.054527124	-5.710893352	–
32	1/3	1/2	1/3	–	-0.29069583	–	–	-0.015614588	1.032769296	5.760080214	–	–	193.9616220
33	1/3	1/2	2	-0.018807273	–	-0.000134816	0.001996301	–	–	-0.049218579	-0.025723379	0.014243209	0.430347911
34	1/3	1	3	8.11248E-05	6.92900E-06	–	–	–	1.67610E-06	0.000193775	0.000125952	0.000151197	-0.001713867
35	1/3	2	0		–	0.007377176	–	–	–	–	-0.002691251	–	0.725363274
36	1/3	2	1/3	0.018872648	-0.007514197	-0.008648184	0.000561973	-0.007775974	-0.054719271	–	-0.024054204	–	-3.065263968
37	1/3	3	0	0.002719157	-2.99538E-05	–	0.002764360	-0.000406371	–	-2.37297E-05	–	0.006042155	0.069087066
38	1/3	3	1/3	-0.005103748	–	-0.001069220	-0.003154042	0.000639660	–	–	–	–	0.002504611
39	1/3	3	1/2	0.001985174	0.000225926	0.000889770	0.001295074	–	0.001748970	–	0.000861132	-0.001671212	0.064193812
40	1/2	0	0	0.072769398	-5.604150947	–	-12.91833009	180.1694818	70.96269449	79.04905394	35.01637327	195.0958416	-409.9154412
41	1/2	0	1/3	-0.064880292	6.381763374	-8.078625842	7.829110200	-510.0190665	-226.4090839	-245.3440866	-102.6309835	-520.0213187	1309.600381
42	1/2	0	1/2	–	–	11.02867858	–	363.6322784	169.3310362	177.4981092	64.86303216	354.4516232	-955.6305107
43	1/2	0	1	0.006109663	-1.373949176	-2.662200029	–	-25.57633917	-13.31164085	-11.52650546	–	-19.99299914	62.87849545
44	1/2	0	2	–	0.075233486	0.124254964	-0.060317890	0.264623868	0.155452352	–	-0.281402090	–	–
45	1/2	0	3	–	-0.001941525	-0.003278130	0.002114272	-0.001965884	-0.001102358	0.003470766	0.009054415	0.004301751	-0.018832291
46	1/2	1/3	0	-0.157843141	0.801038942	0.109010834	-3.742590843	-0.040346250	–	–	20.90150556	-4.464375693	-19.13790349
47	1/2	1/3	2	-0.014155953	–	–	–	–	–	–	–	0.003653487	-0.319682075
48	1/2	1/3	3	0.000215954	7.10900E-06	–	0.000229064	-0.000164645	–	-0.000727262	-0.000892853	0.000801927	0.016699922
49	1/2	1/2	0	1.387602750	-9.468603218	–	3.461746384	–	1.305224198	-9.201836071	-38.99368765	-14.33114167	28.93511436
50	1/2	1/2	1/3	-0.857139387	24.22346633	-0.006271063	0.381620313	–	–	37.96795411	72.07499564	59.07406792	-62.71812853
51	1/2	1/2	1/2	–	-17.27538509	–	–	–	-0.572817096	-29.99605577	-56.82947363	-43.62868441	42.06281945
52	1/2	1/2	1	0.068063834	1.253394711	-0.009874787	-0.038159823	–	0.054841699	2.401100421	4.416686893	3.175312549	1.853727599
53	1/2	1/2	2	0.015791792	-0.016039311	–	–	-5.92686E-05	–	-0.014762850	-0.029411169	-0.037819984	-0.038702724
54	1/2	1/2	3	-0.000332700	0.000161663	–	-0.000195440	0.000146368	-3.58478E-05	0.000425268	0.000606160	–	-0.005256035
55	1/2	1	0	–	–	-0.096847187	–	–	-0.534546754	–	0.705755958	–	–
56	1/2	1	1/3	–	0.031300430	0.023652758	-0.352931053	0.017122595	0.189678093	–	1.579376224	–	0.431205275
57	1/2	1	1	-0.019435990	0.002942045	0.005742715	0.025398792	–	0.012981352	0.001442397	-0.094369396	-0.014264536	-0.662864429
58	1/2	1	3	-1.71375E-05	–	–	1.66244E-05	-1.31558E-05	–	-7.85099E-05	-7.35919E-05	-2.83782E-05	0.000992367
59	1/2	2	0	-0.015355228	0.000833844	0.004159775	-0.025219485		0.016518287	–	–	-0.072182961	-0.530956629
60	1/2	2	1/3	0.012297817	–	–	0.022095670	-0.000780421	-0.003184700	–	-0.034490417	0.031491026	0.356916231
61	1/2	2	1	-0.000717806	-0.000196319	-0.000422623	-0.001488523	-3.05483E-05	-0.000992014	–	0.001824472	–	-0.016747648

Dynamic Trim, Propulsive Coefficients and Power

See Tables 10.30, 10.31 and 10.32.

Table 10.30 Polynomial terms and regression coefficients for θ

$$\theta = \sum_{i=1}^{53} coeff_i \cdot (X_1)^{a_i} \cdot (X_4)^{d_i}$$

			β		
i	**a**	**d**	16	27	38
1	0	0	–634.6364094	18463.00301	1839.125226
2	0	1/3	–	–11497.74607	–
3	0	1	–	–29.06759571	–
4	0	4/3	604.6621485	–	–
5	0	7/3	–65.01351656	–0.073450332	–
6	0	3	–	–	–11.54115526
7	1/3	0	–	–50269.1766	–
8	1/3	1/3	1060.793057	42073.7678	–
9	1/3	2/3	–	–	–9402.757876
10	1/3	1	–	–	8983.269820
11	1/3	4/3	–2162.533542	–	–2788.014903
12	1/3	5/3	–	4.383497838	–
13	1/3	2	–	–	110.0251860
14	1/3	7/3	259.2957698	–	–
15	1/3	3	–	–	57.26172282
16	2/3	0	–460.2054200	–	–
17	2/3	1/3	–	–46581.76531	–
18	2/3	4/3	2487.412038	–	–
19	2/3	2	–	–	–2.172519882
20	2/3	7/3	–377.0172583	–	
21	2/3	3	–	–	–121.0950009
22	1	0	–	127172.7946	–
23	1	4/3	–1265.835277	–	–
24	1	2	180.7786529	–	–
25	1	7/3	162.6137800	–0.186971675	–
26	1	3	0.888384857	–	133.4205720
27	4/3	0	–	–153939.4243	11241.24289
28	4/3	1/3	–	32269.22872	–
29	4/3	2	–	–0.098061479	–
30	4/3	3	–	–	–80.33049354
31	5/3	0	–	61414.01976	–13293.77781
32	5/3	1/3	48.74603199	–17162.13935	–
33	5/3	2	–187.8661802	–	–
34	5/3	7/3	–	–	0.644877162
35	5/3	3	–	–	22.73054976
36	2	5/3	228.0192776	–	–
37	2	7/3	7.954483913	–	–
38	7/3	0		–	7639.734977
39	7/3	4/3	–10.98788879	–	–
40	7/3	7/3	–	–	–0.295338037
41	7/3	8/3	–	0.027083457	–
42	7/3	3	–1.049603559	–	–1.493165472
43	8/3	0	–	–4192.139521	–4391.407319
44	8/3	1/3	–	1460.905499	–
45	8/3	5/3	–58.62118585	–	–
46	8/3	7/3	3.858210486	–	–
47	8/3	8/3	–	–0.012719680	–
48	8/3	3	0.652076570	–	0.242860882
49	3	0	–	1250.279699	778.6221570
50	3	1/3	–	–447.3178958	–
51	3	5/3	14.90568780	–	–
52	3	2	–	–	0.074407280
53	3	8/3	–0.774905725	–	–

Table 10.31 Polynomial terms and regression coefficients for (a) $1 - t_X$, (b) $1 - w_T$, (c) η_H, (d) η_O, (e) η_D

(a)

$$1 - t_X = \sum_{i=1}^{60} coeff_i \cdot (X_1)^{a_i} \cdot (X_4)^{d_i}$$

			β		
i	a	d	16	27	38
1	0	0	582.9743125	3530.929170	–2952.250041
2	0	1/3	–838.4031467	–555.7488296	2915.266617
3	0	2/3	–	–	–651.8353445
4	0	1	339.8417169	–	–
5	0	4/3	–79.55660778	–	–
6	0	7/3	–	1.017677851	–
7	0	8/3	–	–0.401288534	–6.778659381
8	0	3	–	–	0.253753953
9	1/3	0	–	–15889.25018	–8691.700501
10	1/3	1/3	–	1449.326105	32866.94837
11	1/3	2/3	198.4283211	–	–40696.84803
12	1/3	1	–	–	23438.65409
13	1/3	4/3	–168.0668240	–	–7022.837312
14	1/3	5/3	–	–3.878573982	946.5143988
15	1/3	7/3	2.815640455	–	–
16	1/3	8/3	–	–	30.06411250
17	1/3	3	–	0.098489155	
18	2/3	0	–	29154.37403	–3226.456924
19	2/3	1/3	–	–	3163.820203
20	2/3	2/3	–	70.48878096	–
21	2/3	4/3	–45.53842418	–	–62.53581857
22	2/3	5/3	112.8178962	–	–
23	2/3	8/3	–1.141304574	–	–75.67092090
24	2/3	3	–	–0.058671275	0.608771600
25	1	0	18.02983154	–27517.47395	
26	1	1/3	–	–4183.896989	–971.0872708
27	1	2	–31.07238746	–	–
28	1	7/3	–	0.159722886	–
29	1	8/3	–	–	88.74811477
30	1	3	–	–	–0.301893841
31	4/3	0	–	14305.11034	508.4011427
32	4/3	1/3	–81.98165886	5376.565056	–
33	4/3	2/3	–	–	62.38007660
34	4/3	8/3	2.403544674	–	–56.68667611
35	5/3	0	–	–4268.871359	–
36	5/3	1/3	–	–2473.698695	–
37	5/3	2/3	41.15628117	–29.48358228	–
38	5/3	7/3	6.159156761	–	–
39	5/3	8/3	–2.963818778	–	17.06406637
40	2	0	–	979.8949915	–
41	2	1/3	–	–	–42.77319845
42	2	2/3	–	18.58967113	–
43	2	5/3	–4.564213563	–	–
44	2	3	0.248764782	–	–
45	7/3	0	10.14775785	–226.8268865	–
46	7/3	1/3	–12.88819880	335.6204244	–
47	7/3	1	2.543686791	–	–
48	7/3	5/3	–	–0.104261691	–
49	7/3	8/3	–	0.002045042	–1.039991459
50	7/3	3	–0.217205516	–	–
51	8/3	1/3	–	–78.13967182	–
52	8/3	1	–	–	1.916823684
53	8/3	8/3	–0.056086420	–	–
54	8/3	3	0.129735341	–	–
55	3	0	–	13.15790177	–
56	3	5/3	–	–	–0.131867724
57	3	7/3	0.040025677	–	–
58	3	8/3	–	–	0.057518745
59	3	3	–0.024540346	–	–

(b)

$$1 - w_T = \sum_{i=1}^{51} coeff_i \cdot (X_1)^{a_i} \cdot (X_4)^{d_i}$$

			β		
i	a	d	16	27	38
1	0	0	–116.1918809	191.5348352	1351.481200
2	0	1/2	40.67382787	–	–1128.195122
3	0	1	–	–0.217881315	812.3062999
4	0	3/2	–	–	–238.9348394
5	0	2	4.174948152	–	28.17577708
6	0	3	–0.330798043	–	–
7	0	7/2	–	–0.000543264	–
8	0	4	0.004183984	9.72812E–05	–0.019268611
9	1/2	0	–	–1023.839208	–3925.540889
10	1/2	1/2	174.2014292	–	–593.5982772
11	1/2	3/2	–84.37140405	0.354994523	–
12	1/2	5/2	4.530073464	–	–
13	1/2	4	–0.002493759	–	–
14	1/2	9/2	–	–	0.011871410
15	1	0	–	2353.822124	10910.51458
16	1	1/2	–	5.836882114	823.4872384
17	1	1	–216.9905239	–2.835718617	–
18	1	3/2	302.3030098	–	0.012789520
19	1	2	–55.46325021	–	–
20	1	5/2	–1.045376396	–	–
21	1	9/2	–	–	–0.021605961
22	3/2	0	–	–3043.630328	–15843.93045
23	3/2	1/2	–117.5244542	–	–457.9694395
24	3/2	3/2	–188.0468083	–	–
25	3/2	2	45.57921322	–	–
26	3/2	5/2	–	–0.001519587	–
27	3/2	9/2	–	–	0.020584912
28	2	0	204.3949549	2418.832398	12957.03250
29	2	1	221.1227650	1.197713944	–
30	2	2	–12.06524797	–	–
31	2	9/2	–	–	–0.010841703
32	5/2	0	–217.6285779	–1224.646193	–5717.623437
33	5/2	1/2	–	–	99.95627740
34	5/2	1	–113.1188212	–	–
35	5/2	3/2	27.12420570	–	–
36	5/2	2	–	0.001708786	–
37	5/2	9/2	–	–	0.002996035
38	3	0	85.75458413	393.8209279	929.2639667
39	3	1/2	20.11076638	–2.960285905	–30.20825658
40	3	1	5.044755760	–0.382117113	–
41	3	3/2	–2.940023171	–	–
42	3	3	0.002537542	–	–
43	3	9/2	–	–	–0.000339686
44	7/2	0	–16.27160518	–75.26366791	238.7835060
45	7/2	1/2	–	2.370759179	–
46	7/2	1	0.595567253	0.090446610	–
47	4	0	–	6.629323052	–122.8825070
48	4	1/2	–	–0.487424751	–
49	4	1	–	–	–0.001618597
50	9/2	0	–	–	14.59114828
51	9/2	1/2	–	–	0.264191120

(c)

$$\eta_H = \sum_{i=1}^{58} coeff_i \cdot (X_1)^{a_i} \cdot (X_4)^{d_i}$$

i	a	d	β 16	β 27	β 38
1	0	0	−448.9197050	−6.868901638	−74.00031245
2	0	1/2	390.3973700	–	–
3	0	1	−85.75959034	–	−35.99824410
4	0	3/2	–	–	−7.590026496
5	0	3	0.051240292	–	–
6	0	9/2	–	−0.000025439	0.000401350
7	1/2	0	930.1355384	137.8197734	104.4439706
8	1/2	1/2	−874.1077809	−5.412922148	–
9	1/2	1	207.3036743	–	289.9352487
10	1/2	5/2	–	0.000912759	0.722602252
11	1/2	3	−0.127501358	–	–
12	1/2	4	–	–	−0.008513543
13	1/2	9/2	–	0.000004145	
14	1	0	−563.8221973	−438.6536380	−39.57092486
15	1	1/2	609.4217601	–	−19.22645858
16	1	1	−169.6183198	–	−528.8288202
17	1	2	–	–	−1.549456180
18	1	3	0.107077275	–	–
19	1	9/2	–	–	0.001241681
20	3/2	0	–	540.6594967	–
21	3/2	1/2	–	58.27922647	–
22	3/2	1	–	–	308.8733906
23	3/2	3/2	10.05251429	–	–
24	3/2	5/2	−0.295290847	–	–
25	3/2	3	–	0.017719729	–
26	3/2	9/2	–	–	−0.000462956
27	2	0	–	−253.0026209	–
28	2	1/2	−32.22046574	−117.4065953	–
29	2	1	–	–	325.9841841
30	2	3/2	–	–	2.099915340
31	2	2	–	–	−0.394290073
32	2	5/2	–	−0.167056815	–
33	2	3	–	–	0.011639937
34	5/2	0	–	–	3.457922799
35	5/2	1/2	–	80.28435191	–
36	5/2	1	–	–	−694.4983496
37	5/2	2	–	0.681567816	–
38	5/2	7/2	–	–	0.000446629
39	3	0	24.67086115	–	–
40	3	1	–	–	535.9963016
41	3	3/2	−0.138223976	−1.511624376	–
42	3	9/2	–	−0.000000189	–
43	7/2	0	−10.93780000	40.33427999	–
44	7/2	1/2	–	−26.35622515	–
45	7/2	1	–	1.903882320	−221.3579359
46	7/2	3/2	0.085793454	–	–
47	7/2	4	4.09031E−05	–	–
48	7/2	9/2	−6.49539E−06	–	–
49	4	0	1.139998880	−23.00596965	–
50	4	1/2	–	11.02791560	–
51	4	1	0.130195846	–	48.52956223
52	4	3/2	−0.044670575	–	–
53	4	7/2	–	–	−5.14165E−05
54	9/2	0	–	4.140266787	
55	9/2	1/2	–	−1.803444086	−0.034293017
56	9/2	1	–	–	−4.442852827
57	9/2	3/2	–	−0.000126590	–
58	9/2	9/2	–	2.52155E−08	6.35951E−07

(d)

$$\eta_O = \sum_{i=1}^{48} coeff_i \cdot (X_1)^{a_i} \cdot (X_4)^{d_i}$$

i	a	d	β 16	β 27	β 38
1	0	0	−539.9646094	−323.7314017	−981.9975875
2	0	1/2	234.2081809	17.24569041	112.8867187
3	0	1	−53.35882643	–	11.11428039
4	0	1/2	5.841191206	–	–
5	0	2	–	–	0.075115522
6	0	4	–	1.86686E−05	–
7	1/2	0	1948.958149	1691.837213	4677.669316
8	1/2	1/2	−488.9153323	−46.54566534	−244.3071136
9	1/2	1	69.83654546	–	−103.9015400
10	1/2	3/2	−9.520959406	−0.389623973	–
11	1	0	−3234.571716	−3794.580963	−9801.093132
12	1	1/2	272.3105870	23.50016946	143.3682255
13	1	1	55.11461414	–	244.2701654
14	3/2	0	3310.487284	4787.039638	11657.77711
15	3/2	1	−120.2009263	23.61695962	−243.2524091
16	3/2	3/2	5.642199369	–	–
17	3/2	2	–	–	−0.271115211
18	3/2	7/2	–	−0.000126516	
19	2	0	−2201.263462	−3626.708217	−8398.299289
20	2	1	53.70003012	−46.62886027	109.9143262
21	2	3/2	−1.596561157	–	–
22	2	5/2	–	0.005521356	–
23	2	7/2	–	–	−0.000179440
24	5/2	0	883.1813597	1614.474737	3560.798540
25	5/2	1/2	−11.39852858	–	–
26	5/2	1	−7.714356531	40.21291005	−26.84909772
27	5/2	3/2	–	–	2.943272484
28	5/2	5/2	–	–	0.043263618
29	5/2	9/2	7.23797E−06	–	–
30	3	0	−188.2890989	−357.1686594	−749.0763394
31	3	1	–	−19.33755989	–
32	3	2	–	–	−0.380231591
33	3	7/2	−0.000457294	–	–
34	7/2	0	15.82467515	–	–
35	7/2	1/2	–	–	1.094808976
36	7/2	1	–	5.335715856	–
37	7/2	3	–	−0.000717967	–
38	7/2	7/2	0.000204960	9.29276E−05	–
39	4	0	–	16.96658503	31.10439614
40	4	1/2	0.734772168	–	0.187102275
41	4	1	–	−0.935593781	–
42	4	3/2	–	0.044879284	–
43	4	2	–	–	0.060823280
44	9/2	0	–	−2.743963291	−4.399465274
45	9/2	1/2	−0.135893757	0.237815415	–
46	9/2	2	–	–	−0.012714277
47	9/2	4	−1.69973E−06	–	–
48	9/2	9/2	–	–	−6.91935E−08

Note η_O of stock propeller

(e)

$$\eta_D = \sum_{i=1}^{51} coeff_i \cdot (X_1)^{a_i} \cdot (X_4)^{d_i}$$

			β						β		
i	a	d	16	27	38	i	a	d	16	27	38
1	0	0	−1697.527857	−3012.583635	−12558.53639	26	3/4	2	−2.457312631	17.12690239	–
2	0	1/4	−208.6648087	–	−3054.761894	27	1	0	–	−2287.002321	–
3	0	1/3	–	–	5731.260446	28	1	1/4	–	1409.688496	–
4	0	1/2	–	–	−2102.151272	29	1	1/3	–	–	3116.363331
5	0	5/4	3295.227999	−120.5308326	–	30	1	5/4	–	351.9875623	–
6	0	3/2	−4563.078591	172.7186711	–	31	5/4	0	−8532.553146	–	–
7	0	7/4	2302.038967	−93.84143003	–	32	5/4	1/4	–	–	−21171.56317
8	0	2	−361.0569613	37.58570491	–	33	5/4	1/2	–	−1543.092524	–
9	1/4	0	4500.543705	39616.96581	133642.4443	34	5/4	3/2	78.50071312	–	–
10	1/4	5/4	−4414.227614	–	–	35	5/4	2	−5.640828598	–	0.025605665
11	1/4	7/4	−1116.935008	–	–	36	3/2	0	10811.93813	–	
12	1/4	2	238.8224427	−219.1974048	–	37	3/2	1/4	−4334.196865	1012.501108	17622.34797
13	1/3	0	–	−53801.22937	−169070.9407	38	3/2	1/2	–	–	−177.6292385
14	1/3	3/4	–	–	2037.346532	39	3/2	3/4	–	455.7213029	–
15	1/3	5/4	10423.73127	–	−71.30205400	40	3/2	1	–	−211.1959960	–
16	1/3	2	–	272.7511588	–	41	3/2	3/2	−16.75099908	–	–
17	1/2	0	–	18485.76499	47840.95943	42	7/4	0	−3129.261281	–	−32.75621964
18	1/2	1/2	−11771.39585	–	−8233.130491	43	7/4	1/4	1682.420119	−302.9147610	−6141.698763
19	1/2	3/4	8039.352139	–	–	44	7/4	7/4	–	−0.048595877	–
20	1/2	1	−11420.54045	–	–	45	7/4	2	–	–	−0.031617283
21	1/2	7/4	–	114.5430105	–	46	2	0	−28.96904765	5.602032506	−11.98810005
22	1/2	2	–	−109.1135012	0.582095725	47	2	1/4	–	–	852.5390077
23	3/4	1/3	–	–	11729.76897	48	2	5/4	–	2.516913432	
24	3/4	1/2	10338.64064	–	–	49	2	3/2	–	–	0.200206915
25	3/4	3/2	–	−286.0566599	–	50	2	7/4	−2.960994440	–	–
						51	2	2	1.275101545	–	–

Table 10.32 Polynomial terms and regression coefficients for ε_{BTR}

$$\varepsilon_{BTR} = \sum_{i=1}^{41} coeff_i \cdot (X_1)^{a_i} \cdot (X_4)^{d_i}$$

			β						β		
i	a	d	16	27	38	i	a	d	16	27	38
1	0	0	−75.11557898	165.5334799	−96.61876908	21	2	3/2	–	–	0.000186355
2	0	1/2	180.6528927	−32.08809452	57.60677895	22	2	2	1.721986359	–	–
3	0	1	−59.34168690	–	–	23	5/2	0	−122.0236110	−158.1428026	1240.467831
4	0	3/2	7.129496756	−0.070202080	–	24	5/2	1/2	22.00522137	32.66818281	−1035.771938
5	0	2	−0.366894267	–	–	25	5/2	2	0.102483347	–	–
6	1/2	0	–	−791.0198737	787.1852118	26	5/2	5/2	−0.390363802	–	–
7	1/2	1/2	−456.8424596	158.2762657	−408.6449700	27	3	1/2	−2.345849016	–	337.8549472
8	1/2	1	125.1480060	–	–	28	3	3/2	–	−0.000947648	–
9	1/2	3/2	−7.735620595	0.197660404	–	29	3	3	0.034894497	–	–
10	1/2	3	0.003509819	–	–	30	3	7/2	0.001194094	–	–
11	1	0	523.2676054	1509.767793	−2427.530358	31	7/2	0	–	–	−196.5926231
12	1	1/2	342.6528228	−308.5285358	1203.907875	32	7/2	1/2	–	–	−51.55778092
13	1	1	−75.99088211	–	–	33	7/2	7/2	−0.001909626	–	–
14	1	3/2	–	−0.179780824	–	34	7/2	9/2	−1.88073E−05	–	–
15	1	9/2	–	–	1.72793E−07	35	4	0	2.015783256	2.671918520	69.75198786
16	3/2	0	−873.1139870	−1458.647011	3768.579813	36	4	1/2	–	−0.526815762	–
17	3/2	1/2	–	301.4487727	−1923.468882	37	4	1	−0.015124225	–	–
18	3/2	3/2	–	0.057200221	–	38	4	9/2	1.07596E−05	–	–
19	2	0	525.4931424	730.9175464	−3136.854141	39	9/2	0	−0.380457604	−0.529539213	−8.033947560
20	2	1/2	−59.91777426	−151.5398060	1819.309177	40	9/2	1/2	–	0.101831902	0.671969008
						41	9/2	3	4.44381E−05	–	–

10.4.2 Example of Power Prediction

This section contains a comprehensive worked example of a hard-chine hull form catamaran of 44 t. Basic characteristics of the considered catamaran, as well as other data required for the calculations, are given in Table 10.33. Other data are assumed to be the same as in Fig. 10.10. The procedures presented generally correspond to the process typically performed in the estimation of power during the preliminary design stages.

Table 10.33 Input data

Hull characteristics			Other data		
L_{WL}	[m]	23.16	ρ	[t/m^3]	1.02587
B_{WL}	[m]	7.6	ρ_A	[kg/m^3]	1.226
B_{XDH}	[m]	2.01	ν	[m^2/s]	1.19E–06
Δ	[t]	44	g	[m/s^2]	9.81
β_M	[∫]	16	η_M	[-]	0.94
δ_W	[∫]	8	C_{AA}	[-]	0.65
A_V	[m^2]	42	ΔC_F	[-]	$0.3 \cdot 10^{-3}$
			C_F according to ITTC-1957		

Powering prediction examples are done for:

- A stock propeller used for model tests, from now on just *stock propeller* (three bladed, $A_E/A_0 = 1.1$, $P/D = 1.4$, contemporary blade sections with cupped trailing edges), and
- *Installed propeller* of a subject hull (in this example assumed to be a KCA propeller).

Three methods that explain the power prediction are elaborated here:

1. Conventional method (stock propeller)
2. Short method (stock propeller)
3. Conventional method (installed propeller).

(1) Conventional Method (Stock Propeller)

This method is performed in four steps:

- Evaluation of the residual resistance (actually ε_R) and dynamic trim (θ).
- Evaluation of propulsive efficiency (η_D)
- Evaluation of wetted surface (actually *WSC*) and slenderness ratio $\left(L_{WL}/(\nabla/2)^{1/3}\right)$
- Calculation of brake power ($P_B = f(V)$).

Evaluation of Residual Resistance and Dynamic Trim

Residual resistance-to-weight ratio (ε_R) is evaluated for $L_{WL}/B_{XDH} = 11.5$, $\beta_M = 16°$ and $\delta_W = 8°$, according to the speed-independent MM (see Table 10.29). The dynamic trim is evaluated from the MM explained in Table 10.30. Results are given in Table 10.34.

Table 10.34 Residuary resistance-to-weight ratio and dynamic trim

No.	$Fn_{\nabla/2}$	[-]	0.75	1.25	1.50	2.00	3.00	4.00
1	v_S	[m/s]	3.92	6.53	7.83	10.44	15.66	20.88
2	V_S	[kn]	7.61	12.69	15.22	20.30	30.45	40.60
3	$\varepsilon_R\ 10^3$	[-]	12.5958	44.7311	52.7236	49.8696	53.3287	62.9161
4	θ	[°]	0.157	1.085	1.406	1.021	0.303	–0.015

The subject hull's speed, given in row 1, is evaluated from:

$$v_S = Fn_{\nabla/2}\sqrt{g \cdot (\nabla/2)^{1/3}},$$

where volumetric displacement is: $\nabla = \Delta/\rho = 42.89\,\text{m}^3$.

Evaluation of Propulsive Efficiency

Propulsive efficiency (η_D) depends on $Fn_{\nabla/2}$ and L_{WL}/B_{XDH}, is evaluated according to the speed-dependent MM given in Table 10.31e. Results are shown in Table 10.35.

Table 10.35 Propulsive efficiency

No.	$Fn_{\nabla/2}$	[-]	0.75	1.25	1.50	2.00	3.00	4.00
1	V_S	[kn]	7.61	12.69	15.22	20.30	30.45	40.60
2	η_D	[-]	0.5959	0.5888	0.6046	0.6631	0.7238	0.7320

Propulsive efficiency (trial conditions) is valid for a stock propeller on inclined shaft.

Evaluation of Wetted Surface and Slenderness Ratio

Wetted surface coefficient (WSC) and slenderness ratio $\left(L_{WL}\big/(\nabla/2)^{1/3}\right)$ of each demihull depend on β_M and L_{WL}/B_{XDH}, according to the MM's given in Table 10.28. Results are as follows:

$$WSC = S \Big/ (\nabla/2)^{2/3} = 10.1729 \quad \text{and} \quad L_{WL} \Big/ (\nabla/2)^{1/3} = 8.3351.$$

Consequently, an overall wetted surface area (both hulls) is

$$S = 78.523\,\text{m}^2 \quad \text{and} \quad L_{WL} = 23.157\,\text{m}.$$

Evaluation of Brake Power

Evaluation procedure for estimation of brake power for trial conditions (conventional method) is according to Fig. 10.10 and includes estimation of

- resistance for towing tank conditions (R_{TTK})
- air resistance (R_{AA})
- parasitic resistance (R_{PAR}) due to the zink-anodes and various inlet and outlet openings.

These results are given in Table 10.36.

Table 10.36 Brake power corresponding to trial conditions (conventional method, stock propeller)

No.	$Fn_{\nabla/2}$	[-]	0.75	1.25	1.50	2.00	3.00	4.00
1	v_S	[m/s]	3.92	6.53	7.83	10.44	15.66	20.88
2	V_S	[kn]	7.61	12.69	15.22	20.30	30.45	40.60
3	$\varepsilon_R\ 10^3$	[-]	12.5958	44.7311	52.7236	49.8696	53.3287	62.9161
4	η_D	[-]	0.5959	0.5888	0.6046	0.6631	0.7238	0.7320
5	$Rn\ 10^{-8}$	[-]	0.7615	1.2692	1.5230	2.0307	3.0460	4.0613
6	$C_F\ 10^3$	[-]	2.1680	2.0133	1.9620	1.8851	1.7841	1.7172
7	R_{TTK}	[kN]	8.48	27.24	33.93	40.72	64.20	98.02
8	R_{AA}	[kN]	0.88	1.63	2.09	3.18	6.04	9.82
9	R_{PAR}	[kN]	0.19	0.52	0.75	1.33	2.98	5.30
10	R_{TTR}	[kN]	9.72	29.93	37.44	46.03	74.51	115.10
11	P_{BTR}	[kW]	68.0	352.9	515.9	771.1	1715.2	3493.3
12	$P_{BTR}/2$	[kW]	34	176	258	386	858	1747

Values given in rows 1–4 are from Tables 10.34 and 10.35.

5. Reynolds number: $Rn = v_S \cdot L_{WL}/\nu$

6. Specific frictional resistance coefficient: $C_F = 0.075/(\log Rn - 2)^2$
7. Total resistance (with appendages) corresponds to towing tank conditions: $R_{TTK} = \varepsilon_R \cdot \rho \cdot \nabla \cdot g + 0.5 \cdot \rho \cdot v_S^2 \cdot (C_F + \cdot C_F) \cdot 2 \cdot S$, where roughness allowance $\Delta C_F = 0.03 \times 10^{-3}$.
8. Air and wind resistance (see Fig. 10.10):

$$R_{AA} = 0.5 \cdot \rho_A \cdot C_{AA} \cdot (v_S + v_W)^2 \cdot A_V,$$

where air allowance coefficient is $C_{AA} = 0.65$, wind speed $v_W = 3.34$ m/s and above-water transverse area projection $A_V = 42\,\mathrm{m}^2$.
9. Parasitic resistance (see Fig. 10.10): $R_{PAR} = 0.5 \cdot \rho \cdot v_S^2 \cdot [C_{DP}(A_{FZ} + A_X)]$, where structural roughness allowance coefficient is $C_{DP} = 0.15$, area of zink anodes $A_{FZ} = 0.06\,\mathrm{m}^2$ and area of two hull openings is $A_X = 0.098\,\mathrm{m}^2$.
10. Total resistance under trial conditions is: $R_{TTR} = 1.02 \cdot R_{TTK} + R_{AA} + R_{PAR}$
11. Total brake power under trial conditions: $P_{BTR} = R_{TTR} \cdot v_S/(\eta_M \cdot \eta_D)$, where gearbox and shaft losses are $\eta_M = 0.94$.
12. Brake power per propeller.

(2) **Short Method (Stock Propeller)**

This method is suitable for application in the early design phases. It enables a quick estimation of brake power for trial conditions, but is less accurate than the Conventional method. It is based on the assumption that causes for the less important resistance components (i.e. above-water transverse area, size of outlet and inlet openings, size of zinc-anodes, percentage of added resistance due to waves etc.) of a considered catamaran are proportional to those corresponding to the S'89 model (see Müller-Graf et al. 2002).

Specific brake power under trial conditions (ε_{BTR}) is calculated according to the speed-dependent MM which depends on $Fn_{\nabla/2}$ and L_{WL}/B_{XDH}, given in Table 10.32. Note that ε_{BTR} includes the effects the Reynolds number correction of the model propeller. Total required brake power under trial conditions (row 5 of Table 10.37) is estimated from $P_{BTR} = \varepsilon_{BTR} \cdot v_S \cdot \Delta \cdot g$.

Table 10.37 Total brake power corresponding to trial conditions (short method, stock propeller)

No.	$Fn_{\nabla/2}$	[–]	0.75	1.25	1.50	2.00	3.00	4.00
1	v_S	[m/s]	3.92	6.53	7.83	10.44	15.66	20.88
2	V_S	[kn]	7.61	12.69	15.22	20.30	30.45	40.60
3	$\varepsilon_{BTR}\ 10^3$	[–]	39.5927	124.9529	147.1918	161.7375	235.6391	353.6006
4	P_{BTR}	[kW]	66.9	352.0	497.5	728.9	1593.0	3187.3
5	$P_{BTR}/2$	[kW]	33	176	249	364	796	1594

(3) Conventional method (Installed Propeller)

In this example it is assumed that the installed propellers (twin screw installation) are the KCA propellers. Evaluation procedure is performed in three steps:

- Evaluation of propulsive coefficients ($\eta_H, \eta_O, \eta_R, w_T$ and t_Ψ)
- Propeller selection
- Estimation of off-design propeller characteristics.

Evaluation of Propulsive Coefficients

Wake fraction (w_T), hull efficiency (η_H), open water efficiency of stock propeller (η_O) and propulsive efficiency (η_D), depend on $Fn_{\nabla/2}$ and L_{WL}/B_{XDH}, and are estimated according to MM's given in Table 10.31b–e. Results are presented in rows 3 to 6 respectively, in Table 10.38.

Table 10.38 Propulsive coefficients

No.	$Fn_{\nabla/2}$	[–]	0.75	1.25	1.50	2.00	3.00	4.00
1	v_S	[m/s]	3.92	6.53	7.83	10.44	15.66	20.88
2	V_S	[kn]	7.61	12.69	15.22	20.30	30.45	40.60
3	w_T	[–]	–0.0033	–0.0206	–0.0154	0.0241	0.0488	0.0497
4	η_H	[–]	0.9696	0.9521	0.9576	0.9941	1.0390	1.0362
5	η_0	[–]	0.6469	0.6420	0.6618	0.6958	0.7252	0.7346
6	η_D	[–]	0.5959	0.5888	0.6046	0.6631	0.7238	0.7320
7	η_R	[–]	0.9500	0.9633	0.9540	0.9587	0.9606	0.9616
8	t_ψ	[–]	0.0272	0.0283	0.0277	0.0298	0.0117	0.0153

Relative rotative coefficient (η_R) and thrust deduction fraction (t_Ψ), which accounts in shaft inclination, are calculated from:

$$\eta_R = \eta_D/(\eta_O \cdot \eta_H) \quad \text{and} \quad t_\Psi = 1 - \eta_H \cdot (1 - w_T),$$

and are given in rows 7 and 8 respectively, in Table 10.38.

Propeller Selection

In this example the propellers are optimized through few iterations applying the KCA (RA) MM (Sect. 9.2.3, Table 9.4) for V = 30.45 kn corresponding to $Fn_{\nabla/2} = 3.0$. The optimization procedure is explained in Sect. 10.3.3, and therefore there is no need to repeat it here. Note that in this example an assumption is made that each of the two propellers operates under uniform conditions, and hence generates the same thrust. Conditions for which the best propeller is selected are: $V_S = 30.45$ kn, total resistance is $R_{TTR} = 74.5$ kN (see Table 10.36), i.e. $R_{TTR}/2 = 37.25$ kN per propeller. Cavitation and loading criteria are neglected.

Input and output data for selected optimal propeller (three bladed KCA propeller, D = 1.0 m, P/D = 1.5, DAR = 1.1) is given in Table 10.39. Other hydrodynamic characteristics are obtained from the open-water data of a KCA propeller with the abovementioned characteristics (Fig. 10.12), i.e. from the KCA (RA) MM. Thrust and torque of the propeller are determined from:

$$T = K_T \cdot \rho \cdot n^2 \cdot D^4 \quad \text{and} \quad Q_0 = K_Q \cdot \rho \cdot n^2 \cdot D^5.$$

Table 10.39 Input and output data of selected propeller for V = 30.45 kn

Input			Output		
Resistance per propeller	37.25	kN	Propeller rotational speed	783	rpm
Propeller diameter	1.0	m	Pitchadiameter ratio	1.50	–
Speed	30.45	kn	Propeller open–water efficiency	0.6943	–
Number of blades	3	–	Advance coefficient	1.1418	–
Propeller area ratio (DAR)	1.1	–	Thrust ceofficient	0.2160	–
Thrust deduction fraction	0.0117	–	Torque coeffcient	0.0565	–
Wake fraction	0.0488	–	Propeller thrust	37.69	kN
Relative rotative efficiency	0.9606	–	Propeller torque	9.87	kNm

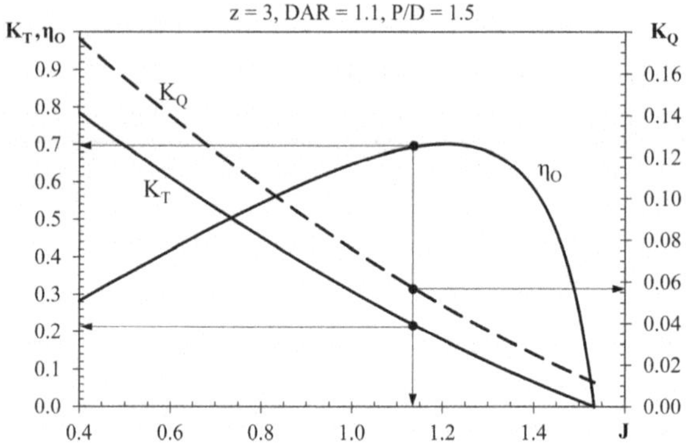

Fig. 10.12 Open-water characteristics of the selected propeller

Off-Design Conditions

The procedure for determining propeller's off-design conditions is explained in Sect. 10.3.3, and the final results only are presented here in Table 10.40, i.e. brake power per propeller (row 3) and total power (row 4).

Table 10.40 Brake power for installed KCA propeller (conventional method, installed propeller)

No.	$Fn_{\nabla/2}$	[–]	0.75	1.25	1.50	2.00	3.00	4.00
1	v_S	[m/s]	3.92	6.53	7.83	10.44	15.66	20.88
2	V_S	[kn]	7.61	12.69	15.22	20.30	30.45	40.60
3	$P_{BTR}/2$	[kW]	33.8	176.2	259.3	390.8	886.6	1812.0
4	P_{BTR}	[kW]	68	352	519	782	1773	3624

A comparison of the final results obtained by the Short and Conventional methods, for both with stock propellers, and for Conventional method only with a KCA propeller, is given in Fig. 10.13. Dynamic trim, evaluated in Table 10.34, is also shown in Fig. 10.13 as it doesn't depend on the method used for power evaluation.

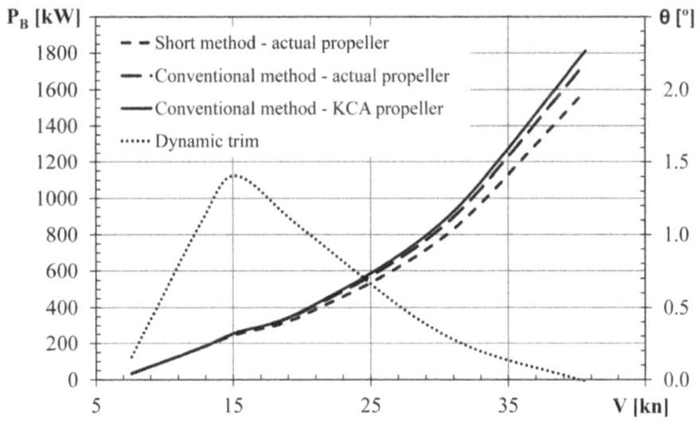

Fig. 10.13 Brake power per propeller obtained by three considered methods

As expected, the Short method underestimates brake power primarily because transverse area exposed to air and wind of a considered catamaran is proportionally larger than that of a corresponding S'89 model. The difference in brake power between stock and installed propellers (both powers determined by the conventional method) is explained by the fact that the stock propeller used for model tests of the Series '89 is a bit more efficient.

References

Bailey D (1982) A Statistical analysis of propulsion data obtained from models of High-speed round bilge hulls. In: RINA symposium on small fast warships and security vessels, London

Blount DL, Bjarne E (1989) Design and selection of propulsors for high-speed craft. In: 7th lips propeller symposium, Nordwijk-on-Sea

Blount DL, Fox DL (1976) Small craft power prediction. Marine Technol 13(1)

Blount DL (2014) Performance by design. ISBN 0-978-9890837-1-3

Gamulin A (1996) A semidisplacement series of ships. Int Shipbuilding Progr 43(43)

Hadler JB (1966) The prediction of Power performance of planing craft. SNAME Trans 74

Hadler JB, Hubble EN (1971) Prediction of the power performance of the series 62 planing hull forms. SNAME Trans, vol 79

Helm K (1961) Optimale Formverhaltnisse Für schnell Fahrgastschiffe auf Binnengessern. Schiff und Hafen, H. 9

Helm K (1963) Systematische Untersuchungen über den Einflüss der Formgebung auf Widerstand and Leistungbedarf von Fahrgastschiffen der Binnenschifart. Schiff und Hafen, H. 2

Hoggard MM (1979) Examining added drag of planing craft operating in the seaway. Hampton Road Section of SNAME

Hoggard MM, Jones MP (1980) Examining pitch, heave and accelerations of planing craft operating in a seaway. In: High-speed surface craft Conference, Brighton

Hofman M, Radojčić D (1997) Resistance and propulsion of high-speed vessels in shallow water. (in Serbian) Masinski fakultet, Beograd

Molland AF, Turnock SR, Hudson DA (2011) Ship resistance and propulsion—practical estimation of ship propulsive power. Cambridge University Press, ISBN 978-0-521-76052-2

Müller-Graf B, Radojčić D, Simić A (2002) Resistance and propulsion characteristics of the VWS hard chine Catamaran hull Series '89. SNAME Trans 110

Radojčić D (1991) An engineering approach to predicting the hydrodynamic performance of planing craft using computer techniques. RINA Trans 133

Radojčić D, Bowles J (2010) On high-speed Monohulls in shallow water. In: SNAME's 2nd chesapeake power boat symposium, Annapolis

The manufacturer's authorised representative in the EU is Springer Nature Customer Service Centre GmbH, Europaplatz 3, 69115 Heidelberg, Germany. If you have any concerns regarding our products, please contact ProductSafety@springernature.com

Printed and bound by CPI Group (UK) Ltd, Croydon, CR0 4YY
05/10/2026
02228131-0006